New Approaches to Synthetic Organic Chemistry

New Approaches to Synthetic Organic Chemistry

Editors

Mircea Darabantu
Alison Rinderspacher
Gloria Proni

Basel • Beijing • Wuhan • Barcelona • Belgrade • Novi Sad • Cluj • Manchester

Editors

Mircea Darabantu
University Babes-Bolyai
Cluj-Napoca
Cluj-Napoca, Romania

Alison Rinderspacher
Columbia University Irving
Medical Center
New York, NY, USA

Gloria Proni
John Jay College of Criminal
Justice
New York, NY, USA

Editorial Office
MDPI
St. Alban-Anlage 66
4052 Basel, Switzerland

This is a reprint of articles from the Special Issue published online in the open access journal *Molecules* (ISSN 1420-3049) (available at: https://www.mdpi.com/journal/molecules/special_issues/Synthetic_OC).

For citation purposes, cite each article independently as indicated on the article page online and as indicated below:

Lastname, A.A.; Lastname, B.B. Article Title. *Journal Name* **Year**, *Volume Number*, Page Range.

ISBN 978-3-0365-9020-2 (Hbk)
ISBN 978-3-0365-9021-9 (PDF)
doi.org/10.3390/books978-3-0365-9021-9

© 2023 by the authors. Articles in this book are Open Access and distributed under the Creative Commons Attribution (CC BY) license. The book as a whole is distributed by MDPI under the terms and conditions of the Creative Commons Attribution-NonCommercial-NoDerivs (CC BY-NC-ND) license.

Contents

Jahyun Koo, Minsu Kim, Kye Jung Shin and Jae Hong Seo
Non-Palladium-Catalyzed Approach to the Synthesis of (*E*)-3-(1,3-Diarylallylidene)Oxindoles
Reprinted from: *Molecules* **2022**, *27*, 5304, doi:10.3390/molecules27165304 1

Yang-Fan Guo, Tao Luo, Guang-Jing Feng, Chun-Yang Liu and Hai Dong
Efficient Synthesis of 2-OH Thioglycosides from Glycals Based on the Reduction of Aryl Disulfides by $NaBH_4$
Reprinted from: *Molecules* **2022**, *27*, 5980, doi:10.3390/molecules27185980 17

Hillary Straub, Pavel Ryabchuk, Marina Rubina and Michael Rubin
Preparation of Chiral Enantioenriched Densely Substituted Cyclopropyl Azoles, Amines, and Ethers via Formal S_N2' Substitution of Bromocylopropanes
Reprinted from: *Molecules* **2022**, *27*, 7069, doi:10.3390/molecules27207069 33

Mihail Lucian Birsa and Laura G. Sarbu
An Improved Synthetic Method for Sensitive Iodine Containing Tricyclic Flavonoids
Reprinted from: *Molecules* **2022**, *27*, 8430, doi:10.3390/molecules27238430 49

Nazar Moshnenko, Alexander Kazantsev, Olga Bakulina, Dmitry Dar'in and Mikhail Krasavin
The Use of Aryl-Substituted Homophthalic Anhydrides in the Castagnoli–Cushman Reaction Provides Access to Novel Tetrahydroisoquinolone Carboxylic Acid Bearing an All-Carbon Quaternary Stereogenic Center
Reprinted from: *Molecules* **2022**, *27*, 8462, doi:10.3390/molecules27238462 57

Lu Yang, Hongwei Su, Yue Sun, Sen Zhang, Maosheng Cheng and Yongxiang Liu
Recent Advances in Gold(I)-Catalyzed Approaches to Three-Type Small-Molecule Scaffolds via Arylalkyne Activation
Reprinted from: *Molecules* **2022**, *27*, 8956, doi:10.3390/molecules27248956 71

Kosuke Yamamoto, Keisuke Miyamoto, Mizuki Ueno, Yuki Takemoto, Masami Kuriyama and Osamu Onomura
Copper-Catalyzed Asymmetric Sulfonylative Desymmetrization of Glycerol
Reprinted from: *Molecules* **2022**, *27*, 9025, doi:10.3390/molecules27249025 97

Mohamed S. H. Salem, Ahmed Sabri, Md. Imrul Khalid, Hiroaki Sasai and Shinobu Takizawa
Two-Step Synthesis, Structure, and Optical Features of a Double Hetero[7]helicene
Reprinted from: *Molecules* **2022**, *27*, 9068, doi:10.3390/molecules27249068 109

Denis N. Tomilin, Lyubov N. Sobenina, Alexandra M. Belogolova, Alexander B. Trofimov, Igor A. Ushakov and Boris A. Trofimov
Unexpected Decarbonylation of Acylethynylpyrroles under the Action of Cyanomethyl Carbanion: A Robust Access to Ethynylpyrroles
Reprinted from: *Molecules* **2023**, *28*, 1389, doi:10.3390/molecules28031389 121

Luka Ciber, Franc Požgan, Helena Brodnik, Bogdan Štefane, Jurij Svete, Mario Waser and Uroš Grošelj
Synthesis and Catalytic Activity of Bifunctional Phase-Transfer Organocatalysts Based on Camphor
Reprinted from: *Molecules* **2023**, *28*, 1515, doi:10.3390/molecules28031515 135

Caiyun Yang, Sirou Hu, Xinhui Pan, Ke Yang, Ke Zhang, Qingguang Liu, et al.
Novel Synthesis of Dihydroisoxazoles by p-TsOH-Participated 1,3-Dipolar Cycloaddition of Dipolarophiles with α-Nitroketones
Reprinted from: *Molecules* **2023**, *28*, 2565, doi:10.3390/molecules28062565 . . . . **153**

Patricia Camarero González, Sergio Rossi, Miguel Sanz, Francesca Vasile and Maurizio Benaglia
Synthesis of Tetrasubstituted Nitroalkenes and Preliminary Studies of Their Enantioselective Organocatalytic Reduction
Reprinted from: *Molecules* **2023**, *28*, 3156, doi:10.3390/molecules28073156 . . . . **163**

Article

Non-Palladium-Catalyzed Approach to the Synthesis of (*E*)-3-(1,3-Diarylallylidene)Oxindoles

Jahyun Koo, Minsu Kim, Kye Jung Shin and Jae Hong Seo *

Integrated Research Institute of Pharmaceutical Sciences, College of Pharmacy, The Catholic University of Korea, Bucheon-si 420-743, Korea
* Correspondence: jaehongseo@catholic.ac.kr; Tel.: +82-2-2164-6531; Fax: +82-2-2164-4059

Abstract: Two novel synthetic approaches for synthesizing (*E*)-3-(1,3-diarylallylidene)oxindoles from oxindole were developed. All previously reported methods for synthesizing 3-(1,3-diarylallylidene) oxindoles utilized palladium-catalyzed reactions as a key step to form this unique skeleton. Despite high efficiency, palladium-catalyzed reactions have limitations in terms of substrate scope. Especially, an iodoaryl moiety cannot be introduced by the previous methods due to its high reactivity toward the palladium catalyst. Our Knoevenagel/allylic oxidation/Wittig and Knoevenagel/aldol/dehydration strategies complement each other and show broad substrate scope, including substrates with iodoaryl groups. The current methods utilized acetophenones, benzylidene phosphonium ylides, and benzaldehydes that are commercially available or easily accessible. Thus, the current synthetic approaches to (*E*)-3-(1,3-diarylallylidene)oxindoles are readily amendable for variety of oxindole derivatives.

Keywords: 3-(1,3-diarylallylidene)oxindole; knoevenagel condensation; allylic oxidation; wittig reaction; aldol reaction; non-palladium-catalyzed

1. Introduction

3-(Diarylmethylene)oxindoles belong to a major oxindole family that has recently been reported to have novel biological activities, such as AMPK activation [1] and estrogen receptor-related anticancer activity against breast cancer [2]. As valuable derivatives of 3-(diarylmethylene)oxindoles in the field of medicinal chemistry, 3-(1,3-diarylallylidene) oxindoles, which have a vinyl linker at the 3-methylene position, have attracted considerable attention from synthetic chemists, and several synthetic methods have been reported (Scheme 1) [3–6]. In 2005, a 3-(1,3-diarylallylidene)oxindole was first synthesized by Takemoto et al., utilizing double Heck reactions [3]. Recently, Sekar et al., improved this approach using palladium binaphthyl nanoparticles (Pd-BNPs) as a catalyst to broaden the substrate scope and allow easy separation [4]. In 2008, Murakami et al., developed another synthetic method featuring palladium-catalyzed oxidative cyclization of 2-(alkynyl)isocyanate, followed by the Suzuki-Miyaura reaction with styrylboronic acid [5]. As part of our ongoing efforts to identify novel synthetic methods for 3-methyleneoxindole derivatives [6–10], we recently reported a palladium-catalyzed multicomponent tandem reaction, which allowed a stereoselective approach to (*E*)- and (*Z*)-isomers of 3-(1,3-diarylallylidene)oxindoles by changing phosphine ligands, reaction temperature, and time [6]. Although several synthetic methods for 3-(1,3-diarylallylidene)oxindoles have already been developed, as described above, the narrow substrate scope and/or limited accessibility of reagents when using these methods necessitate the development of a more general approach to this unique skeleton. In all previous methods palladium-catalyzed reactions were the key reactions, which greatly limited the range of products to which these procedures could be applied. A few non-palladium-catalyzed approaches to 3-allylideneoxindoles have been reported, but they cannot be applied to the synthesis of 3-(1,3-diarylallylidene)oxindoles [11,12]. Therefore, we attempted to develop a novel synthetic method for 3-(1,3-diarylallylidene)oxindoles with a wide substrate scope using commercially available or easily accessible reagents, and not involving palladium-catalyzed reactions.

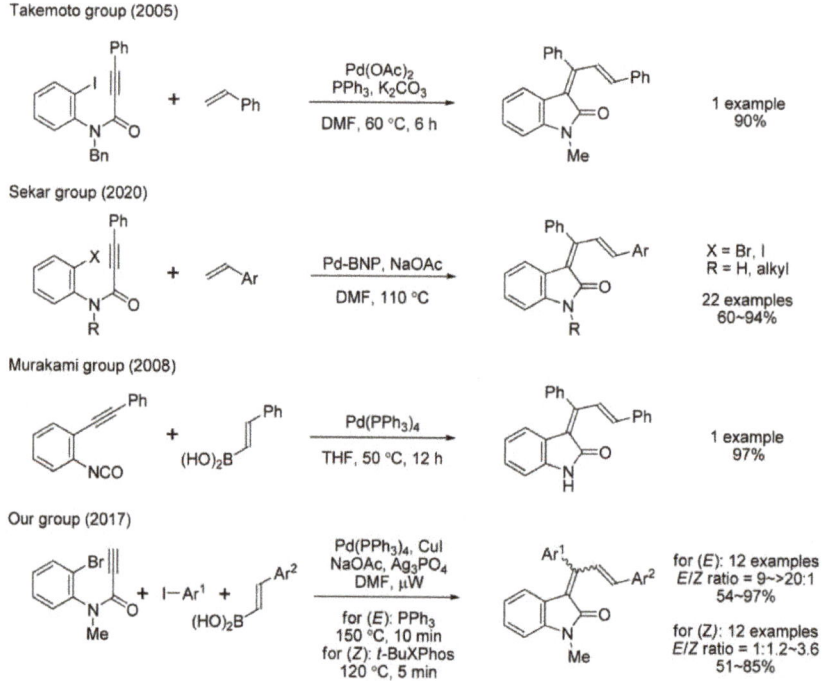

Scheme 1. Previous synthetic approaches to 3-(1,3-diarylallylidene)oxindoles.

2. Results and Discussion

2.1. Unsuccessful Direct Knoevenagel Approach

To develop a novel and efficient approach for synthesizing 3-(1,3-diarylallylidene)oxindoles, we first examined the feasibility of direct Knoevenagel condensation between oxindole **1** and chalcone (**2**) (Scheme 2). Several Knoevenagel condensations of oxindoles with α,β-unsaturated aldehydes have been reported [13–17], but there is no precedent for using α,β-unsaturated ketones, such as chalcone. Despite intensive efforts, the desired 3-(1,3-diphenylallylidene)oxindole **3** could not be formed, as in previous studies. Only a small amount of 1,4-addition adduct **4** was obtained under Ti(O^{i}Pr)$_4$/pyridine-mediated conditions [14,18].

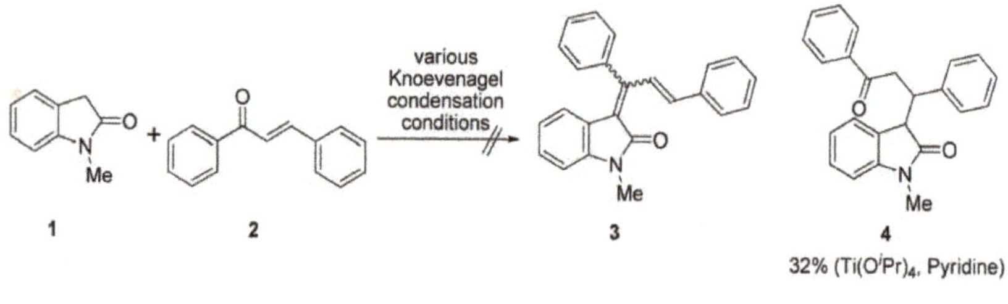

Scheme 2. Knoevenagel condensation of oxindole **1** with chalcone (**2**).

2.2. Stepwise Approach 1 (Knoevenagel/Allylic Oxidation/Wittig)

The disappointing results of the direct Knoevenagel strategy prompted us to apply a stepwise approach (Scheme 3). Using an acetophenone as a "partner" of Ti(O^{i}Pr)$_4$/pyridine-mediated Knoevenagel condensation, 3-methyleneoxindole **5** was easily obtained from oxindole **1** in 93% yield with good Z-stereoselectivity (Z:E = 5:1). The preference for the Z-isomer could be explained by a chelation-controlled transition state [14]. The geometry of each isomer was confirmed by comparing ^{1}H NMR data for (**E**)-**5a** [19] and the chemical shift of H$_4$ [6.84 ppm (Z-isomer), 6.14 ppm (E-isomer)]. On ^{1}H NMR analysis of known 3-arylmethylenoxindoles, the chemical shift of H$_4$ is upfield (generally 6.50–6.00 ppm) compared to the usual aromatic area when the aryl group attached to the 3-methylene position of oxindole is located close to H$_4$ [6,9]. Next, the methyl group should be transformed into a proper functional group to introduce the second olefin. Under radical bromination conditions [20], allylic bromide **6** was obtained as a single geometric isomer, regardless of the geometry of **5a**. The structure of **6** was elucidated by intensive NMR studies, including HSQC, HMBC, COSY, and ROESY. In addition, the chemical shift of H$_4$ was 6.11 ppm, which supported the (Z)-geometry of **6**. The Krische group reported similar scrambling of olefin geometry during radical allylic bromination [21]. Unfortunately, the Wittig reaction of the corresponding ylide derived from **6** did not afford the desired **3aa**. Given these disappointing results, we exchanged the positions of the functional groups in the Wittig reaction. Therefore, the second functionalization of the methyl group was allylic oxidation. After analyzing several oxidation conditions, we found that SeO$_2$ oxidation [22] afforded aldehyde **7a** in 84% yield, including as a single geometric isomer from both (Z)- and (E)-**5a**. Interestingly, the olefin geometry of **7a** had the Z-configuration, which was confirmed by comparison with previously reported ^{1}H NMR data for **7a** [23]. In addition, the chemical shift of H$_4$ for **7a** also appeared at 6.26 ppm. The Z-stereoselectivity of allylic oxidation may have been due to coordination of the oxindole carbonyl group to selenium [24]. However, considering the high reaction temperature, the possibility of isomerization of (**E**)-**7a** to the more stable (**Z**)-**7a** during the reaction could not be excluded. Then, the Wittig reaction of aldehyde **7a** with the ylide proceeded smoothly and provided the desired **3aa** in 84% yield. The ^{1}H and ^{13}C NMR data of **3aa** exactly matched the results obtained in our previous study [6].

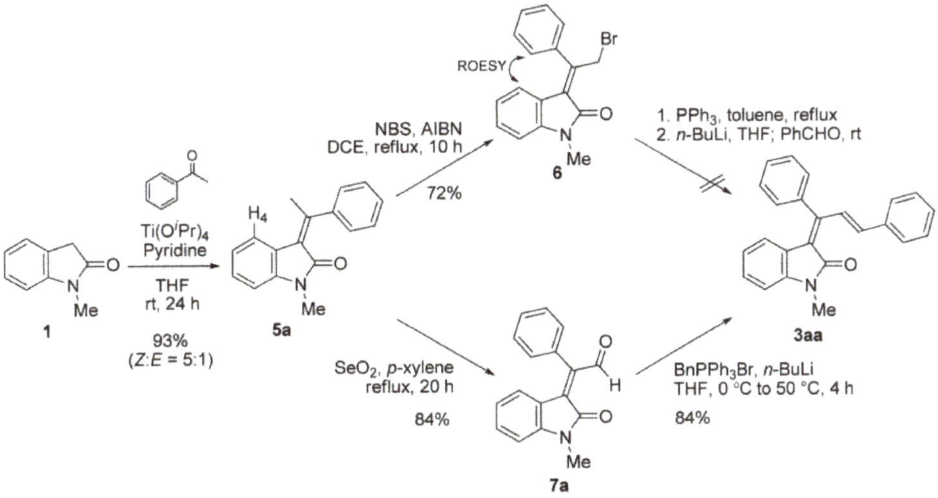

Scheme 3. Stepwise approach to 3-(1,3-diphenylallylidene)oxindole **3aa**.

By applying the successful stepwise approach to 3-(1,3-diphenylallylidene)oxindole, we investigated the substrate scope of aldehyde **7** (Table 1). The Knoevenagel condensation of oxindole **1** and acetophenones with chloro, nitro, and methoxy substituents proceeded well, affording **5b–d** in good yield (77–95%) with Z-stereoselectivity (Z:E = 4–10:1) (Entries 1–3). SeO$_2$-mediated allylic oxidation of **5b** and **5c** also proceeded smoothly to afford the corresponding **7** in 85% and 87% yield, respectively. However, the oxidation of **5d**, bearing a methoxy substituent at the aryl group, was unsuccessful and resulted in complete decomposition. Unfortunately, neither a lower reaction temperature nor other oxidation conditions overcame this decomposition problem.

Table 1. Substrate scope for preparation of aldehyde **7**.

Entry	5	R	Yield [1] (%)	Z:E [2]	7	Yield [3] (%)
1	5b	Cl	91	4.6:1	7b	85
2	5c	NO$_2$	95	10:1 [4]	7c	87
3	5d	OMe	77	4:1	7d	0

[1] Sum of isolated yields of (Z)- and (E)-isomers, [2] Ratio of isolated yield of (Z)- and (E)-isomers, [3] Isolated yield. [4] Ratio in ^{1}H NMR of mixture of two isomers, which could not be isolated.

Setting aside the problematic **5d**, we assessed the substrate scope of the final Wittig reaction of **5a–c** with ylides bearing various substituents on the aryl group (Table 2). Fortunately, all reactions afforded 3-(1,3-diarylallylidene)oxindoles **3**, regardless of the substituent combination, in moderate to good yield (53–95%) (Entries 1–12).

Table 2. Substrate scope of the Wittig reaction.

Entry	7	3	R	R^1	Yield [1] (%)
1	7a	3aa	H	H	84
2	7a	3ab	H	Cl	67
3	7a	3ac	H	NO$_2$	64
4	7a	3ad	H	OMe	53
5	7b	3ba	Cl	H	95
6	7b	3bb	Cl	Cl	79
7	7b	3bc	Cl	NO$_2$	82
8	7b	3bd	Cl	OMe	68
9	7c	3ca	NO$_2$	H	80
10	7c	3cb	NO$_2$	Cl	79
11	7c	3cc	NO$_2$	NO$_2$	68
12	7c	3cd	NO$_2$	OMe	82

[1] Isolated yield.

2.3. Stepwise Approach 2 (Knoevenagel/Aldol/Dehydration)

As shown above, the allylic oxidation/Wittig reaction strategy allowed synthesis of various 3-(1,3-diarylallylidene)oxindoles **3** from Knoevenagel adducts **5**. However, decomposition of **5d** in SeO$_2$ oxidation limited the application scope of this strategy. Therefore, we investigated another stepwise approach, which could be applied to **5d** and overcome the limitation of the first strategy. Based on the fact that the functional handle of the methyl group was located at the γ-position of the α,β-unsaturated carbonyl moiety, we assumed that aldol reaction may be feasible. In addition, several examples of similar aldol reactions were found in the literature [24–26]. According to the results of base screening, only *n*-BuLi could provide the desired aldol product, **8aa**, in good yield (Scheme 4). The olefin geometry of **8aa** was assigned as *E*, as the chemical shift of H$_4$ appeared at 6.14 ppm. Notably, unlike bromination and allylic oxidation, the olefin geometry of **5a** significantly affected the aldol reaction rate. Under optimized conditions, (*Z*)-**5a** was rapidly converted to **8aa** in 91% yield, while the reaction of (*E*)-**5a** afforded **8aa** in 55% yield. A longer reaction time and/or elevated reaction temperature did not increase product yield. The difference in reaction rate may have been caused by the lithiated intermediate from (*E*)-**5a** assuming a stable chelated form via coordination of the oxindole carbonyl group. Dehydration of **8aa** proceeded smoothly under acidic conditions to provide **3aa** in 95% yield [27,28].

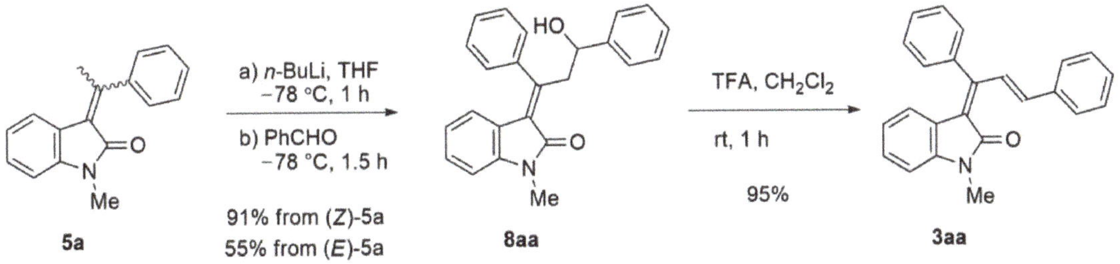

Scheme 4. The second stepwise approach utilizing aldol reaction/dehydration.

Next, we examined whether the second stepwise approach (aldol/dehydration) was applicable to **5d** (Table 3). The first aldol reaction of **5d** proceeded well with various benzaldehydes, giving aldol adduct **8** in moderate to good yields (Entries 1–3 and 5). With the exception of **8dc**, TFA-mediated dehydration of **8** also proceeded well to afford **3** in excellent yields (Entries 1, 2, and 5). The strong electron-withdrawing action of the nitro group in **8dc** may hamper dehydration under acidic conditions. Even under reflux conditions, the desired **3dc** was produced in only 45% yield (Entry 3). After several tests, we found that, under basic conditions (TsCl, DMAP, NEt$_3$, CH$_2$Cl$_2$, room temperature, 3 h), **3dc** formed in 79% yield (Entry 4). The aldol/dehydration approach could serve as an additional option for synthesis of 3-(1,3-diarylallylidene)oxindoles **3** with the previous allylic oxidation/Wittig strategy.

Table 3. Substrate scope of the aldol/dehydration strategy from **(Z)-5d**.

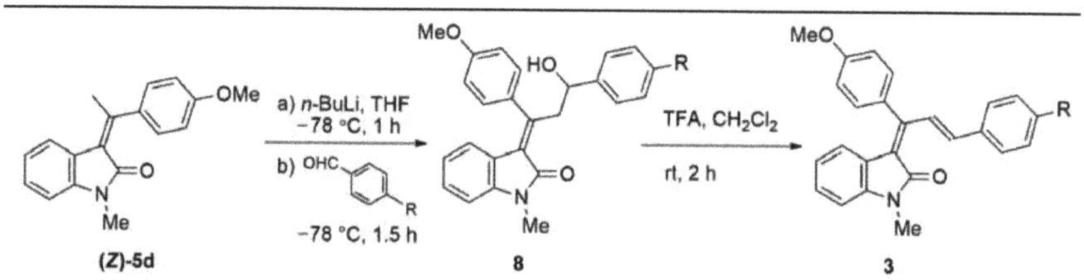

Entry	R	8	Yield [1] (%)	3	Yield [1] (%)
1	H	8da	93	3da	90
2	Cl	8db	75	3db	93
3	NO$_2$	8dc	86	3dc	45 [2]
4	NO$_2$			3dc	79 [3]
5	OMe	8dd	63	3dd	91

[1] Isolated yield, [2] Reflux, 20 h, [3] TsCl, DMAP, NEt$_3$, CH$_2$Cl$_2$, rt, 3 h.

2.4. Application of Stepwise Approach to An Iodoaryl Compound

To demonstrate that our stepwise approach is useful for synthesis of 3-(1,3-diarylallylidene)oxindoles **3**, which are not accessible by previous palladium-catalyzed methods, we synthesized **3** with an iodoaryl moiety using a Knoevenagel/allylic oxidation/Wittig strategy (Scheme 5). Due to its high reactivity toward the palladium catalyst, the iodoaryl group was not compatible with palladium-catalyzed reactions. Ti(O^{i}Pr)$_4$/pyridine-mediated Knoevenagel condensation of oxindole **1** with *p*-iodoacetophenone gave 3-methyleneoxindole **5e** in 89% yield, with a preference for the Z-isomer (Z:E=10:1). Allylic oxidation of **5e** afforded aldehyde **7e** in 78% yield. In addition, the last Wittig reaction of **7e** produced **3ea** (80% yield), which could not be formed using previous methods. The iodoaryl group of **3ea** could be used as a functional handle for further molecular modifications. For example, Suzuki-Miyaura reaction of **3ea** introduced another phenyl group to give **9** in 84% yield.

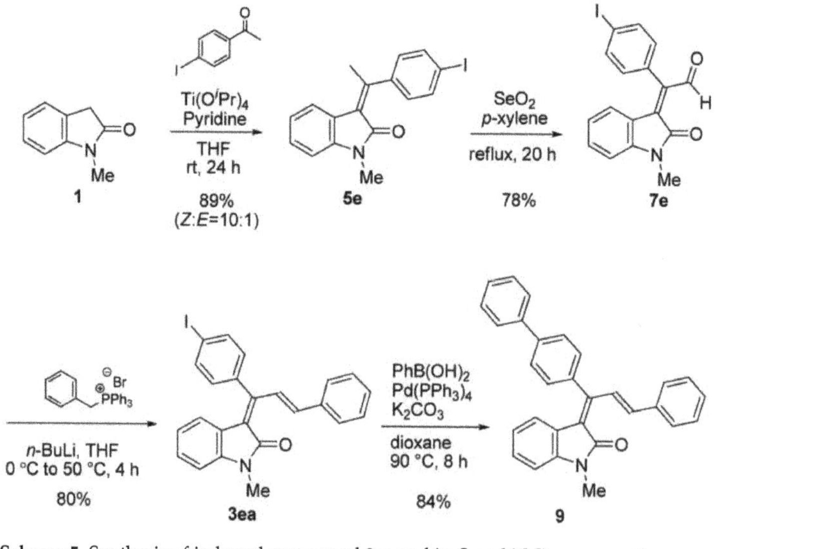

Scheme 5. Synthesis of iodoaryl compound **3ea** and its Suzuki-Miyaura reaction.

In conclusion, we have developed two complementary stepwise approaches to synthesize (E)-3-(1,3-diarylallylidene)oxindoles **3** from oxindole **1** (Knoevenagel/allylic oxidation/Wittig and Knoevenagel/aldol/dehydration). These strategies enable the synthesis of various **3**, regardless of substituents on the aryl moiety. Especially, **3** with palladium-sensitive functional groups, such as iodoaryl groups, could be obtained by these stepwise methods, which could help to expand the applications of (E)-3-(1,3-diarylallylidene)oxindoles.

3. Experimentals

3.1. General Information

All reactions were performed under an argon atmosphere with dry solvents, unless otherwise stated. Dry tetrahydrofuran (THF) and methylene chloride (CH_2Cl_2) were obtained from Ultimate Solvent Purification System (JC Meyer Solvent System, Laguna Beach, CA, USA). Other dry solvents were purchased as anhydrous grade. All glassware is oven-dried and/or flame-dried before use. All commercially available reagents were purchased and used without further purification. Reactions were monitored by thin-layer chromatography (TLC) on silica gel plates (Merck TLC Silica Gel 60 F254, Darmstadt, Germany) using UV light, PMA (an ethanolic solution of phosphomolybdic acid) or ANIS (an ethanolic solution of para-anisaldehyde) as visualizing agent. Purification of products was conducted by column chromatography through silica gel 60 (0.060−0.200 mm). Melting points of all solid compounds were determined by Buchi M-565. NMR spectra were obtained on Bruker AVANCE III 500 MHz (Bruker Corporation, Billerica, MA, USA) at 20 °C using residual undeuterated solvent or TMS (tetramethylsilane) as an internal reference. High-resolution mass spectra (HR-MS) were recorded on a JEOL JMS-700 (JEOL, Tokyo, Japan) using EI (electron impact).

3.2. General Procedure of $Ti(O^iPr)_4$/pyridine Mediated Knoevenagel Condensation (**4**, **5a-e**)

To a stirred solution of 1-methyl-2-oxindole (**1**) (1.0 mmol) and the corresponding acetophenone or chalcone (**2**) (1.0 mmol, 1.0 equiv.) in THF (10 mL) were added pyridine (0.17 mL, 2.0 mmol, 2.0 equiv.) and $Ti(O^iPr)_4$ (0.90 mL, 3.0 mmol, 3.0 equiv.). The reaction mixture was stirred at rt for 24 h, Then, the mixture was diluted with EtOAc (100 mL) and 1 N aq. HCl solution (30 mL). The organic layer was separated and washed with sat. aq. $NaHCO_3$ (30 mL) and brine (30 mL). The remained organic layer was dried over Na_2SO_4, filtered, and concentrated under reduced pressure. The crude residue was purified by column chromatography to afford Knoevenagel product **5** or 1,4-addition adduct **4**.

3.2.1. 1-Methyl-3-(3-oxo-1,3-diphenylpropyl)indolin-2-one (**4**)

32% Yield; white solid; mp = 122.9–125.6 °C; R_f = 0.24 (silica gel, hexane:EtOAc = 4:1); ^{1}H NMR (500 MHz, CDCl$_3$): δ 8.09–8.01 (m, 2H), 7.57 (td, J = 7.4, 1.2 Hz 1H), 7.47 (dd, J = 10.6, 4.7 Hz, 2H), 7.33 (d, J = 7.3 Hz, 1H), 7.17 (t, J = 7.7 Hz, 1H), 7.11–7.05 (m, 5H), 7.03 (td, J = 7.7, 0.7 Hz, 1H), 6.57 (d, J = 7.7 Hz, 1H), 4.33–4.17 (m, 2H), 3.84 (d, J = 3.6 Hz, 1H), 3.54 (dd, J = 17.0, 5.1 Hz, 1H), 2.99 (s, 3H) ppm; ^{13}C NMR (125 MHz, CDCl$_3$) δ 199.1, 176.6, 144.1, 139.6, 137.2, 133.3, 128.7, 128.3, 127.98, 127.96, 127.8, 126.9, 124.4, 122.3, 107.8, 49.8, 42.2, 39.7, 25.9 ppm; HRMS (EI): calcd for $C_{24}H_{21}NO_2$ [M$^+$] : 355.1572, found 355.1573.

3.2.2. (Z)-1-Methyl-3-(1-phenylethylidene)indolin-2-one ((**Z**)-**5a**)

80% Yield; yellow solid; mp = 128.9–130.2 °C; R_f = 0.45 (silica gel, hexane:EtOAc = 4:1); ^{1}H NMR (500 MHz, CDCl$_3$): δ 7.67 (d, J = 7.6 Hz, 1H), 7.46–7.36 (m, 3H), 7.35–7.27 (m, 3H), 7.10 (td, J = 7.7, 1.0 Hz, 1H), 6.84 (d, J = 7.7 Hz, 1H), 3.15 (s, 3H), 2.65 (s, 3H) ppm; ^{13}C NMR (125 MHz, CDCl$_3$) δ 166.4, 153.2, 143.4, 142.6, 128.6, 128.2, 128.1, 127.4, 124.0, 123.7, 123.6, 121.8, 107.8, 25.8, 25.6 ppm; HRMS (EI): calcd for $C_{17}H_{15}NO$ [M$^+$] : 249.1154, found 249.1151.

3.2.3. (E)-1-Methyl-3-(1-phenylethylidene)indolin-2-one ((E)-5a) [19]

13% Yield; yellow solid; mp = 140.4–142.2 °C (lit. [19] 142.7-144.1 °C); R_f = 0.55 (silica gel, hexane:EtOAc = 4:1); ^{1}H NMR (500 MHz, CDCl$_3$): δ 7.51–7.41 (m, 3H), 7.28 (dt, J = 3.6, 2.0 Hz, 2H), 7.14 (dd, J = 11.2, 4.2 Hz, 1H), 6.76 (d, J = 7.8 Hz, 1H), 6.64 (td, J = 7.7, 0.8 Hz, 1H), 6.14 (d, J = 7.7 Hz, 1H), 3.28 (s, 3H), 2.81 (s, 3H) ppm; ^{13}C NMR (125 MHz, CDCl$_3$) δ 168.3, 155.0, 143.1, 142.4, 129.3, 128.4, 128.2, 126.6, 123.5, 122.9, 122.8, 121.5, 107.6, 29.8, 25.9, 23.0 ppm.

3.2.4. (Z)-1-Methyl-3-(1-(4-chlorophenyl)ethylidene)indolin-2-one ((Z)-5b)

75% Yield; yellow solid; mp = 141.5–143.1 °C; R_f = 0.42 (silica gel, hexane:EtOAc = 4:1); ^{1}H NMR (500 MHz, CDCl$_3$): δ 7.67 (d, J = 7.6 Hz, 1H), 7.44–7.38 (m, 2H), 7.34 (td, J = 7.7, 0.9 Hz, 1H), 7.29–7.26 (m, 2H), 7.11 (td, J = 7.7, 0.9 Hz, 1H), 6.86 (d, J = 7.7 Hz, 1H), 3.17 (s, 3H), 2.64 (s, 3H) ppm; ^{13}C NMR (125 MHz, CDCl$_3$) δ 166.3, 151.5, 143.5, 140.9, 134.1, 129.0, 128.8, 128.5, 124.2, 124.1, 123.4, 122.0, 108.0, 25.8, 25.4 ppm; HRMS (EI): calcd for C$_{17}$H$_{14}$ClNO [M$^+$]: 283.0764, found 283.0763.

3.2.5. (E)-1-Methyl-3-(1-(4-chlorophenyl)ethylidene)indolin-2-one ((E)-5b)

16% Yield; yellow solid; mp = 132.1–133.9 °C; R_f = 0.43 (silica gel, hexane:EtOAc = 4:1); ^{1}H NMR (500 MHz, CDCl$_3$): δ 7.50–7.42 (m, 2H), 7.25–7.21 (m, 2H), 7.16 (t, J = 7.2 Hz, 1H), 6.77 (d, J = 7.8 Hz, 1H), 6.69 (td, J = 7.7, 0.9 Hz, 1H), 6.22 (d, J = 7.7 Hz, 1H), 3.27 (s, 3H), 2.78 (s, 3H) ppm; ^{13}C NMR (125 MHz, CDCl$_3$) δ 166.3, 151.5, 143.5, 140.9, 134.1, 129.0, 128.8, 128.5, 124.2, 124.1, 123.4, 122.0, 108.0, 25.8, 25.4 ppm; HRMS (EI): calcd for C$_{17}$H$_{14}$ClNO [M$^+$]: 283.0764, found 283.0761.

3.2.6. (Z)-1-Methyl-3-(1-(4-nitrophenyl)ethylidene)indolin-2-one ((Z)-5c)

86% Yield; orange solid; mp = 189.8–192.2 °C; R_f = 0.26 (silica gel, hexane:EtOAc = 3:1); ^{1}H NMR (500 MHz, CDCl$_3$): δ 8.28 (d, J = 8.5 Hz, 2H), 7.67 (d, J = 7.6 Hz, 1H), 7.44 (d, J = 8.5 Hz, 2H), 7.35 (t, J = 7.8 Hz, 1H), 7.12 (t, J = 7.7 Hz, 1H), 6.85 (d, J = 7.8 Hz, 1H), 3.14 (s, 3H), 2.63 (s, 3H) ppm; ^{13}C NMR (125 MHz, CDCl$_3$) δ 166.2, 149.8, 149.1, 147.4, 143.8, 129.5, 128.4, 125.1, 124.3, 123.8, 122.8, 122.3, 108.3, 25.9, 24.8 ppm; HRMS (EI): calcd for C$_{17}$H$_{14}$N$_2$O$_3$ [M$^+$]: 294.1004, found 294.1001.

3.2.7. (Z)-1-Methyl-3-(1-(4-methoxyphenyl)ethylidene)indolin-2-one ((Z)-5d)

62% Yield; yellow solid; mp = 125.0–127.3 °C; R_f = 0.26 (silica gel, hexane:EtOAc: CH$_2$Cl$_2$ = 4:1:1); ^{1}H NMR (500 MHz, CDCl$_3$): δ 7.61 (d, J = 7.6 Hz, 1H), 7.31 (dd, J = 6.4, 4.8 Hz, 2H), 7.25 (t, J = 7.7 Hz, 1H), 7.05 (t, J = 7.6 Hz, 1H), 6.92 (t, J = 5.6 Hz, 2H), 6.79 (d, J = 7.7 Hz, 1H), 3.82 (s, 3H), 3.13 (s, 3H), 2.61 (s, 3H) ppm; ^{13}C NMR (125 MHz, CDCl$_3$) δ 166.5, 160.0, 153.4, 143.1, 134.2, 129.6, 128.2, 124.0, 123.9, 123.2, 121.8, 113.5, 107.8, 55.4, 25.8, 25.7 ppm; HRMS (EI): calcd for C$_{18}$H$_{17}$NO$_2$ [M$^+$]: 279.1259, found 279.1255.

3.2.8. (E)-1-Methyl-3-(1-(4-methoxyphenyl)ethylidene)indolin-2-one ((E)-5d)

15% Yield; gummy solid; R_f = 0.34 (silica gel, hexanes:EtOAc:CH$_2$Cl$_2$ = 4:1:1); ^{1}H NMR (500 MHz, CDCl$_3$): δ 7.26–7.22 (m, 2H), 7.14 (dd, J = 7.7, 7.0 Hz, 1H), 7.02–6.97 (m, 2H), 6.76 (d, J = 7.7 Hz, 1H), 6.67 (td, J = 7.7, 0.8 Hz, 1H), 6.36 (d, J = 7.7 Hz, 1H), 3.89 (s, 3H), 3.27 (s, 3H), 2.79 (s, 3H) ppm; ^{13}C NMR (125 MHz, CDCl$_3$) δ 168.4, 159.9, 155.2, 142.3, 135.3, 128.4, 128.0, 123.4, 123.0, 122.8, 121.4, 114.5, 107.6, 55.5, 25.9, 23.1 ppm; HRMS (EI): calcd for C$_{18}$H$_{17}$NO$_2$ [M$^+$]: 279.1259, found 279.1257.

3.2.9. (Z)-1-Methyl-3-(1-(4-iodophenyl)ethylidene)indolin-2-one ((Z)-5e)

81% Yield; yellow solid; mp = 132.5–134.0 °C; R_f = 0.64 (silica gel, hexane:EtOAc:CH$_2$Cl$_2$ = 4:1:1); ^{1}H NMR (500 MHz, CDCl$_3$): δ 7.78–7.71 (m, 2H), 7.65 (d, J = 7.6 Hz, 1H), 7.32 (td, J = 7.7, 0.9 Hz, 1H), 7.13–7.04 (m, 3H), 6.84 (d, J = 7.7 Hz, 1H), 3.15 (s, 3H), 2.61 (s, 3H) ppm; ^{13}C NMR (125 MHz, CDCl$_3$) δ 166.3, 151.4, 143.5, 142.1, 137.3, 129.4, 128.9, 124.12, 124.11,

123.4, 122.0, 108.0, 94.1, 25.8, 25.3 ppm; HRMS (EI): calcd for $C_{17}H_{14}INO$ [M$^+$] : 375.0120, found 375.0119.

3.2.10. (E)-1-Methyl-3-(1-(4-iodoxyphenyl)ethylidene)indolin-2-one ((**E**)-**5e**)

8% Yield; yellow solid; mp = 117.1–119.5 °C R_f = 0.61 (silica gel, hexane:EtOAc:CH$_2$Cl$_2$ = 4:1:1); ^{1}H NMR (500 MHz, CDCl$_3$): δ 7.82 (d, J = 8.4 Hz, 2H), 7.16 (td, J = 7.7, 1.0 Hz, 1H), 7.08–7.01 (m, 2H), 6.77 (d, J = 7.7 Hz, 1H), 6.69 (td, J = 7.7, 1.0 Hz, 1H), 6.24 (d, J = 7.7 Hz, 1H), 3.27 (s, 3H), 2.77 (s, 3H) ppm; ^{13}C NMR (125 MHz, CDCl$_3$) δ 168.1, 153.1, 142.5, 142.5, 138.5, 128.7, 128.5, 123.8, 122.9, 122.4, 121.6, 107.7, 94.2, 25.9, 22.8 ppm; HRMS (EI): calcd for $C_{17}H_{14}INO$ [M$^+$] : 375.0120, found 375.0120.

3.2.11. (Z)-3-(2-Bromo-1-phenylethylidene)-1-methylindolin-2-one (**6**)

To a stirred solution of 3-methyleneoxindole **5a** (40.4 mg, 0.162 mmol) were added NBS (34.6 mg, 0.194 mmol, 1.2 equiv.) and AIBN (16 μL, 8.1 μmol, 5 mol%) in anhydrous 1,2-dichloroethane(DCE) (6 mL). The reaction mixture was refluxed under argon atmosphere for 10 h, then cooled to rt and diluted with EtOAc (50 mL) and water (50 mL). The organic layer was separated, dried over Na$_2$SO$_4$, filtered, and concentrated under reduced pressure. The crude residue was purified by column chromatography to afford **6** (38.1 mg, 72% yield) as a yellow solid (mp = 161.5–163.1 °C). R_f = 0.53 (silica gel, hexane:EtOAc = 3:1); ^{1}H NMR (500 MHz, CDCl$_3$): δ 7.51 (dt, J = 4.8, 2.5 Hz, 3H), 7.40–7.35 (m, 2H), 7.18 (td, J = 7.7, 1.1 Hz, 1H), 6.75 (d, J = 7.8 Hz, 1H), 6.64 (td, J = 7.7, 0.9 Hz, 1H), 6.11 (d, J = 7.7 Hz, 1H), 5.20 (s, 2H), 3.27 (s, 3H) ppm; ^{13}C NMR (125 MHz, CDCl$_3$) δ 167.2, 149.5, 143.4, 139.1, 129.9, 129.4, 129.3, 129.2, 127.8, 123.9, 122.0, 121.9, 108.0, 31.8, 26.0 ppm; HRMS (EI): calcd for $C_{17}H_{14}BrNO$ [M$^+$] : 327.0259, found 327.0254.

3.3. *General Procedure of SeO2 Mediated Allylic Oxidation* (**7a–c, 7e**)

To a stirred solution of the corresponding 3-methyleneoxindole **5** (0.256 mmol) in *p*-xylene (3.0 mL) was added SeO$_2$ (170.4 mg, 2.0 mmol, 2.0 equiv.). The reaction mixture was refluxe for 24 h, then, the mixture was diluted with EtOAc (50 mL) and brine (30 mL). The organic layer was separated, dried over Na$_2$SO$_4$, filtered, and concentrated under reduced pressure. The crude residue was purified by column chromatography to afford aldehyde **7**.

3.3.1. (Z)-2-(1-Methyl-2-oxoindolin-3-ylidene)-2-phenylacetaldehyde (**7a**) [23]

84% Yield; orange solid; mp = 187.8–190.7 °C (lit. [23] 174-176 °C); R_f = 0.35 (silica gel, hexanes:EtOAc = 3:1); ^{1}H NMR (500 MHz, CDCl$_3$): δ 11.40 (s, 1H), 7.49 (dd, J = 3.7, 2.7 Hz, 3H), 7.28–7.19 (m, 3H), 6.77 (d, J = 7.8 Hz, 1H), 6.68 (td, J = 7.7, 0.9 Hz, 1H), 6.26 (d, J = 7.7 Hz, 1H), 3.25 (s, 3H) ppm; ^{13}C NMR (125 MHz, CDCl$_3$) δ 193.3, 167.3, 145.8, 145.6, 135.5, 133.3, 132.2, 129.4, 129.2, 128.9, 126.2, 122.6, 121.4, 108.6, 26.2 ppm; HRMS (EI): calcd for $C_{17}H_{13}NO_2$ [M$^+$] : 263.0946, found 263.0942.

3.3.2. (Z)-2-(4-Chlorophenyl)-2-(1-methyl-2-oxoindolin-3-ylidene)acetaldehyde (**7b**)

85% Yield; orange solid; mp = 186.8–189.1 °C; R_f = 0.32 (silica gel, hexane:EtOAc = 3:1); ^{1}H NMR (500 MHz, CDCl$_3$): δ 11.38 (s, 1H), 7.52–7.45 (m, 2H), 7.29 (td, J = 7.8, 1.1 Hz, 1H), 7.20–7.16 (m, 2H), 6.79 (d, J = 7.8 Hz, 1H), 6.74 (td, J = 7.7, 0.9 Hz, 1H), 6.36 (d, J = 7.4 Hz, 1H), 3.27 (s, 3H) ppm; ^{13}C NMR (125 MHz, CDCl$_3$) δ 192.9, 167.1, 145.7, 144.4, 135.9, 135.6, 132.6, 131.6, 130.6, 129.6, 126.1, 122.7, 121.1, 108.8, 26.3 ppm; HRMS (EI): calcd for $C_{17}H_{12}ClNO_2$ [M$^+$] : 297.0557, found 297.0556.

3.3.3. (Z)-2-(1-Methyl-2-oxoindolin-3-ylidene)-2-(4-nitrophenyl)acetaldehyde (**7c**)

87% Yield; red solid; mp = 194.2–196.8 °C; R_f = 0.34 (silica gel, hexane:EtOAc = 4:1); ^{1}H NMR (500 MHz, CDCl$_3$): δ 11.41 (s, 1H), 8.36 (d, J = 8.5 Hz, 2H), 7.43 (d, J = 8.5 Hz, 2H), 7.31 (t, J = 7.7 Hz, 1H), 6.81 (d, J = 7.8 Hz, 1H), 6.71 (t, J = 7.7 Hz, 1H), 6.20 (d, J = 7.7 Hz, 1H), 3.27 (s, 3H) ppm; ^{13}C NMR (125 MHz, CDCl$_3$) δ 192.1, 166.8, 148.4, 146.1, 142.9, 140.2, 136.4,

133.2, 130.4, 126.1, 124.4, 122.9, 120.6, 109.1, 26.3 ppm; HRMS (EI): calcd for $C_{17}H_{12}N_2O_4$ [M$^+$] : 308.0797, found 308.0797.

3.3.4. (Z)-2-(4-Iodophenyl)-2-(1-methyl-2-oxoindolin-3-ylidene)acetaldehyde (7e)

78% Yield; orange solid; mp = 157.2–159.5 °C; R_f = 0.20 (silica gel, hexane:EtOAc = 3:1); ^{1}H NMR (500 MHz, CDCl$_3$): δ 11.36 (s, 1H), 7.83 (d, J = 8.2 Hz, 2H), 7.29 (td, J = 7.8, 0.8 Hz, 1H), 6.98 (d, J = 8.2 Hz, 2H), 6.78 (d, J = 7.8 Hz, 1H), 6.74 (t, J = 7.7 Hz, 1H), 6.38 (d, J = 7.7 Hz, 1H), 3.25 (s, 3H) ppm; ^{13}C NMR (125 MHz, CDCl$_3$) δ 192.7, 167.0, 145.7, 144.3, 138.3, 135.7, 132.7, 132.6, 130.9, 126.1, 122.7, 121.0, 108.7, 95.7, 26.2 ppm; HRMS (EI): calcd for $C_{17}H_{12}INO_2$ [M$^+$] : 388.9913, found 388.9908.

3.4. General Procedure of Wittig Reaction (3aa-ad, 3ba-bd, 3ca-cd, 3ea)

To a stirred solution of the corresponding benzyltriphenylphosphonium bromide (0.098 mmol, 1.5 equiv.) in THF (2.0 mL) was slowly added *n*-butyllithium (2.5 M in hexane, 0.036 mL, 0.091 mmol, 1.4 equiv.) at 0 °C under argon atmosphere. After stirring for 1 h at the same temperature, a solution of the corresponding aldehyde 7 (0.065 mmol, 1.0 equiv.) in THF (1.0 mL) was added. Then, the reaction temperature was raised to 50 °C. After 4 h, the reaction mixture was cooled to rt and diluted with sat. aq. NH$_4$Cl (30 mL) and EtOAc (70 mL). The organic layer was separated, dried over Na$_2$SO$_4$, filtered, and concentrated under reduced pressure. The crude residue was purified by column chromatography to afford 3-(1,3-diarylallyliden)oxindole 3.

3.4.1. (E)-3-((E)-1,3-Diphenylallylidene)-1-methylindolin-2-one (3aa)

84% Yield; yellow solid; mp = 125.7 °C; R_f = 0.4 (silica gel, hexane:EtOAc = 4:1); ^{1}H NMR (500 MHz, CDCl$_3$): δ 9.38 (d, J = 16.0 Hz, 1H), 7.55–7.51 (m, 5H), 7.33–7.25 (m, 5H), 7.08 (td, J = 8.2, 7.6 Hz, 1H), 6.74 (d, J = 7.8 Hz, 1H), 6.60 (t, J = 7.7 Hz, 1H), 6.47 (d, J = 16.0 Hz, 1H), 5.72 (d, J = 7.8 Hz, 1H), 3.3 (s, 3H) ppm; ^{13}C NMR (125 MHz, CDCl$_3$): δ 168.1, 151.2, 142.9, 141.6, 137.8, 136.9, 129.3, 129.2, 128.8, 128.7, 128.6, 128.3, 128.0, 127.7, 123.7, 123.4, 122.6, 121.6, 107.6, 25.9 ppm; HRMS (EI): calcd for $C_{24}H_{19}NO$ [M$^+$]: 337.1467, found 337.1466.

3.4.2. (E)-3-((E)-3-(4-Chlorophenyl)-1-phenylallylidene)-1-methylindolin-2-one (3ab)

67% Yield; yellow solid; mp = 123.2 °C; R_f = 0.44 (silica gel, hexane:EtOAc = 3:1); ^{1}H NMR (500 MHz, CDCl$_3$): δ 9.36 (d, J = 16 Hz, 1H), 7.57–7.53 (m, 3H), 7.44 (dt, J = 13.3, 2.3 Hz, 2H), 7.29–7.26 (m, 4H), 7.10 (td, J = 7.7, 1.1 Hz, 1H), 6.75 (d, J = 7.7 Hz, 1H), 6.59 (td, J = 7.7, 1.0 Hz, 1H), 6.40 (d, J = 16.0 Hz, 1H), 5.72 (d, J = 7.8 Hz, 1H), 3.30 (s, 3H) ppm; ^{13}C NMR (125 MHz, CDCl$_3$): δ 168.1, 150.8, 143.0, 139.9, 137.5, 135.5, 134.8, 129.4, 129.1, 129.0, 128.7, 128.6, 128.5, 128.2, 123.8, 123.3, 123.0, 121.7, 107.7, 25.9 ppm; HRMS (EI): calcd for $C_{24}H_{18}ClNO$ [M$^+$]: 371.1077, found 371.1077.

3.4.3. (E)-1-Methyl-3-((E)-3-(4-nitrophenyl)-1-phenylallylidene)indolin-2-one (3ac)

64% Yield; red solid; mp = 183.3–184.7 °C; R_f = 0.32 (silica gel, hexane:EtOAc = 8:1); ^{1}H NMR (500 MHz, CDCl$_3$): δ 9.52 (d, J = 16.0 Hz, 1H), 8.22–8.10 (m, 2H), 7.64 (d, J = 8.8 Hz, 2H), 7.61–7.53 (m, 3H), 7.31–7.26 (m, 2H), 7.15 (td, J = 7.7, 1.1 Hz, 1H), 6.76 (d, J = 7.7 Hz, 1H), 6.62 (td, J = 7.7, 0.9 Hz, 1H), 6.46 (d, J = 16.0 Hz, 1H), 5.76 (d, J = 7.5 Hz, 1H), 3.30 (s, 3H) ppm; ^{13}C NMR (125 MHz, CDCl$_3$): δ 168.0, 149.5, 147.5, 143.4, 137.9, 137.0, 131.8, 129.6, 129.3, 128.9, 128.6, 128.3, 124.9, 124.21, 124.16, 123.0, 121.9, 107.9, 25.9 ppm; HRMS (EI): calcd for $C_{24}H_{18}N_2O_3$ [M$^+$] : 382.1317, found 382.1316.

3.4.4. (E)-3-((E)-3-(4-Methoxyphenyl)-1-phenylallylidene)-1-methylindolin-2-one (3ad)

53% Yield; yellow solid; mp = 135.5 °C; R_f = 0.2 (silica gel, hexane:EtOAc = 5:1); ^{1}H NMR (500 MHz, CDCl$_3$): δ 9.27 (d, J = 16.0 Hz, 1H), 7.55–7.52 (m, 3H), 7.48 (d, J = 8.7 Hz, 2H), 7.28–7.27 (m, 2H), 7.08 (td, J = 7.7, 1.1 Hz, 1H), 6.84 (dt, J = 14.3, 2.9 Hz, 2H), 6.75 (d, J = 7.7 Hz, 1H), 6.59 (td, J = 7.7, 1.0 Hz, 1H), 6.44 (d, J = 16 Hz, 1H), 5.70 (d, J = 7.7 Hz, 1H),

3.82 (s, 3H), 3.30(s, 3H) ppm; ^{13}C NMR (125 MHz, CDCl$_3$): δ 168.2, 160.7, 151.9, 142.7, 141.5, 137.9, 129.9, 129.6, 129.3, 128.7, 128.5, 128.0, 125.7, 123.6, 123.5, 121.5, 114.3, 107.5, 55.5, 25.9 ppm; HRMS (EI): calcd for C$_{25}$H$_{21}$NO$_2$ [M$^+$]: 367.1572, found 367.1572.

3.4.5. (E)-3-((E)-1-(4-Chlorophenyl)-3-phenylallylidene)-1-methylindolin-2-one (**3ba**)

95% Yield; yellow solid; mp = 178.2 °C; R_f = 0.3 (silica gel, hexane:EtOAc = 6:1); ^{1}H NMR (500 MHz, CDCl$_3$): δ 9.37 (d, J = 16.1 Hz, 1H), 7.56–7.53 (m, 4H), 7.33 (t, J = 7.3 Hz, 2H), 7.29 (d, J = 7.1 Hz, 1H), 7.24 (dt, J = 12.8, 2.2 Hz, 2H), 7.13 (td, J = 7.7, 1.0 Hz, 1H), 6.76 (d, J = 7.7 Hz, 1H), 6.66 (td, J = 7.7, 0.9 Hz, 1H), 6.43 (d, J = 16.1 Hz, 1H), 5.86 (d, J = 7.7 Hz, 1H), 3.29 (s, 3H) ppm; ^{13}C NMR (125 MHz, CDCl$_3$): δ 167.9, 149.6, 143.0, 141.4, 136.8, 136.1, 134.6, 130.3, 129.7, 129.3, 128.9, 128.6, 128.0, 127.5, 123.5, 123.1, 122.8, 121.7, 107.8, 25.9 ppm; HRMS (EI): calcd for C$_{24}$H$_{18}$ClNO [M$^+$]: 371.1077, found 371.1078.

3.4.6. (E)-3-((E)-1,3-Bis(4-chlorophenyl)allylidene)-1-methylindolin-2-one (**3bb**)

79% Yield; yellow solid; mp = 193.0 °C; R_f = 0.27 (silica gel, hexane:EtOAc = 8:1); ^{1}H NMR (500 MHz, CDCl$_3$): δ 9.33 (d, J = 16.1 Hz, 1H), 7.55 (d, J = 8.3 Hz, 2H), 7.44 (d, J = 8.4 Hz, 2H), 7.28 (d, J = 8.4 Hz, 2H), 7.23 (d, J = 8.3 Hz, 2H), 7.15 (t, J = 7.7 Hz, 1H), 6.76 (d, J = 7.8 Hz, 1H), 6.67 (t, J = 7.7 Hz, 1H), 6.35 (d, J = 16.1 Hz, 1H), 5.86 (d, J = 7.7 Hz, 1H), 3.28 (s, 3H) ppm; ^{13}C NMR (125 MHz, CDCl$_3$): δ 167.9, 167.2, 163.1, 139.7, 135.9, 135.3, 135.0, 134.8, 130.3, 129.8, 129.1, 128.8, 128.0, 123.7, 123.2, 123.0, 121.8, 107.8, 25.9 ppm; HRMS (EI): calcd for C$_{24}$H$_{17}$Cl$_2$NO [M$^+$]: 405.0687, found 405.0689.

3.4.7. (E)-3-((E)-1-(4-Chlorophenyl)-3-(4-nitrophenyl)allylidene)-1-methylindolin-2-one (**3bc**)

82% Yield; red solid; mp = >250 °C; R_f = 0.50 (silica gel, hexane:EtOAc:CH$_2$Cl$_2$ = 4:1:1); ^{1}H NMR (500 MHz, CDCl$_3$): δ 9.53 (d, J = 16.2 Hz, 1H), 8.47 (d, J = 8.1 Hz, 2H), 8.18 (d, J = 8.3 Hz, 2H), 7.63 (d, J = 8.3 Hz, 2H), 7.53 (d, J = 8.1 Hz, 2H), 7.20 (t, J = 7.6 Hz, 1H), 6.80 (d, J = 7.6 Hz, 1H), 6.66 (t, J = 7.6 Hz, 1H), 6.30 (d, J = 16.2 Hz, 1H), 5.74 (d, J = 7.6 Hz, 1H), 3.31 (s, 3H) ppm; ^{13}C NMR (125 MHz, CDCl$_3$): δ 167.5, 148.3, 147.8, 146.3, 143.9, 143.7, 142.8, 137.8, 130.8, 130.1, 128.3, 125.2, 125.0, 124.3, 123.8, 122.19, 122.15, 108.4, 26.1 ppm; HRMS (EI): calcd for C$_{24}$H$_{17}$N$_3$O$_5$ [M$^+$] : 427.1168, found 427.1169.

3.4.8. (E)-3-((E)-1-(4-Chlorophenyl)-3-(4-methoxyphenyl)allylidene)-1-methylindolin-2-one (**3bd**)

68% Yield; yellow solid; mp = 129.0 °C; R_f = 0.38 (silica gel, CH$_2$Cl$_2$:hexane 5:1); ^{1}H NMR (500 MHz, CDCl$_3$): δ 9.24 (d, J = 16 Hz, 1H), 7.53 (dt, J = 13.1, 2.2 Hz, 2H), 7.48 (d, J = 8.7 Hz, 2H), 7.22 (dt, J = 12.6, 2.1 Hz, 2H), 7.11 (td, J = 7.4, 1.1 Hz, 1H), 6.85 (dt, J = 14.4, 2.4 Hz, 2H), 6.76 (d, J = 7.7 Hz, 1H), 6.64 (td, J = 7.7, 1.0, 1H), 6.38 (d, J = 16 Hz, 1H), 5.83 (d, J = 7.5 Hz, 1H), 3.82 (s, 3H), 3.30 (s, 3H) ppm; ^{13}C NMR (125 MHz, CDCl$_3$): δ 168.0, 160.8, 150.2, 142.8, 141.4, 136.4, 134.5, 130.3, 129.73, 129.65, 129.6, 128.3, 125.5, 123.3, 121.6, 114.4, 107.7, 55.5, 25.9. ppm; HRMS (EI): calcd for C$_{25}$H$_{20}$ClNO$_2$ [M$^+$]: 401.1183, found 401.1185.

3.4.9. (E)-1-Methyl-3-((E)-1-(4-nitrophenyl)-3-phenylallylidene)indolin-2-one (**3ca**)

80% Yield; yellow solid; mp = 228.2 °C; R_f = 0.3 (silica gel, hexanes:EtOAc = 3:1); ^{1}H NMR (500 MHz, CDCl$_3$): δ 9.39 (d, J = 16.2 Hz, 1H), 8.45 (d, J = 8.8 Hz, 2H), 7.53–7.50 (m, 4H), 7.35–7.30 (m, 3H), 7.15 (td, J = 7.7, 1.1 Hz, 1H), 6.78 (d, J = 7.7 Hz, 1H), 6.62 (td, J = 7.7, 1.0 Hz, 1H), 6.30 (d, J = 16.2 Hz, 1H), 5.70 (d, J = 7.5 Hz, 1H), 3.31 (s, 3H) ppm; ^{13}C NMR (125 MHz, CDCl$_3$): δ 167.7, 148.1, 148.0, 144.7, 143.2, 141.5, 136.5, 130.2, 129.6, 129.2, 123.0, 128.0, 126.8, 124.8, 123.3, 122.9, 122.6, 121.9, 108.1, 26.0 ppm; HRMS (EI): calcd for C$_{24}$H$_{18}$N$_2$O$_3$ [M$^+$]: 382.1317, found 382.1318.

3.4.10. (E)-3-((E)-3-(4-Chlorophenyl)-1-(4-nitrophenyl)allylidene)-1-methylindolin-2-one (**3cb**)

79% Yield; orange solid; mp = 249.5 °C; R_f = 0.34 (silica gel, hexanes:EtOAc = 5:1); ^{1}H NMR (500 MHz, CDCl$_3$): δ 9.37 (d, J = 16.2 Hz, 1H), 8.44 (dt, J = 12.9, 2.3 Hz, 2H), 7.50 (dt, J = 13.0, 2.3 Hz, 2H), 7.43 (d, J = 8.5 Hz, 2H), 7.30 (d, J = 8.6 Hz, 2H), 7.15 (td, J = 7.7, 0.9 Hz, 1H), 6.78 (d, J = 7.8 Hz, 1H), 6.62 (td, J = 7.7, 0.8 Hz, 1H), 6.23 (d, J = 16.2 Hz, 1H),

5.71 (d, J = 7.7 Hz, 1H), 3.30 (s, 3H) ppm; ^{13}C NMR (125 MHz, CDCl$_3$): δ 167.6, 148.2, 147.5, 144.4, 143.3, 139.8, 135.4, 135.0, 130.2, 129.4, 129.3, 129.2, 129.1, 128.3, 127.3, 127.1, 124.8, 124.3, 123.4, 123.3, 122.5, 122.0, 108.2, 26.0 ppm; HRMS (EI): calcd for C$_{24}$H$_{17}$ClN$_2$O$_3$ [M$^+$]: 416.0928, found 416.0928.

3.4.11. (E)-3-((E)-1,3-Bis(4-nitrophenyl)allylidene)-1-methylindolin-2-one (**3cc**)

68% Yield; orange solid; mp = 243.5–245.2 °C; R_f = 0.38 (silica gel, hexanes:EtOAc = 5:1); ^{1}H NMR (500 MHz, CDCl$_3$): δ 9.49 (d, J = 16.1 Hz, 1H), 8.16 (d, J = 8.7 Hz, 2H), 7.63 (d, J = 8.7 Hz, 2H), 7.57 (d, J = 8.2 Hz, 2H), 7.24 (d, J = 8.2 Hz, 2H), 7.17 (d, J = 7.6 Hz, 1H), 6.77 (d, J = 7.7 Hz, 1H), 6.68 (t, J = 7.6 Hz, 1H), 6.41 (d, J = 16.1 Hz, 1H), 5.88 (d, J = 7.7 Hz, 1H), 3.29 (s, 3H) ppm; ^{13}C NMR (125 MHz, CDCl$_3$): δ 167.8, 148.0, 147.6, 143.5, 143.1, 137.8, 135.4, 135.0, 131.5, 130.2, 130.0, 129.6, 128.3, 125.1, 124.2, 124.1, 122.7, 122.1, 108.1, 26.0 ppm; HRMS (EI): calcd for C$_{24}$H$_{17}$ClN$_2$O$_3$ [M$^+$]: 416.0928, found 416.0923.

3.4.12. (E)-3-((E)-3-(4-Methoxyphenyl)-1-(4-nitrophenyl)allylidene)-1-methylindolin-2-one (**3cd**)

82% Yield; orange solid; mp = 197.0 °C; R_f = 0.23 (silica gel, hexanes:EtOAc = 4:1); ^{1}H NMR (500 MHz, CDCl$_3$): δ 9.27 (d, J = 16.2 Hz, 1H), 8.43 (dt, J = 13.0, 2.3 Hz, 2H), 7.50 (dt, J = 13, 2.3 Hz, 2H), 7.46 (d, J = 8.8 Hz, 2H), 7.13 (td, J = 7.7, 0.9 Hz, 1H), 6.9 (d, J = 8.8 Hz, 2H), 6.78 (d, J = 7.8 Hz, 1H), 6.60 (td, J = 7.7, 0.8 Hz, 1H), 6.26 (d, J = 16.2 Hz, 1H), 5.68 (,d J = 7.7 Hz, 1H), 3.82 (s, 3H), 3.30 (s, 3H) ppm; ^{13}C NMR (125 MHz, CDCl$_3$): δ 167.8, 161.1, 148.6, 148.1, 144.9, 143.0, 141.4, 130.2, 129.7, 129.4, 128.8, 124.8, 124.7, 123.1, 122.8, 121.8, 121.7, 114.5, 108.0, 55.5, 25.9 ppm; HRMS (EI): calcd for C$_{25}$H$_{20}$N$_2$O$_4$ [M$^+$]: 412.1423, found 412.1422.

3.4.13. (E)-3-((E)-1-(4-Iodophenyl)-3-phenylallylidene)-1-methylindolin-2-one (**3ea**)

80% Yield; orange solid; mp = 182.7–184.0 °C; R_f = 0.31 (silica gel, hexanes:EtOAc = 6:1); ^{1}H NMR (500 MHz, CDCl$_3$): δ 9.35 (d, J = 16.1 Hz, 1H), 7.93–7.85 (m, 2H), 7.56–7.50 (m, 2H), 7.36–7.31 (m, 2H), 7.31–7.27 (m, 1H), 7.15 (td, J = 7.7, 1.1 Hz, 1H), 7.07–7.02 (m, 2H), 6.76 (d, J = 7.7 Hz, 1H), 6.68 (td, J = 7.7, 1.0 Hz, 1H), 6.42 (d, J = 16.1 Hz, 1H), 5.86 (d, J = 7.4 Hz, 1H), 3.30 (s, 3H) ppm; ^{13}C NMR (125 MHz, CDCl$_3$): δ 167.9, 149.7, 143.0, 141.5, 138.6, 137.2, 136.8, 130.8, 129.3, 128.9, 128.7, 128.0, 127.3, 123.6, 123.1, 122.7, 121.8, 107.8, 94.5, 25.9 ppm; HRMS (EI): calcd for C$_{24}$H$_{18}$INO [M$^+$] : 463.0433, found 463.0430.

3.5. General Procedure of Aldol Reaction (**8aa**, **8da-dd**)

To a stirred solution of the corresponding 3-methyleneoxindole **5** (0.20 mmol) in THF (2.0 mL) was slowly added *n*-butyllithium (2.5 M in hexane, 0.088 mL, 0.22 mmol, 1.1 equiv.) at −78 °C under argon atmosphere. After stirring for 1 h at the same temperature, the corresponding benzaldehyde (0.26 mmol, 1.3 equiv.) was added. After stirring for additional 1.5 h at −78 °C, the reaction mixture was diluted with sat. aq. NH$_4$Cl (10 mL) and EtOAc (50 mL). The organic layer was separated, dried over Na$_2$SO$_4$, filtered, and concentrated under reduced pressure. The crude residue was purified by column chromatography to afford aldol adduct **8**.

3.5.1. (E)-3-(3-Hydroxy-1,3-diphenylpropylidene)-1-methylindolin-2-one (**8aa**)

91% Yield; yellow foam; R_f = 0.37 (silica gel, hexane:EtOAc = 2:1); ^{1}H NMR (500 MHz, CDCl$_3$): δ 7.57–7.43 (m, 4H), 7.40 (d, J = 7.4 Hz, 2H), 7.31 (t, J = 7.5 Hz, 2H), 7.27–7.19 (m, 2H), 7.15 (t, J = 7.6 Hz, 1H), 6.79 (d, J = 7.7 Hz, 1H), 6.66 (t, J = 7.6 Hz, 1H), 6.14 (d, J = 7.7 Hz, 1H), 4.82 (d, J = 9.8 Hz, 1H), 4.50 (s, 1H), 4.42 (dd, J = 12.6, 11.0 Hz, 1H), 3.30 (s, 3H), 2.88 (dd, J = 12.8, 2.8 Hz, 1H) ppm; ^{13}C NMR (125 MHz, CDCl$_3$): δ 169.6, 155.6, 145.5, 142.3, 141.3, 129.6, 129.1, 128.9, 128.7, 128.5, 127.3, 126.5, 126.3, 125.7, 123.0, 122.8, 122.1, 108.0, 73.7, 46.0, 26.2 ppm; HRMS (EI): calcd for C$_{24}$H$_{21}$NO$_2$ [M$^+$] : 355.1572, found 355.1568.

3.5.2. (E)-3-(3-Hydroxy-1-(4-methoxyphenyl)-3-phenylpropylidene)-1-methylindolin-2-one (**8da**)

93% Yield; yellow foam; R_f = 0.18 (silica gel, hexane:EtOAc:CH$_2$Cl$_2$ = 4:1:1); ^{1}H NMR (500 MHz, CDCl$_3$): δ 7.42 (dd, J = 13.8, 7.3 Hz, 3H), 7.33 (dd, J = 10.4, 4.8 Hz, 2H), 7.26–7.19 (m, 2H), 7.19–7.13 (m, 1H), 7.09–6.96 (m, 2H), 6.81 (d, J = 7.7 Hz, 1H), 6.71 (td, J = 7.7, 1.0 Hz, 1H), 6.38 (d, J = 7.4 Hz, 1H), 4.82–4.73 (m, 1H), 4.65 (d, J = 7.7 Hz, 1H), 4.39 (dd, J = 12.8, 10.8 Hz, 1H), 3.92 (s, 3H), 3.32 (s, 3H), 2.88 (dd, J = 12.8, 2.9 Hz, 1H) ppm; ^{13}C NMR (125 MHz, CDCl$_3$): δ 169.8, 160.4, 156.1, 145.7, 142.2, 133.2, 130.7, 128.53, 128.48, 128.0, 127.3, 126.3, 125.7, 123.1, 122.8, 122.1, 115.1, 114.3, 108.1, 74.1, 55.5, 46.3, 26.2 ppm; HRMS (EI): calcd for C$_{25}$H$_{21}$NO$_2$ [M-H$_2$O]$^+$: 367.1572, found 367.1567.

3.5.3. (E)-3-(3-(4-Chlorophenyl)-3-hydroxy-1-(4-methoxyphenyl)propylidene)-1-methylindolin-2-one (**8db**)

75% Yield; orange foam; R_f = 0.38 (silica gel, hexane:EtOAc:CH$_2$Cl$_2$ = 4:1:1); ^{1}H NMR (500 MHz, CDCl$_3$): δ 7.39 (d, J = 7.8 Hz, 1H), 7.35–7.22 (m, 4H), 7.18 (dd, J = 15.7, 8.0 Hz, 2H), 7.00 (dd, J = 22.8, 8.2 Hz, 2H), 6.80 (d, J = 7.7 Hz, 1H), 6.71 (t, J = 7.6 Hz, 1H), 6.38 (d, J = 7.7 Hz, 1H), 4.78 (s, 2H), 4.32–4.19 (m, 1H), 3.90 (s, 3H), 3.30 (s, 3H), 2.97–2.88 (m, 1H) ppm; ^{13}C NMR (125 MHz, CDCl$_3$): δ 169.7, 160.4, 155.6, 144.1, 142.2, 133.0, 132.8, 130.5, 128.6, 128.5, 128.1, 127.1, 126.4, 123.0, 122.8, 122.2, 115.0, 114.3, 108.1, 73.4, 55.5, 46.2, 26.2 ppm; HRMS (EI): calcd for C$_{25}$H$_{20}$ClNO$_2$ [M-H$_2$O]$^+$: 401.1183, found 401.1183.

3.5.4. (E)-3-(3-Hydroxy-1-(4-methoxyphenyl)-3-(4-nitrophenyl)propylidene)-1-methylindolin-2-one (**8dc**)

86% Yield; orange foam; R_f = 0.20 (silica gel, hexane:EtOAc:CH$_2$Cl$_2$ = 4:1:1); ^{1}H NMR (500 MHz, CDCl$_3$): δ 8.14 (d, J = 8.5 Hz, 2H), 7.54 (d, J = 8.4 Hz, 2H), 7.39 (d, J = 7.7 Hz, 1H), 7.19 (t, J = 7.5 Hz, 2H), 7.00 (t, J = 9.6 Hz, 2H), 6.82 (d, J = 7.7 Hz, 1H), 6.73 (t, J = 7.6 Hz, 1H), 6.41 (d, J = 7.7 Hz, 1H), 5.18 (d, J = 6.7 Hz, 1H), 5.00–4.86 (m, 1H), 4.17 (dd, J = 12.7, 10.4 Hz, 1H), 3.90 (s, 3H), 3.32 (s, 3H), 3.05 (dd, J = 12.8, 2.5 Hz, 1H) ppm; ^{13}C NMR (125 MHz, CDCl$_3$): δ 169.8, 160.6, 154.7, 153.0, 147.1, 142.2, 132.8, 130.5, 128.8, 128.3, 126.8, 126.5, 123.7, 122.83, 122.81, 122.4, 115.0, 114.5, 108.3, 73.4, 55.5, 45.9, 26.3 ppm; HRMS (EI): calcd for C$_{25}$H$_{20}$N$_2$O$_4$ [M-H$_2$O]$^+$: 412.1423, found 412.1421.

3.5.5. (E)-3-(3-Hydroxy-1,3-bis(4-methoxyphenyl)propylidene)-1-methylindolin-2-one (**8dd**)

63% Yield; orange foam; R_f = 0.14 (silica gel, hexane:EtOAc:CH$_2$Cl$_2$ = 4:1:1); ^{1}H NMR (500 MHz, CDCl$_3$): δ 7.41 (d, J = 8.0 Hz, 1H), 7.33 (d, J = 8.6 Hz, 2H), 7.21 (d, J = 7.7 Hz, 1H), 7.16 (t, J = 7.4 Hz, 1H), 7.02 (dd, J = 25.1, 8.3 Hz, 2H), 6.86 (d, J = 8.6 Hz, 2H), 6.79 (d, J = 7.7 Hz, 1H), 6.71 (t, J = 7.6 Hz, 1H), 6.37 (d, J = 7.7 Hz, 1H), 4.73 (d, J = 9.8 Hz, 1H), 4.48 (s, 1H), 4.37 (dd, J = 12.7, 10.7 Hz, 1H), 3.91 (s, 3H), 3.79 (s, 3H), 3.31 (s, 3H), 2.90 (dd, J = 12.8, 3.0 Hz, 1H) ppm; ^{13}C NMR (125 MHz, CDCl$_3$): δ 169.7, 160.3, 158.9, 156.2, 142.2, 137.9, 133.2, 130.6, 128.5, 128.1, 126.9, 126.2, 123.1, 122.8, 122.1, 115.0, 114.3, 113.9, 108.0, 73.6, 55.5, 55.4, 46.2, 26.2 ppm; HRMS (EI): calcd for C$_{26}$H$_{23}$NO$_3$ [M-H$_2$O]$^+$: 397.1678, found 397.1678.

3.6. General Procedure of Dehydration under Acidic Conditions (**3aa, 3da-dd**)

To a stirred solution of the corresponding alcohol **8** (0.20 mmol) in CH$_2$Cl$_2$ (2.4 mL) was added trifluoroacetic acid (TFA) (0.3 mL) at rt. After 1 h, volatile material was distilled off under reduced pressure. The crude residue was purified by column chromatography to afford 3-(1,3-diarylallyliden)oxindole **3**.

3.6.1. (E)-3-((E)-1,3-Diphenylallylidene)-1-methylindolin-2-one (**3aa**)

91% Yield; yellow foam; R_f = 0.37 (silica gel, hexane:EtOAc = 2:1); ^{1}H NMR (500 MHz, CDCl$_3$): δ 7.57–7.43 (m, 4H), 7.40 (d, J = 7.4 Hz, 2H), 7.31 (t, J = 7.5 Hz, 2H), 7.27–7.19 (m, 2H), 7.15 (t, J = 7.6 Hz, 1H), 6.79 (d, J = 7.7 Hz, 1H), 6.66 (t, J = 7.6 Hz, 1H), 6.14 (d, J = 7.7 Hz, 1H), 4.82 (d, J = 9.8 Hz, 1H), 4.50 (s, 1H), 4.42 (dd, J = 12.6, 11.0 Hz, 1H), 3.30 (s, 3H), 2.88

(dd, *J* = 12.8, 2.8 Hz, 1H) ppm; ^{13}C NMR (125 MHz, CDCl$_3$): δ 169.6, 155.6, 145.5, 142.3, 141.3, 129.6, 129.1, 128.9, 128.7, 128.5, 127.3, 126.5, 126.3, 125.7, 123.0, 122.8, 122.1, 108.0, 73.7, 46.0, 26.2 ppm; HRMS (EI): calcd for C$_{24}$H$_{21}$NO$_2$ [M$^+$] : 355.1572, found 355.1568.

3.6.2. (E)-3-((E)-1-(4-Methoxyphenyl)-3-phenylallylidene)-1-methylindolin-2-one (**3da**)

90% Yield; yellow solid; mp = 153.6 °C; R_f = 0.3 (silica gel, hexane:EtOAc = 4:1); ^{1}H NMR (500 MHz, CDCl$_3$): δ 9.36 (d, *J* = 16 Hz, 1H), 7.55 (d, *J* = 7.4 Hz, 2H), 7.32 (t, *J* = 7.4 Hz, 2H), 7.27 (d, *J* = 5.8 Hz, 1H), 7.21 (d, *J* = 8.5 Hz, 2H), 7.12 (t, *J* = 7.6 Hz, 1H), 7.08 (d, *J* = 8.5 Hz, 2H), 6.75 (d, *J* = 7.7 Hz, 1H), 6.65 (t, *J* = 7.6 Hz, 1H), 6.53 (d, *J* = 16 Hz, 1H), 5.92 (d, *J* = 7.7 Hz, 1H), 3.94 (s, 3H), 3.30 (s, 3H) ppm; ^{13}C NMR (125 MHz, CDCl$_3$): δ 168.1, 159.8, 151.2, 142.8, 141.4, 137.0, 130.1, 129.9, 129.1, 128.8, 128.21, 128.19, 128.0, 123.64, 123.58, 122.8, 121.6, 114.7, 107.6, 55.5, 25.8 ppm; HRMS (EI): calcd for C$_{25}$H$_{21}$NO$_2$ [M$^+$]: 367.1572, found 367.1571.

3.6.3. (E)-3-((E)-3-(4-Chlorophenyl)-1-(4-methoxyphenyl)allylidene)-1-methylindolin-2-one (**3db**)

93% Yield; yellow solid; mp = 143.1 °C; R_f = 0.36 (silica gel, hexane:EtOAc = 5:1); ^{1}H NMR (500 MHz, CDCl$_3$): δ 9.31 (d, *J* = 16.0 Hz, 1H), 7.45 (d, *J* = 8.5 Hz, 2H), 7.27 (d, *J* = 8.6 Hz, 2H), 7.19 (d, *J* = 8.6 Hz, 2H), 7.12 (t, *J* = 7.7 Hz, 1H), 7.07 (d, *J* = 8.6 Hz, 2H), 6.74 (d, *J* = 7.8 Hz, 1H), 6.65 (t, *J* = 7.8 Hz, 1H), 6.45 (d, *J* = 16.0 Hz, 1H), 5.92 (d, *J* = 7.7 Hz, 1H), 3.93 (s, 3H), 3.28 (s, 3H) ppm; ^{13}C NMR (125 MHz, CDCl$_3$): δ 168.1, 159.9, 150.7, 142.9, 139.7, 135.6, 134.7, 130.1, 129.6, 129.04, 128.97, 128.7, 128.4, 123.7, 123.5, 123.2, 121.6, 114.7, 107.6, 55.5, 25.8 ppm; HRMS (EI): calcd for C$_{25}$H$_{20}$ClNO$_2$ [M$^+$]: 401.1185, found 401.1185.

3.6.4. (E)-3-((E)-1-(4-Methoxyphenyl)-3-(4-nitrophenyl)allylidene)-1-methylindolin-2-one (**3dc**)

Reflux for 20 h; 45% yield; red solid; mp = 143.1–144.6 °C; R_f = 0.54 (silica gel, hexane:EtOAc:CH$_2$Cl$_2$ = 4:1:2); ^{1}H NMR (500 MHz, CDCl$_3$): δ 9.48 (d, *J* = 16.0 Hz, 1H), 8.19–8.15 (m, 2H), 7.65 (d, *J* = 8.7 Hz, 2H), 7.22–7.18 (m, 2H), 7.16 (td, *J* = 7.7, 1.1 Hz, 1H), 7.12–7.06 (m, 2H), 6.77 (d, *J* = 7.7 Hz, 1H), 6.67 (td, *J* = 7.7, 1.0 Hz, 1H), 6.51 (d, *J* = 16.0 Hz, 1H), 5.95 (d, *J* = 7.4 Hz, 1H), 3.95 (s, 3H), 3.30 (s, 3H) ppm; ^{13}C NMR (125 MHz, CDCl$_3$): δ 168.0, 160.1, 149.6, 147.5, 143.5, 143.3, 137.8, 132.3, 130.1, 129.2, 129.1 128.3, 125.1, 124.2, 123.2, 121.9, 114.9, 107.9, 55.6, 26.0 ppm; HRMS (EI): calcd for C$_{25}$H$_{20}$N$_2$O$_4$ [M$^+$] : 412.1423, found 412.1422.

3.6.5. (E)-3-((E)-1,3-Bis(4-methoxyphenyl)allylidene)-1-methylindolin-2-one (**3dd**)

91% yield; yellow solid; mp = 131.7 °C; R_f = 0.21 (silica gel, hexane:EtOAc = 5:1); ^{1}H NMR (500 MHz, CDCl$_3$): δ 9.23 (d, *J* = 15.9 Hz, 1H), 7.49 (d, *J* = 8.8 Hz, 2H), 7.19 (dt, *J* = 13.9, 2.3 Hz, 2H), 7.09 (td, *J* = 7.7, 0.8 Hz, 1H), 7.06 (dt, *J* = 14.0, 2.4 Hz, 2H), 6.85 (d, *J* = 8.8 Hz, 2H), 6.74 (d, *J* = 7.7 Hz, 1H), 6.63 (td, *J* = 7.7, 0.8 Hz, 1H), 6.48 (d, *J* = 15.9 Hz, 1H), 5.89 (d, *J* = 7.7 Hz, 1H), 3.9 (s, 3H), 3.82 (s, 3H), 3.30 (s, 3H) ppm; ^{13}C NMR (125 MHz, CDCl$_3$): δ 168.2, 160.6, 159.8, 151.8, 142.6, 141.3, 130.1, 130.0, 129.6, 127.9, 126.2, 123.8, 123.4, 121.8, 121.5, 114.6, 114.3, 107.5, 55.49, 55.46, 25.8 ppm; HRMS (EI): calcd for C$_{26}$H$_{23}$NO$_3$ [M$^+$]: 397.1678, found 397.1677.

3.7. Dehydration under Basic Conditions (**3dc**)

To a stirred solution of the alcohol **8dc** (15.3 mg, 0.0355 mmol) in CH$_2$Cl$_2$ (1.0 mL) was added NEt$_3$ (40 µL, 0.29 mmol, 8 equiv.), MsCl (10 µL, 0.13 mmol, 3.7 equiv.), and DMAP (1.1 mg, 9.0 µmol, 0.25 equiv.) at rt. After stirring for 1 h, the reaction mixture was diluted with sat. aq. NH$_4$Cl (5 mL) and EtOAc (30 mL). The organic layer was separated, dried over Na$_2$SO$_4$, filtered, and concentrated under reduced pressure. The crude residue was purified by column chromatography (silica gel, Hexane:EtOAc:CH$_2$Cl$_2$ = 4:1:1) to afford aldol adduct **3dc** (11.6 mg, 0.0281 mmol, 79% yield).

3.8. (E)-3-((E)-1-(Biphenyl-4-yl)-3-phenylallylidene)-1-methylindolin-2-one (**9**)

To a solution of **3ea** (16.4 mg, 0.0354 mmol) in dioxane (1.0 mL) were added phenylboronic acid (5.2 mg, 0.043 mmol, 1.2 equiv.), K$_2$CO$_3$ (14.7 mg, 0.106 mmol, 3 equiv.) and

Pd(PPh$_3$)$_4$ (2.0 mg, 1.7 µmol, 5 mol%). The reaction mixture was stirred at 90 °C for 8 h, then cooled to rt and diluted with EtOAc (50 mL) and water (20 mL). The organic layer was separated, dried over Na$_2$SO$_4$, filtered, and concentrated under reduced pressure. The crude residue was purified by column chromatography (silica gel, hexane:EtOAc = 8:1) to afford **9** (12.2 mg, 81% yield) as orange gum. R_f = 0.43 (silica gel, hexane:EtOAc = 5:1); ^{1}H NMR (500 MHz, CDCl$_3$): δ 9.41 (d, J = 16.0 Hz, 1H), 7.85–7.79 (m, 2H), 7.77 (dd, J = 8.2, 1.1 Hz, 2H), 7.56 (d, J = 7.3 Hz, 2H), 7.53 (t, J = 7.7 Hz, 2H), 7.45–7.40 (m, 1H), 7.39–7.35 (m, 2H), 7.33 (t, J = 7.3 Hz, 2H), 7.31–7.27 (m, 1H), 7.13 (td, J = 7.7, 1.0 Hz, 1H), 6.77 (d, J = 7.7 Hz, 1H), 6.63 (td, J = 7.7, 0.9 Hz, 1H), 6.56 (d, J = 16.0 Hz, 1H), 5.92 (d, J = 7.7 Hz, 1H), 3.32 (s, 3H) ppm; ^{13}C NMR (125 MHz, CDCl$_3$) δ 168.1, 151.0, 142.9, 141.6, 141.2, 140.4, 137.0, 136.7, 129.3, 129.2, 129.1, 128.8, 128.4, 128.0, 127.9, 127.7, 127.2, 123.7, 123.4, 122.7, 121.7, 107.7, 25.9 ppm; HRMS (EI): calcd for C$_{30}$H$_{23}$NO [M$^+$]: 413.1780, found 413.1776.

Supplementary Materials: The following supporting information can be downloaded at: https://www.mdpi.com/article/10.3390/molecules27165304/s1, copies of NMR spectra of **4–9**, and **3** (**3ac**, **3bc**, **3cc**, **3dc**, **3ea**).

Author Contributions: Conceptualization, J.H.S.; methodology, J.K. and M.K.; software, J.K. and M.K.; validation, J.K. and M.K.; formal analysis, J.K. and M.K.; investigation, J.K. and M.K.; resources, J.H.S. and K.J.S.; data curation, J.K. and M.K.; writing—original draft preparation, J.H.S. and J.K.; writing—review and editing, J.H.S. and K.J.S.; visualization, J.H.S. and J.K.; supervision, J.H.S. and K.J.S.; project administration, J.H.S.; funding acquisition, J.H.S. and K.J.S. All authors have read and agreed to the published version of the manuscript.

Funding: This research was supported by Basic Science Research Program through the National Research Foundation of Korea (NRF) funded by the Ministry of Education (2018R1A6A1A03025108) and Research Fund of 2019 of The Catholic University of Korea.

Institutional Review Board Statement: Not applicable.

Informed Consent Statement: Not applicable.

Data Availability Statement: The data presented in this study are available in the Supplementary Materials.

Acknowledgments: We thank Dain Lee and Boyoung Kim for their assistance in data collection.

Conflicts of Interest: The authors declare no conflict of interest.

Sample Availability: Samples of the compounds **3–9** are available from the authors.

References

1. Yu, L.F.; Li, Y.Y.; Su, M.B.; Zhang, M.; Zhang, W.; Zhang, L.N.; Pang, T.; Zhang, R.T.; Liu, B.; Li, J.Y.; et al. Development of Novel Alkene Oxindole Derivatives as Orally Efficacious AMP-Activated Protein Kinase Activators. *ACS Med. Chem. Lett.* **2013**, *4*, 475–480. [CrossRef] [PubMed]
2. Pal, A.; Ganguly, A.; Ghosh, A.; Yousuf, M.; Rathore, B.; Banerjee, R.; Adhikari, S. Bis-arylidene Oxindoles as Anti-Breast-Cancer Agents Acting via the Estrogen Receptor. *ChemMedChem* **2014**, *9*, 727–732. [CrossRef] [PubMed]
3. Yanada, R.; Obika, S.; Inokuma, T.; Yanada, K.; Yamashita, M.; Ohta, S.; Takemoto, Y. Stereoselective Synthesis of 3-Alkylideneoxindoles via Palladium-Catalyzed Domino Reactions. *J. Org. Chem.* **2005**, *70*, 6972–6975. [CrossRef] [PubMed]
4. Praveen, N.; Sekar, G. Palladium Nanoparticle-Catalyzed Stereoselective Domino Synthesis of 3-Allylidene-2(3*H*)-oxindole and 3-Allylidene-2(3*H*)-benzofuranones. *J. Org. Chem.* **2020**, *85*, 4682–4694. [CrossRef] [PubMed]
5. Miura, T.; Toyoshima, T.; Takahashi, Y.; Murakami, M. Stereoselective Synhthesis of 3-Alkylideneoxindoles by Palladium-Catalyzed Cyclization Reaction of 2-(Alkynyl)aryl Isocyanates with Organoboron Reagents. *Org. Lett.* **2008**, *10*, 4887–4889. [CrossRef] [PubMed]
6. Yu, Y.; Shin, K.J.; Seo, J.H. Stereoselective Synthesis of 3-(1,3-Diarylallylidene)oxindoles via a Palladium-Catalyzed Tandem Reaction. *J. Org. Chem.* **2017**, *82*, 1864–1871. [CrossRef] [PubMed]
7. Dong, G.R.; Park, S.; Lee, D.; Shin, K.J.; Seo, J.H. Synthesis of 3-(Diarylmethylene)oxindoles via a Palladium-Catalyzed One-Pot Reaction: Sonogashira-Heck-Suzuki-Miyaura Combined Reaction. *Synlett* **2013**, *24*, 1993–1997. [CrossRef]
8. Park, S.; Shin, K.J.; Seo, J.H. Palladium-Catalyzed Tandem Approach to 3-(Diarylmethylene)oxindoles Using Microwave Irradiation. *Synlett* **2015**, *26*, 2296–2300. [CrossRef]
9. Park, S.; Lee, J.; Shin, K.J.; Seo, J.H. Consecutive One-Pot versus Domino Multicomponent Approaches to 3-(Diarylmethylene)oxindoles. *Molecules* **2017**, *22*, 503. [CrossRef]

10. Lee, J.; Park, S.; Shin, K.J.; Seo, J.H. Palladium-Catalyzed One-Pot Approach to 3-(1,3-Diarylprop-2-yn-1-ylidene)oxindoles. *Heterocycles* **2018**, *96*, 1795–1807.
11. Samineni, R.; Madapa, J.; Pabbaraja, S.; Mehta, G. Stitching Oxindoles and Ynones in a Domino Process: Access to Spirooxindoles and Application to a Short Synthesis of Spindomycin, B. *Org. Lett.* **2017**, *19*, 6152–6155. [CrossRef]
12. Kumar, S.; Pratap, R.; Kumar, A.; Kumar, B.; Tandon, V.K.; Ji Ram, V. Direct Alkenylation of Indolin-2-ones by 6-Aryl-4-methylthio-2H-pyran-2-one-3-carbonitriles: A Novel Approach. *Beilstein J. Org. Chem.* **2013**, *9*, 809–817. [CrossRef]
13. Asahara, H.; Kida, T.; Hinoue, T.; Akashi, M. Cyclodextrin Host as a Supramolecular Catalyst in Nonpolar Solvents: Stereoselective Synthesis of (E)-3-alkylideneoxindoles. *Tetrahedron* **2013**, *69*, 9428–9433. [CrossRef]
14. Lee, H.J.; Lim, J.W.; Yu, J.; Kim, J.N. An Expedient Synthesis of 3-Alkylideneoxindoles by Ti(O^{i}Pr)$_4$/pyridine-Mediated Knoevenagel Condensation. *Tetrahedron Lett.* **2014**, *55*, 1183–1187. [CrossRef]
15. Chu, W.; Zhou, D.; Gabe, V.; Liu, J.; Li, S.; Peng, X.; Xu, J.; Dhavale, D.; Bagchi, D.P.; d'Avianon, A.; et al. Design, Synthesis, and Characterization of 3-(Benzylidene)indolin-2-one Derivatives as Ligands for α–Synuclein Fibrils. *J. Med. Chem.* **2015**, *58*, 6002–6017. [CrossRef]
16. Furuta, K.; Kawai, Y.; Mizuno, Y.; Hattori, Y.; Koyama, H.; Hirata, Y. Synthesis of 3-[4-(Dimethylamino)phenyl]alkyl-2-oxindole Derivatives and Their Effects on Neuronal Cell Death. *Bioorg. Med. Chem. Lett.* **2017**, *27*, 4457–4461. [CrossRef]
17. Suthar, S.K.; Bansal, S.; Narkhede, N.; Guleria, M.; Alex, A.T.; Joseph, A. Design, Synthesis and Biological Evaluation of Oxindole-Based Chalcones as Small-Molecule Inhibitors of Melanogenic Tyrosinase. *Chem. Pharm. Bull.* **2017**, *65*, 833–839. [CrossRef]
18. Robichaud, B.A.; Liu, K.G. Titanium isopropoxide/pyridine mediated Knoevenagel reactions. *Tetrahedron Lett.* **2011**, *52*, 6935–6938. [CrossRef]
19. Tang, S.; Peng, P.; Zhong, P.; Li, J.-H. Palladium-Catalyzed C-H Functionalization of N-Arylpropiolamides with Aryliodonium Salts: Selective Synthesis of 3-(1-Arylmethylene)oxindoles. *J. Org. Chem.* **2008**, *73*, 5476–5480. [CrossRef]
20. Cheng, C.; Ge, L.; Lu, X.; Huang, J.; Huang, H.; Chen, J.; Cao, W.; Wu, X. Cu-Pybox Catalyzed Synthesis of 2,3-Disubstituted imidazo[1,2-a]pyridines from 2-Aminopyridines and Propargyl Alcohol Derivatives. *Tetrahedron* **2016**, *72*, 6866–6874. [CrossRef]
21. Sam, B.; Montgomery, P.; Krische, M.J. Ruthenium Catalyzed Reductive Coupling of Paraformaldehyde to Trifluoromethyl Allenes: CF$_3$-Bearing All-Carbon Quaternary Centers. *Org. Lett.* **2013**, *15*, 3790–3793. [CrossRef] [PubMed]
22. Ito, K.; Nakajima, K. Selenium Dioxide Oxidation of Alkylcoumarines and Related Methy-Substituted Heteroarmatics. *J. Heterocycl. Chem.* **1988**, *25*, 511–515. [CrossRef]
23. Buynak, J.D.; Rao, M.N.; Pajouhesh, H.; Chandrasekaran, R.Y.; Finn, K.; de Meester, P.; Chu, S.C. Useful Chemistry of 3-(1-Methylethylidene)-4-acetoxy-2-azetidinone: A Formal Synthesis of (±)-Asparenomycin, C. *J. Org. Chem.* **1985**, *50*, 4245–4252. [CrossRef]
24. Singh, K.; Singh, K.; Balzarini, J. Regioselective Synthesis of 6-Substituted-2-amino-5-bromo-4(3H)-pyrimidnones and Evaluation of Their Antiviral Activity. *Eur. J. Med. Chem.* **2013**, *67*, 428–433. [CrossRef]
25. Peixoto, P.A.; Boulangé, A.; Leleu, S.; Franck, X. Versitile Synthesis of Acylfuranones by Reaction of Acylketenes with α-Hydroxy Ketones: Application to the One-Step Multicomponent Synthesis of Cadiolide B and Its Analogues. *Eur. J. Org. Chem.* **2013**, *2013*, 3316–3327. [CrossRef]
26. Diaz Ropero, B.P.F.; Elsegood, M.R.J.; Fairley, G.; Pritchard, G.J.; Weaver, G.W. Pyridone Functionalization: Regioselective Deprotonation of 6-Methylpyridin-2(1H)- and -4(1H)-one Derivatives. *Eur. J. Org. Chem.* **2016**, *2016*, 5238–5242. [CrossRef]
27. Badejo, I.T.; Karaman, R.; Fry, J.L. Unstable Compounds. Synthesis and Experimental and Computational Study of the Chemical Behavior of 9-[1-(2,4,6-Cycloheptatrienyl)]-9-xanthydrol. *J. Org. Chem.* **1989**, *54*, 4591–4596. [CrossRef]
28. Hu, S.; Han, X.; Xie, X.; Fang, F.; Wang, Y.; Saidahmatov, A.; Liu, H.; Wang, J. Synthesis of Pyrazolo[1,2-a]cinnolines via Rhodium(III)-Catalyzed [4+2] Annulation Reactions of Pyrazolidinones with Sulfoxonium Ylides. *Adv. Synth. Catal.* **2021**, *363*, 3311–3317. [CrossRef]

Article

Efficient Synthesis of 2-OH Thioglycosides from Glycals Based on the Reduction of Aryl Disulfides by NaBH$_4$

Yang-Fan Guo [†], Tao Luo [†], Guang-Jing Feng, Chun-Yang Liu and Hai Dong *

Key Laboratory of Material Chemistry for Energy Conversion and Storage, Ministry of Education, Hubei Key Laboratory of Material Chemistry and Service Failure, School of Chemistry & Chemical Engineering, Huazhong University of Science & Technology, Luoyu Road 1037, Wuhan 430074, China
* Correspondence: hdong@mail.hust.edu.cn
† These authors contributed equally to this work.

Abstract: An improved method to efficiently synthesize 2-OH thioaryl glycosides starting from corresponding per-protected glycals was developed, where 1,2-anhydro sugars were prepared by the oxidation of glycals with oxone, followed by reaction of crude crystalline 1,2-anhydro sugars with NaBH$_4$ and aryl disulfides. This method has been further used in a one-pot reaction to synthesize glycosyl donors having both "armed" and "NGP (neighboring group participation)" effects.

Keywords: glycals; 2-OH thioglycosides; glycosyl donors; glycosylation; one-pot reaction

Citation: Guo, Y.-F.; Luo, T.; Feng, G.-J.; Liu, C.-Y.; Dong, H. Efficient Synthesis of 2-OH Thioglycosides from Glycals Based on the Reduction of Aryl Disulfides by NaBH$_4$. *Molecules* 2022, 27, 5980. https://doi.org/10.3390/molecules27185980

Academic Editors: Mircea Darabantu, Alison Rinderspacher and Gloria Proni

Received: 28 August 2022
Accepted: 9 September 2022
Published: 14 September 2022

Publisher's Note: MDPI stays neutral with regard to jurisdictional claims in published maps and institutional affiliations.

Copyright: © 2022 by the authors. Licensee MDPI, Basel, Switzerland. This article is an open access article distributed under the terms and conditions of the Creative Commons Attribution (CC BY) license (https:// creativecommons.org/licenses/by/ 4.0/).

1. Introduction

The chemical synthesis of complex carbohydrates is an important research topic in carbohydrate chemistry due to their crucial roles in biological processes [1–5]. Thioglycoside donors are widely used in these syntheses due to their advantages, such as easy preparation, stable chemical properties, and various activation methods [6–12]. Generally, each glycosylation reaction between the donor and acceptor may produce a product in the α or β configuration, which will lead to adverse effects, such as reduced yield and difficult purification. Therefore, many studies have been devoted to solving the problem of stereoselectivity in glycosylation. Among them, the use of neighboring group participation (NGP) from 2-positions of glycoside donors to control stereoselectivity is a very effective method. The acyl group at 2-position is the most commonly used "NGP" group, which leads to the 1,2-*trans*-configuration of the products in glycosylation [13–17]. However, the "disarmed" effect of acyl groups often leads to low reactivity of glycosyl donors in glycosylation. Therefore, several methods using NGP from 2-ether groups of thioglycoside donors to control stereoselectivity have been developed recently, and phenyl-3,4,6-tri-*O*-benzyl-1-thio-β-D-glucopyranoside **1** is often the precursor required for the synthesis of these donors (Figure 1) [18–22].

Although excellent stereoselectivity and high reactivity were shown in these methods, the synthesis of **1** is a challenge, which reduces the practicality of these methods. For example, the traditional "orthoester method" requires multiple-step protection and deprotection, leading to low synthesis efficiency (Figure 2a) [23–25]. More efficient methods involved the formation of 1,2-anhydro sugars by oxidation of glycals with oxone in acetone and the installation of the 1-thiophenyl group through ring-opening reactions of these 1,2-anhydro sugars. For example, the oxidation of glucal **2** by oxone yielded 1,2-anhydro glucose **3** (crude crystalline product): (a) the ring opening of **3** led to 43% (the use of TBASPh) [26] or 47% (the use of NaSPh) [27] yield of **1** in the presence of phenylthiolate at room temperature for overnight, and (b) the ring opening of **3** led to 37–55% yield [19,20] of **1** in the presence of PhSH and ZnCl$_2$ at room temperature for overnight (Figure 2b). In this study, an improved method for the efficient synthesis of 2-OH, 1-thioaryl glycosides was developed, in which the ring opening reaction of 1,2-anhydro sugars occurred in the presence of NaBH$_4$ and alkyl disulfides at room temperature. As a result, 73% yield of **1** could be efficiently prepared from **2** (two-step reaction was completed within 90 min) under very mild conditions (Figure 2c). Furthermore, the glycosyl donors having both "armed" and "NGP" effects can be efficiently synthesized in a one-pot reaction based on this method.

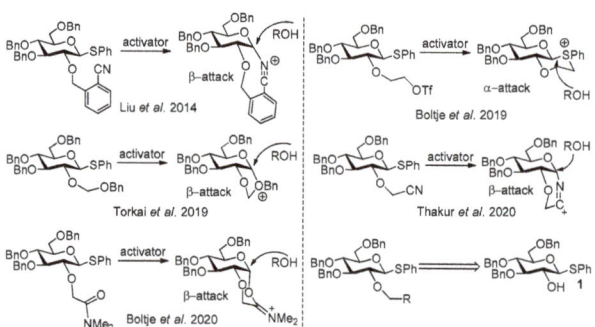

Figure 1. Using NGP from 2-ether groups to control stereoselectivity.

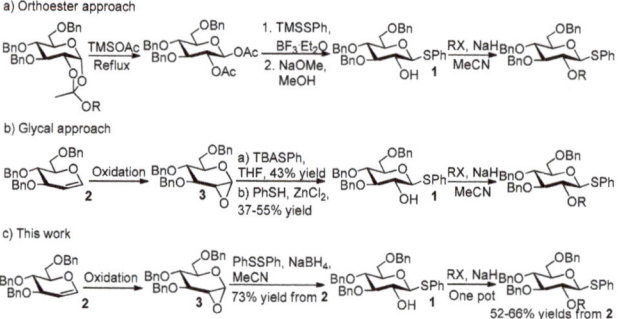

Figure 2. Comparison of this method with previous methods.

2. Results

In our lab, we were working on developing relative green methods for 1-thioglycosides by avoiding the use of odorous thioacetic acids and alkylthiols [28–30]. We noticed a report where NaBH$_4$ and disulfides were used instead of sodium arylthiolates in the synthesis of 1-thioglycosides [31]. It was observed that phenylselenolate and phenylthiolate were quickly generated by mixing diselenide or disulfide precursors with a stoichiometric amount of NaBH$_4$ in acetonitrile (Reaction formula shown in Figure 3a,b). This inspired us to explore whether a system of disulfides and NaBH$_4$ could be used to improve the synthesis of **1** [26,27] starting from glucal **2**.

Thus, 1,2-anhydro glucose **3** was first prepared by oxidation of glycal **2** with oxone in acetone, and then, its crude crystals were directly reacted with 0.7 equiv of phenyl disulfide and 1.5 equiv of NaBH$_4$ (equivalent to 1.4 equiv of NaBH$_3$SPh) at rt in acetonitrile for 1 h to yield **1** in 72% yield, yielding **1** in 75% yield when the reaction was performed at 0 °C for 4 h (entry 1 in Table 1). Due to concerns about direct hydrogenation reduction of NaBH$_4$ to **3**, we first allowed NaBH$_4$ to react with phenyl disulfide at 50 °C in acetonitrile for 1 h and then added crude crystalline **3** to the reaction mixture (entry 2). Yielding **1** in 73% yield indicated that we had been overly concerned about the possible side effects caused by NaBH$_4$. Reducing the amount of NaBH$_4$ to 1.0 equiv (equivalent to 1.0 equiv of NaBH$_3$SPh and 0.4 equiv of HSPh) resulted in a decrease in the yield of **1** to 68%, and reducing the amount of NaBH$_4$ to 0.7 equiv (equivalent to 0.7 equiv of NaBH$_3$SPh and 0.7 equiv of

HSPh) resulted in a decrease in the yield of **1** to 60% (entry 3). The use of 1.0/2.0 equiv of NaBH$_4$ and 0.5/1.0 equiv of phenyl disulfide (equivalent to 1.0/2.0 equiv of NaBH$_3$SPh) led to 55%/65% yield of **1** (entry 4). The use of 1.2 equiv of NaBH$_4$ and 0.6/0.8 equiv of phenyl disulfide led to 65%/69% yield of **1** (entry 5). These results suggested that 1.5 equiv of NaBH$_4$ and 0.7 equiv of phenyl disulfide should be the optimal conditions. We also examined the effect of solvents (acetone, DMF, MeOH, and DCM) on the reaction (entries 6 and 7). These results suggested that acetonitrile should be the optimal solvent. As a comparison, we allowed **3** to react with 1.5 equiv of NaSPh at rt in acetonitrile, which gave **1** in 43% yield after 36 h, indicating the low reactivity of this reaction (entry 8).

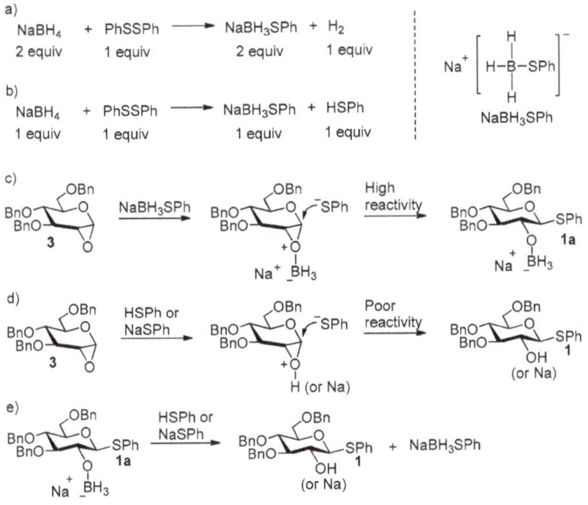

Figure 3. Proposed reaction mechanism.

Table 1. Comparison of results by variation of reaction conditions [a].

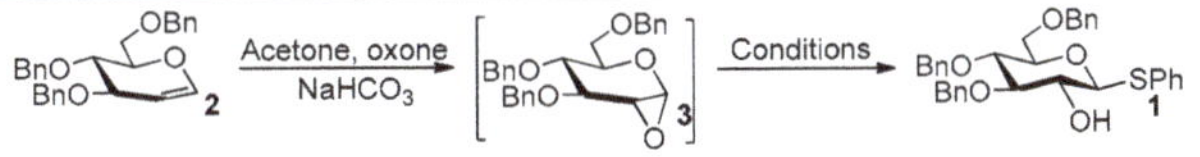

Entry	NaBH$_4$ (equiv)	PhSSPh (equiv)	Reaction Conditions (equiv)	Yields (%)
1	1.5	0.7	ACN, rt/0 °C, 1/4 h	72/75
2	1.5	0.7	ACN, rt, 1 h	73 [b]/70 [c]
3	1.0/0.7	0.7	ACN, rt, 1 h	68/60
4	1.0/2.0	0.5/1.0	ACN, rt, 1 h	55/65
5	1.2	0.6/0.8	ACN, rt, 1 h	65/69
6	1.5	0.7	Acetone/DMF, rt, 1 h	18/<5
7	1.5	0.7	MeOH/DCM, rt, 1 h	21/-
8	-	-	ACN, NaSPh (1.5), rt, 36 h	43
9	0.1	0.1	ACN, NaSPh (1.2), rt, 4 h	55
10	0.3	0.3	ACN, NaSPh (0.8), rt, 4 h	72
11	0.5	0.5	ACN, NaSPh (0.4), rt, 4 h	65

[a] Reagents and conditions: substrate **2** (0.1 mmol), solvents (1 mL), yields based on **2**. [b] Treatment of PhSSPh with NaBH$_4$ in acetonitrile at 50 °C for 1 h, then cooling to rt, and adding crude substrate **3**. [c] Large scale.

We proposed the mechanism of the reaction between **3** and NaBH$_3$SPh in Figure 3c and the mechanism of the reaction between **3** and NaSPh in Figure 3d. The coordination of the boron atom of borane instead of Na$^+$ (or H$^+$) to the 1,2-anhydro oxygen atom may greatly enhance the nucleophilic attack activity of $^-$SPh towards the 1-position of **3**, which explains the results shown in entries 1, 2, and 8 in Table 1. The results shown in entry 3 in Table 1 suggested that NaBH₃SPh may catalyze the reaction of **3** with HSPh. A possible catalytic mechanism is shown in Figure 3c,e where NaBH₃SPh reacts with **3** to form **1a** (Figure 3c) and regenerates (Figure 3e) from the exchange of [NaBH₃]⁺ and H⁺ (Na⁺) between **1a** and HSPh (or NaSPh). Compared to the result (**1**, 42%, 36 h) shown in entry 8, the result (**1**, 55%, 4 h) shown in entry 9 showed that the use of 0.1 equiv of NaBH₄ and 0.1 equiv of phenyl disulfide (equivalent to 0.1 equiv of NaBH₃SPh and 0.1 equiv of HSPh) in the presence of 1.2 equiv of NaSPh (1.4 equiv of ⁻SPh existing in the system) led to higher reactivity. The optimal conditions were the use of 0.3 equiv of NaBH₄, 0.3 equiv of phenyl disulfide and 0.8 equiv of NaSPh (1.4 equiv of ⁻SPh existing in the system also), by which 72% yield of **1** was obtained after 4 h' reaction (entry 10); continuing to increase the amount of NaBH₄ and phenyl disulfide to 0.5 equiv (using 0.4 equiv of NaSPh in order to keep 1.4 equiv of ⁻SPh present in the system) instead reduced the yield of **1** to 65% (entry 11). Since aryl disulfides are generally more commercially available reagents than sodium arylthiolates, the conditions shown in entries 1–2 are obviously more practical than that shown in entry 10.

With the optimized conditions in hand, we next set out to evaluate this method using phenyl disulfide with various glycals as substrates (Figure 4). As can be seen, phenyl-2-OH-1-thio-β-D-glucopyranosides **4–11** and phenyl-2-OH-1-thio-β-D-galactopyranosides **12–14** were efficiently synthesized in 50–70% yields starting from the corresponding glucals and galactals with various protecting groups. For compounds **4, 5, 6, 13**, and **14**, the TBS, acetyl, or benzoyl can be removed orthogonally in the presence of benzyl-protecting group under corresponding acid–base conditions. Thus, these compounds can be used as building blocks for the elongation of sugar chains and the synthesis of branched oligosaccharides. Phenyl-2-OH-3,4-di-OBn-1-thio-β-D-xylopyranoside **15** was synthesized in 65% yield from 3,4-di-OBn xylal, and phenyl-2-OH-1-thio-β-D-lactoside **16** was synthesized in 56% yield from per-benzylated lactal. These results suggested that this method should be applicable to various glycals.

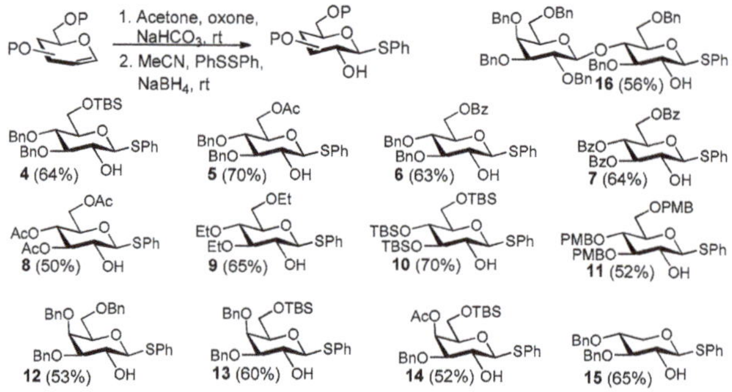

Figure 4. Synthesis of 1-thiophenyl glycosides with 2-OH starting from corresponding glycals.

We next evaluated this method using various disulfides with glucal **2** as the substrate (Figure 5). As can be seen, aryl disulfides worked well in this method, leading to 2-OH, β-D-thioglucosides **17–22** in 70–76% yields, but non-aryl disulfides did not. The 2-OH, β-D-thioglucosides **23–25** could not be obtained by this method. Hydroreduction product **27** was isolated in 26–37% yield in the reaction with non-aryl disulfides, indicating that NaBH₄ had not been consumed by the reaction with non-aryl disulfides. Further experiments indicated that NaBH₄ could not reduce non-aryl disulfides even at 50 °C.

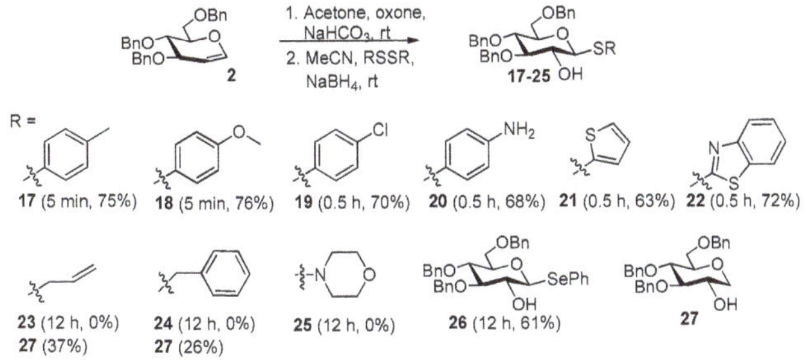

Figure 5. Evaluation of reactions between various disulfides and glucal 2.

Phenyl diselenide also worked well in this method, leading to 2-OH, β-D-selenoglucoside **26** in 61% yield. However, this reaction took a long time due to the low reactivity for reduction of diselenide by NaBH₄. In light of the mechanism shown in Figure 3c, we speculated that **1a** should be able to react directly with RX (X represents $^-$Cl or $^-$Br) in the present of NaH to form various thioglycoside donors containing "NGP" group at their 2-positions. This speculation was supported by further experiments and a one-pot method was developed by us (Figure 6). Once the TLC plate showed complete consumption of 1,2-anhydro sugar, NaH and RX were added to the reaction mixture, and the reaction proceeded at rt for 1–4 h, leading to thioglycoside donors **28–34** in 48–68% yields based on glycals, respectively. It has been reported that 2-Pic STaz-donors exhibited good reactivity and steroselectivity in glycosylation with Cu(OTf)₂ as promoter (2-Pic glucoside STaz-donor was obtained in 60% yield over four steps from orthoester) [10], while 2-Pic glucoside SEt-donor exhibited no reactivity with NIS/TfOH as promoter [10b]. We then evaluated the glycosylation between 2-Pic SPh-donors **28/29** and various acceptors with NIS/TfOH as promoter (Figure 7). As can be seen, disaccharides **35–41** with absolute β-configuration were obtained in 50–86% yields.

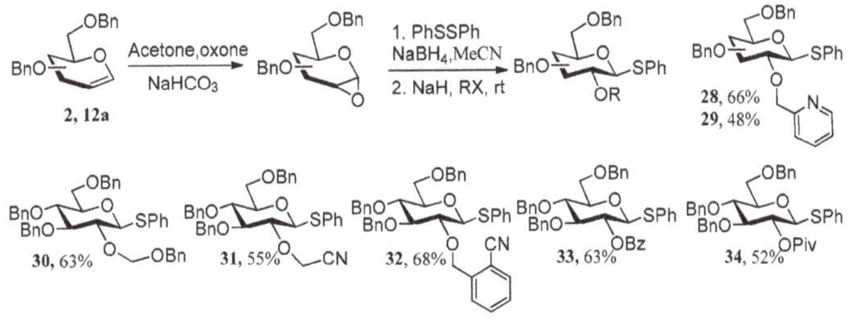

Figure 6. One-pot synthesis of various thioglycoside donors.

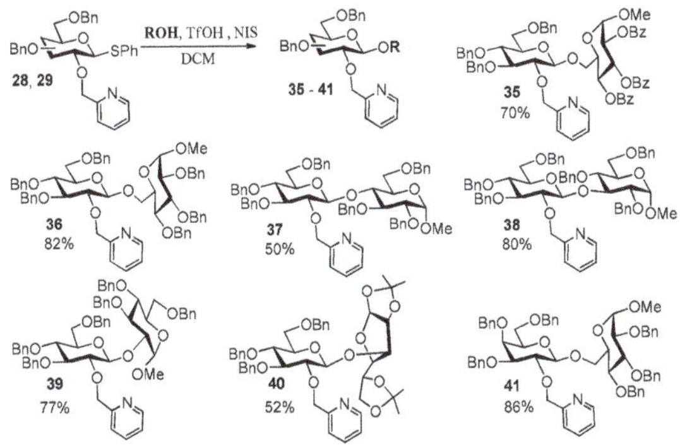

Figure 7. Application of this method in glycosylation.

3. Conclusions

In conclusion, in order to efficiently obtain thioglycoside donors whose protecting groups at 2-position have both "armed" and "NGP" effects, we developed an efficient method for the synthesis of 2-OH thioaryl glycosides starting from their corresponding glycals. In this method, the oxidation of glycals with oxone led to 1,2-anhydro sugars, which are easily isolated by crystallization, and the obtained crude crystalline 1,2-anhydro sugars were then treated with 1.5 equiv of $NaBH_4$ and 0.7 equiv of aryl disulfides in acetonitrile at mild conditions to yield the corresponding 1-thioaryl glycosides with 2-OH in 50–75% total yields. Based on this method, thioglycoside donors having both "armed" and "NGP" effects can be efficiently synthesized in a one-pot reaction. Compared with previous methods [19,20,26,27], this method shows three outstanding advantages: good yields, high synthesis efficiency, and the use of relatively green reagents (avoiding the use of foul-smelling aryl thiol reagents).

4. Materials and Methods

General Methods. All commercially available starting materials and solvents were of reagent grade and used without further purification. Chemical reactions were monitored with thin-layer chromatography using precoated silica gel 60 (0.25 mm thickness) plates. Flash column chromatography was performed on silica gel 60 (SDS 0.040–0.063 mm). ^1H NMR spectra were recorded at 298 K in $CDCl_3$ using the residual signals from $CHCl_3$ (^1H: = 7.26 ppm) as internal standard. ^1H peak assignments were made by first order analysis of the spectra, supported by standard ^1H-^1H correlation spectroscopy (COSY) (see Supplementary Materials).

General process A for synthesis of 2-OH 1-thioaryl glycosides from glycals. *Step 1.* To a cooled (0 °C) solution of a per-protected glycal (1 mmol) in DCM (4 mL) were added acetone (0.4 mL) and saturated aqueous $NaHCO_3$ (7 mL). The mixture was stirred vigorously, and a solution of oxone (2 mmol) in H_2O (2.5 mL) was added dropwise over 10 min. The mixture was stirred vigorously at 0 °C for 30 min and then at rt until TLC indicated consumption of the starting material. The organic phase was separated, and the aqueous phase was extracted with DCM (2 × 10 mL). The combined organic phases were dried ($MgSO_4$) and concentrated in vacuo to obtain the crude 1,2-anhydro sugar. *Step 2.* To a mixture of phenyl disulfide (or phenyl diselenide) (0.7 mmol) and $NaBH_4$ (53 mg, 1.4 mmol) was added acetonitrile (5 mL). The mixture was stirred at rt for 30 min to 2 h until TLC indicated full conversion of the phenyl disulfide (or phenyl diselenide). The mixture was then added to the crude α-1,2-anhydro sugars. The reaction was stirred at rt for 5–60 min until TLC indicated full conversion of the starting material. The mixture was diluted with DCM and washed with water. The aqueous phase was re-extracted with DCM, and collected organic phases were dried and evaporated under vacuum. The residue was purified by silica gel flash chromatography.

General process B for one-pot synthesis of thioglycoside donors containing a "NGP" group at the 2-position.

Step 1. Same as *step 1* in general process A.

Step 2. To a mixture of phenyl disulfide (145 mg, 0.7 mmol) and $NaBH_4$ (53 mg, 1.4 mmol) was added acetonitrile (5 mL). The mixture was stirred at rt for 30 min to 2 h until TLC indicated full conversion of the phenyl disulfide. The mixture was then added to the crude α-1,2-anhydro sugars. The reaction was stirred at rt for 5–60 min until TLC indicated full conversion of the starting material. The reaction mixture was then cooled to 0 °C, followed the slow addition of sodium hydride (6.0 mmol, 6 equiv, 60% oil dispersion), and allowed to stir at 0 °C for 10 min. After that, alkylation/acylation reagents (2–3 equiv) were added to the reaction mixture. The reaction mixture was allowed to warm to rt and then stirred for 1–4 h. Upon completion, the reaction was quenched by adding crushed ice (10 g), stirred until cessation of H_2 evolution, and then extracted with ethyl acetate (3 × 80 mL). The combined organic phase was washed with water (3 × 40 mL), separated, dried with MgSO₄, and evaporated in vacuo. The residue was purified by column chromatography.

General process C for typical NIS/TfOH-promoted glycosylation procedure. A mixture of a glycosyl donor (0.13 mmol), a glycosyl acceptor (0.10 mmol), and freshly activated molecular sieves (4 Å, 200 mg) in CH_2Cl_2 (1.6 mL) was stirred under an atmosphere of argon for 1 h. After NIS (0.26 mmol) and TfOH (0.013 mmol) were added at −25 °C, the reaction mixture was allowed to warm to rt over 1 h and then was quenched with TEA and stirred for 30 min. The mixture was then diluted with CH_2Cl_2, the solid was filtered-off, and the residue was washed with CH_2Cl_2. After the combined filtrate (30 mL) was washed with water (4 × 10 mL), the organic phase was separated, dried with MgSO₄, and concentrated in vacuo. The residue was purified by silica gel flash chromatography.

Phenyl 3,4,6-tri-O-benzyl-1-thio-β-D-glucopyranoside (1) [20]. Following general process A, starting from **2** (100 mg, 0.24 mmol), after 1 h of ring-opening reaction for the crude anhydro sugar, purification by silica gel flash column chromatography afforded **1** as a white solid (94 mg, 72%). Rf = 0.43 (petroleum ether/ethyl acetate 4:1); ^{1}H NMR (600 MHz, chloroform-*d*) δ 7.54–7.64 (m, 2H), 7.42–7.25 (m, 16H), 7.24–7.19 (m, 2H), 4.94 (d, *J* = 11.2 Hz, 1H), 4.90–4.81 (m, 2H), 4.67–4.55 (m, 3H), 4.53 (d, *J* = 9.6 Hz, 1H), 3.83 (dd, *J* = 11.0, 2.0 Hz, 1H), 3.77 (dd, *J* = 11.0, 4.5 Hz, 1H), 3.67–3.59 (m, 2H), 3.58–3.44 (m, 2H), 2.43 (s, 1H) ppm.

Phenyl 3,4-di-O-benzyl-6-O-tert-butyl-dimethylsily-1-thio-β-D-glucopyranoside (4). Following general process A, starting from **4a** (50 mg, 0.114 mmol), after 1 h of ring-opening reaction for the crude anhydro sugar, purification by silica gel flash column chromatography afforded **4** as a colorless oil (40 mg, 64%). Rf = 0.41 (petroleum ether/ethyl acetate 8:1); ^{1}H NMR (400 MHz, chloroform-*d*) δ 7.51–7.43 (m, 2H), 7.31–7.17 (m, 13H), 4.85–4.73 (m, 3H), 4.60 (d, *J* = 10.8 Hz, 1H), 4.39 (d, *J* = 9.6 Hz, 1H), 3.86–3.74 (m, 2H), 3.57–3.45 (m, 2H), 3.41–3.32 (m, 1H), 3.28 (ddd, *J* = 9.2, 3.6, 1.8 Hz, 1H), 2.31 (d, *J* = 2.1 Hz, 1H), 0.83 (s, 9H), 0.01 (s, 6H) ppm. ^{13}C NMR (100 MHz, chloroform-*d*) δ 138.49, 138.32, 133.03, 131.69, 128.92, 128.54, 128.47, 128.09, 128.04, 127.95, 127.84, 127.81, 87.89, 85.98, 80.44, 75.43, 75.07, 72.45, 62.14, 25.93, 18.31, −5.11, −5.34 ppm. $[α]^{20}_D$ = −20.3 (c 0.32, CH_2Cl_2); HRMS (ESI-TOF) (m/z): [M + Na]⁺ calculated for $C_{31}H_{32}O_5S_2Na^+$, 589.2420; found, 589.2379.

Phenyl 3,4-di-O-benzyl-6-O-acetyl-1-thio-β-D-glucopyranoside (5). Following general process A, starting from **5a** (100 mg, 0.271 mmol), after 0.5 h of ring-opening reaction for the crude anhydro sugar, purification by silica gel flash column chromatography afforded **5** as colorless syrup (93.6 mg, 70%). Rf = 0.58 (petroleum ether/ethyl acetate 4:1); ^{1}H NMR (400 MHz, chloroform-*d*) δ 7.61–7.50 (m, 2H), 7.43–7.26 (m, 13H), 4.98 (d, *J* = 11.1 Hz, 1H), 4.93–4.84 (m, 2H), 4.60 (d, *J* = 10.9 Hz, 1H), 4.52 (d, *J* = 9.7 Hz, 1H), 4.43 (dd, *J* = 11.9, 2.1 Hz, 1H), 4.23 (dd, *J* = 11.9, 5.2 Hz, 1H), 3.70–3.56 (m, 2H), 3.55–3.45 (m, 2H), 2.50 (d, *J* = 2.2 Hz, 1H), 2.08 (s, 3H) ppm. ^{13}C NMR (100 MHz, chloroform-*d*) δ 170.67, 138.31, 137.64, 133.04, 131.59, 128.95, 128.57, 128.53, 128.24, 128.10, 128.04, 127.91, 127.00, 88.03, 85.91, 77.19, 75.41, 75.13, 72.72, 66.32, 63.18, 20.86 ppm. $[α]^{20}_D$ = −21.6 (c 0.25, CH_2Cl_2); HRMS (ESI-TOF) (m/z): [M + Na]⁺ calculated for $C_{28}H_{30}O_6SNa^+$, 517.1661; found, 517.1640.

Phenyl 3,4-di-O-benzyl-6-O-benzoyl-1-thio-β-D-glucopyranoside (6). Following general process A, starting from **6a** (100 mg, 0.232 mmol), after 0.5 h of ring-opening reaction for the crude anhydro sugar, purification by silica gel flash column chromatography afforded **6** as a colorless oil (81.5 mg, 63%). Rf = 0.36 (petroleum ether/ethyl acetate 6:1); ^{1}H NMR (400 MHz, chloroform-*d*) δ 8.07–7.93 (m, 2H), 7.60 (t, *J* = 7.4 Hz, 1H), 7.54–7.05 (m, 17H), 4.96 (d, *J* = 11.0 Hz, 1H), 4.90–4.80 (m, 2H), 4.69 (dd, *J* = 12.0, 2.2 Hz, 1H), 4.61 (d, *J* = 10.8 Hz, 1H), 4.52 (d, *J* = 9.7 Hz, 1H), 4.44 (dd, *J* = 11.9, 4.7 Hz, 1H), 3.74–3.63 (m, 2H), 3.59 (t, *J* = 9.2 Hz, 1H), 3.53–3.44 (m, 1H), 2.46 (d, *J* = 2.2 Hz, 1H) ppm. ^{13}C NMR (100 MHz, chloroform-*d*) δ 166.10, 138.25, 137.56, 133.34, 133.16, 131.07, 129.93, 129.77, 128.91, 128.60, 128.54, 128.42, 128.26, 128.16, 128.05, 127.98, 87.78, 85.93, 77.25, 77.01, 75.55, 75.25, 72.55, 63.37 ppm. $[α]^{20}_D$ = −31.3 (c 0.15, CH_2Cl_2); HRMS (ESI-TOF) (m/z): [M + Na]⁺ calculated for $C_{31}H_{32}O_5SNa^+$, 579.1817; found, 579.1803.

Phenyl 3,4,6-tri-O-benzoyl-1-thio-β-D-glucopyranoside (7). Following general process A, starting from **7a** (100 mg, 0.218 mmol), after 20 min of ring-opening reaction for the crude anhydro sugar, purification by silica gel flash column chromatography afforded **7** as colorless syrup (81.6 mg, 64%). Rf = 0.43 (petroleum ether/ethyl acetate 4:1); ^{1}H NMR (400 MHz, chloroform-*d*) δ 8.07–7.90 (m, 6H), 7.64–7.29 (m, 12H), 7.23–7.18 (m, 2H), 5.66–5.51 (m, 2H), 4.80 (d, *J* = 9.7 Hz, 1H), 4.67 (dd, *J* = 12.2, 2.8 Hz, 1H), 4.47 (dd, *J* = 12.2, 5.8 Hz, 1H), 4.13 (ddd, *J* = 9.7, 5.7, 2.8 Hz, 1H), 3.77 (t, *J* = 9.3 Hz, 1H), 2.92–2.88 (m, 1H) ppm. ^{13}C NMR (100 MHz, chloroform-*d*) δ 166.65, 166.06, 165.35, 133.51, 133.45, 133.41, 133.18, 129.91, 129.86, 129.82, 129.80, 129.67, 129.06, 129.01, 128.73, 128.53, 128.45, 128.41, 128.38, 88.30, 76.52, 76.20, 70.93, 68.93, 63.16 ppm. $[\alpha]^{20}_D = -20.7$ (c 0.058, CH_2Cl_2); HRMS (ESI-TOF) (*m/z*): $[M + Na]^+$ calculated for $C_{33}H_{28}O_8SNa^+$, 607.1367; found, 607.1403.

Phenyl 3,4,6-tri-O-acetyl-1-thio-β-D-glucopyranoside (8). Following general process A, starting from **8a** (100 mg, 0.367 mmol), after 15 min of ring-opening reaction for the crude anhydro sugar, purification by silica gel flash column chromatography afforded **8** as colorless syrup (73.2 mg, 50%). Rf = 0.44 (petroleum ether/ethyl acetate 2:1); ^{1}H NMR (400 MHz, chloroform-*d*) δ 7.59–7.53 (m, 2H), 7.39–7.28 (m, 3H), 5.13 (t, *J* = 9.3 Hz, 1H), 4.98 (t, *J* = 9.8 Hz, 1H), 4.57 (d, *J* = 9.7 Hz, 1H), 4.25–4.12 (m, 2H), 3.73 (ddd, *J* = 10.1, 5.0, 2.5 Hz, 1H), 3.50 (td, *J* = 9.4, 2.8 Hz, 1H), 2.53 (d, *J* = 2.9 Hz, 1H), 2.09 (s, 3H), 2.07 (s, 3H), 2.03 (s, 3H) ppm. ^{13}C NMR (100 MHz, chloroform-*d*) δ 170.77, 170.61, 133.59, 130.60, 129.10, 128.72, 88.08, 75.89, 75.77, 70.31, 68.12, 62.24, 20.80, 20.76, 20.62 ppm. $[\alpha]^{20}_D = -70.0$ (c 0.05, CH_2Cl_2); HRMS (ESI-TOF) (*m/z*): $[M + Na]^+$ calculated for $C_{31}H_{32}O_5SNa^+$, 421.0933; found, 421.0950.

Phenyl 3,4,6-tri-O-ethyl-1-thio-β-D-glucopyranoside (9). Following general process A, starting from **9a** (100 mg, 0.435 mmol), after 1 h of ring-opening reaction for the crude anhydro sugar, purification by silica gel flash column chromatography afforded **9** as a white solid (99.7 mg, 65%): mp 83.3–84.5 °C; Rf = 0.31 (petroleum ether/ethyl acetate 8:1); ^{1}H NMR (400 MHz, chloroform-*d*) δ 7.62–7.53 (m, 2H), 7.34–7.26 (m, 3H), 4.49 (d, *J* = 9.4 Hz, 1H), 3.96–3.77 (m, 3H), 3.73 (dd, *J* = 11.0, 2.0 Hz, 1H), 3.70–3.48 (m, 4H), 4.44–4.21 (m, 4H), 2.49 (d, *J* = 2.1 Hz, 1H), 1.27–1.17 (m, 9H) ppm. ^{13}C NMR (100 MHz, chloroform-*d*) δ 132.73, 128.88, 127.92, 88.04, 85.83, 79.64, 77.58, 72.25, 69.43, 68.73, 68.27, 66.97, 15.79, 15.72, 15.25 ppm. $[\alpha]^{20}_D = -75.7$ (c 0.14, CH_2Cl_2); HRMS (ESI-TOF) (*m/z*): $[M + Na]^+$ calculated for $C_{31}H_{32}O_5S_2Na^+$, 379.1555; found, 379.1558.

Phenyl 3,4,6-tri-O-tert-butyl-dimethylsily-1-thio-β-D-glucopyranoside (10). Following general process A, starting from **10a** (100 mg, 0.205 mmol), after 1 h of ring-opening reaction for the crude anhydro sugar, purification by silica gel flash column chromatography afforded **10** as a colorless oil (88 mg, 70%); Rf = 0.55 (petroleum ether/ethyl acetate 50:1); ^{1}H NMR (400 MHz, chloroform-*d*) δ 7.56–7.49 (m, 2H), 7.32–7.19 (m, 3H), 4.59 (d, *J* = 8.9 Hz, 1H), 3.92 (dd, *J* = 11.3, 1.9 Hz, 1H), 3.74 (dd, *J* = 11.3, 5.6 Hz, 1H), 3.58–3.49 (m, 1H), 3.49–3.40 (m, 2H), 3.29 (ddd, *J* = 9.2, 5.7, 1.9 Hz, 1H), 2.17 (d, *J* = 2.7 Hz, 1H), 0.97–0.88 (m, 27H), 0.26–0.03 (m, 18H) ppm. ^{13}C NMR (100 MHz, chloroform-*d*) δ 135.53, 130.74, 128.75, 126.78, 89.09, 81.04, 80.17, 74.40, 70.87, 62.82, 26.14, 25.96, 18.43, 18.24, −3.67, −3.85, −4.08, −4.87, −5.09, −5.33 ppm. $[\alpha]^{20}_D = -63.8$ (c 0.16, CH_2Cl_2); HRMS (ESI-TOF) (*m/z*): $[M + Na]^+$ calculated for $C_{30}H_{58}O_5Si_3SNa^+$, 637.3210; found, 637.3174.

Phenyl 3,4,6-tri-O-p-methoxybenzyl-1-thio-β-D-glucopyranoside (11). Following general process A, starting from **11a** (50 mg, 0.1 mmol), after 1 h of ring-opening reaction for the crude anhydro sugar, purification by silica gel flash column chromatography afforded **11** as a white solid (32 mg, 52%): mp 102.3–104.6 °C; Rf = 0.33 (petroleum ether/ethyl acetate 6:1); ^{1}H NMR (400 MHz, chloroform-*d*) δ 7.60–7.54 (m, 2H), 7.36–7.23 (m, 7H), 7.18–7.08 (m, 2H), 6.93–6.79 (m, 6H), 4.83 (s, 2H), 4.76 (d, *J* = 10.4 Hz, 1H), 4.57 (d, *J* = 11.6 Hz, 1H), 4.54–4.46 (m, 3H), 3.83 (s, 9H), 3.79–3.67 (m, 2H), 3.60–3.43 (m, 4H), 2.40 (d, *J* = 2.1 Hz, 1H) ppm. ^{13}C NMR (100 MHz, chloroform-*d*) δ 159.34, 159.32, 159.19, 132.81, 131.96, 130.66, 130.35, 130.24, 129.65, 129.37, 128.95, 128.00, 113.97, 113.82, 113.76, 88.00, 85.64, 79.47, 77.12, 74.96, 74.69, 73.09, 68.62, 55.28, 43.68, 29.71, 14.63 ppm. $[\alpha]^{20}_D = -83.3$ (c 0.09, CH_2Cl_2); HRMS (ESI-TOF) (*m/z*): $[M + Na]^+$ calculated for $C_{31}H_{32}O_5S_2Na^+$, 655.2342; found, 655.2326.

Phenyl 3,4,6-tri-O-benzyl-1-thio-β-D-galactopyranoside (12). Following general process A, starting from **12a** (50 mg, 0.12 mmol), after 1 h of ring-opening reaction for the crude anhydro sugar, purification by silica gel flash column chromatography afforded **12** as a white solid (34.2 mg, 53%): mp 89.8–90.4 °C; Rf = 0.33 (petroleum ether/ethyl acetate 4:1); ^{1}H NMR (400 MHz, chloroform-d) δ 7.59–7.48 (m, 2H), 7.41–7.07 (m, 18H), 4.89 (d, J = 11.5 Hz, 1H), 4.78–4.61 (m, 2H), 4.61–4.40 (m, 4H,H-1, ArCH_2), 4.07–3.90 (m, 2H, H-2, H-4), 3.66 (s, 3H, H-5, H-6a and H-6b), 3.48 (dd, J = 9.3, 2.7 Hz, 1H, H-3), 2.46 (d, J = 2.2 Hz, 1H, OH) ppm. ^{13}C NMR (100 MHz, chloroform-d) δ 138.64, 137.99, 137.85, 132.60, 132.21, 128.84, 128.56, 128.46, 128.19, 127.94, 127.88, 127.85, 127.75, 127.71, 127.58, 127.47, 88.51, 83.22, 77.62, 74.41, 73.61, 73.20, 72.43, 69.07, 68.70 ppm. $[α]^{20}_D$ = −39.2 (c 0.13, CH_2Cl_2); HRMS (ESI-TOF) (m/z): [M + Na]$^+$ calculated for $C_{31}H_{32}O_5S_2Na^+$, 565.2025; found, 565.2014.

Phenyl 3,4-di-O-benzyl-6-O-tert-butyl-dimethylsily-1-thio-β-D-galactopyranoside (13). Following general process A, starting from **13a** (50 mg, 0.114 mmol), after 1 h of ring-opening reaction for the crude anhydro sugar, purification by silica gel flash column chromatography afforded **13** as a colorless oil (38.6 mg, 60%). Rf = 0.45 (petroleum ether/ethyl acetate 8:1); ^{1}H NMR (400 MHz, chloroform-d) δ 7.54–7.47 (m, 2H), 7.36–7.09 (m, 13H), 4.87 (d, J = 11.4 Hz, 1H), 4.68 (s, 2H), 4.57 (d, J = 11.4 Hz, 1H), 4.48 (d, J = 9.6 Hz, 1H), 4.01–3.93 (m, 1H), 3.91 (d, J = 2.7 Hz, 1H), 3.77–3.64 (m, 2H), 3.50–3.41 (m, 2H), 2.45–2.40 (m, 1H), 0.85 (s, 9H), 0.00 (s, 6H) ppm. ^{13}C NMR (100 MHz, chloroform-d) δ 138.85, 138.11, 132.67, 132.14, 128.83, 128.56, 128.16, 127.89, 127.76, 127.59, 127.52, 127.38, 88.52, 83.31, 79.31, 74.43, 73.08, 72.53, 69.10, 61.52, 25.93, 18.23, −5.32, −5.42 ppm. $[α]^{20}_D$ = +35.0 (c 0.1, CH_2Cl_2); HRMS (ESI-TOF) (m/z): [M + Na]$^+$ calculated for $C_{32}H_{42}O_5SiSNa^+$, 589.2420; found, 589.2396.

Phenyl 3-O-benzyl-4-O-acetyl-6-O-tert-butyl-dimethylsily-1-thio-β-D-galactopyranoside (14). Following general process A, starting from **14a** (50 mg, 0.127 mmol), after 0.5 h of ring-opening reaction for the crude anhydro sugar, purification by silica gel flash column chromatography afforded **14** as a white solid (34.5 mg, 52%): mp 87.1–89.3 °C; Rf = 0.53 (petroleum ether/ethyl acetate 4:1); ^{1}H NMR (400 MHz, chloroform-d) δ 7.55–7.48 (m, 2H), 7.32–7.21 (m, 8H), 5.56 (d, J = 3.0 Hz, 1H), 4.77 (d, J = 11.2 Hz, 1H), 4.55 (d, J = 9.7 Hz, 1H), 4.43 (d, J = 11.2 Hz, 1H), 3.73–3.65 (m, 2H), 3.64–3.54 (m, 2H), 3.46 (dd, J = 9.2, 3.1 Hz, 1H), 2.43 (d, J = 2.0 Hz, 1H), 2.04 (s, 3H), 0.84 (s, 9H), 0.00 (s, 6H) ppm. ^{13}C NMR (100 MHz, chloroform-d) δ 170.05, 137.40, 132.64, 132.34, 128.87, 128.55, 128.27, 128.03, 127.82, 88.52, 80.45, 77.77, 71.73, 68.65, 65.98, 61.28, 25.81, 20.83, 18.21, −5.51, −5.61 ppm. $[α]^{20}_D$ = −110 (c 0.03, CH_2Cl_2); HRMS (ESI-TOF) (m/z): [M + Na]$^+$ calculated for $C_{27}H_{38}O_6SiSNa^+$, 541.2056; found, 541.2047.

Phenyl 3,4-di-O-benzyl-1-thio-β-D-xyloside (15). Following general process A, starting from **15a** (50 mg, 0.169 mmol), after 1 h of ring-opening reaction for the crude anhydro sugar, purification by silica gel flash column chromatography afforded **15** as colorless syrup (46.3 mg, 65%). Rf = 0.53 (petroleum ether/ethyl acetate 4:1); ^{1}H NMR (400 MHz, chloroform-d) δ 7.54–7.48 (m, 2H), 7.41–7.19 (m, 13H), 4.92 (d, J = 5.9 Hz, 1H, H-1), 4.83 (d, J = 11.6 Hz, 1H), 4.74 (d, J = 11.6 Hz, 1H), 4.63 (s, 2H), 4.29 (dd, J = 11.7, 3.0 Hz, 1H, H-5b), 3.72 (q, J = 6.1 Hz, 1H, H-2), 3.63 (t, J = 6.1 Hz, 1H, H-3), 3.59–3.45 (m, 2H, H-4 and H-5a), 3.25 (d, J = 6.3 Hz, 1H, OH) ppm. ^{13}C NMR (100 MHz, chloroform-d) δ 138.08, 137.56, 134.18, 131.91, 128.98, 128.57, 128.53, 128.05, 127.89, 127.83, 127.57, 88.95, 79.32, 77.27, 75.92, 73.89, 72.41, 70.83, 63.55 ppm. $[α]^{20}_D$ = +85.0 (c 0.02, CH_2Cl_2); HRMS (ESI-TOF) (m/z): [M + Na]$^+$ calculated for $C_{25}H_{26}O_4SNa^+$, 445.1449; found, 445.1467.

Phenyl 2,3,3′,4,6,6′-hexa-O-benzyl-D-1-thio-β-lactoside (16) [32]. Following general process A, starting from **16a** (100 mg, 0.118 mmol), after 1 h of ring-opening reaction for the crude anhydro sugar, purification by silica gel flash column chromatography afforded **16** as colorless syrup (64.2 mg, 56%). Rf = 0.61 (petroleum ether/ethyl acetate 3:1); ^{1}H NMR (400 MHz, chloroform-d) δ 7.60 –7.55 (m, 2H), 7.41–7.15 (m, 33H), 5.07 (d, J = 11.0 Hz, 1H), 4.96 (d, J = 11.5 Hz, 1H), 4.85–4.73 (m, 2H), 4.73–4.62 (m, 3H), 4.58–4.47 (m, 3H), 4.48–4.37 (m, 2H), 4.34 (d, J = 11.7 Hz, 1H), 4.26 (d, J = 11.8 Hz, 1H), 3.98–3.89 (m, 2H), 3.88–3.73 (m, 3H), 3.62–3.54 (m, 1H), 3.53–3.35 (m, 6H), 2.50 (s, 1H) ppm. ^{13}C NMR (100 MHz, chloroform-d) δ 139.00, 138.80, 138.71, 138.47, 132.97, 132.00, 128.87, 128.41, 128.38, 128.27, 128.23, 128.17, 128.13, 127.91, 127.87, 127.75, 127.67, 127.57, 127.49, 127.45, 127.35, 102.86, 87.40, 84.18, 82.51, 79.98, 79.76, 76.06, 75.37, 75.03, 74.67, 73.60, 73.46, 73.10, 73.02, 72.65,

71.59, 68.34, 68.14 ppm. $[\alpha]^{20}_D$ = −46.3 (c 0.08, CH_2Cl_2); HRMS (ESI-TOF) (m/z): $[M + Na]^+$ calculated for $C_{60}H_{62}O_{10}SNa^+$, 997.3961; found, 997.3990.

4-Methylphenyl 3,4,6-tri-O-benzyl-1-thio-β-D-glucopyranoside (17) [33]. Following general process A, starting from **2** (100 mg, 0.24 mmol), after 5 min of ring-opening reaction for the crude anhydro sugar, purification by silica gel flash column chromatography afforded **17** as a white solid (99.2 mg, 74%). Rf = 0.51 (petroleum ether/ethyl acetate 6:1); ^{1}H NMR (400 MHz, chloroform-d) δ 7.50–7.41 (m, 2H), 7.40–7.15 (m, 15H), 7.10–6.97 (m, 2H), 4.91 (d, J = 11.2 Hz, 1H), 4.87–4.78 (m, 2H), 4.65–4.50 (m, 3H), 4.43 (d, J = 9.6 Hz, 1H), 3.81–3.68 (m, 2H), 3.61–3.55 (m, 2H), 3.54–3.49 (m, 1H), 3.49–3.40 (m, 1H), 2.40 (d, J = 2.0 Hz, 1H), 2.31 (s, 3H) ppm.

4-Methoxyphenyl 3,4,6-tri-O-benzyl-1-thio-β-D-glucopyranoside (18) [20]. Following general process A, starting from **2** (100 mg, 0.24 mmol), after 5 min of ring-opening reaction for the crude anhydro sugar, purification by silica gel flash column chromatography afforded **18** as colorless syrup (105 mg, 76%). Rf = 0.63 (petroleum ether/ethyl acetate 8:1); ^{1}H NMR (400 MHz, chloroform-d) δ 7.60–7.46 (m, 2H), 7.41–7.29 (m, 13H), 7.25–7.13 (m, 2H), 6.82–6.70 (m, 2H), 4.96–4.80 (m, 3H), 4.68–4.53 (m, 3H), 4.39 (d, J = 9.6 Hz, 1H), 3.80–3.77 (m, 5H), 3.62–3.50 (m, 3H), 3.42 (dd, J = 9.6, 8.4 Hz, 1H), 2.43 (s, 1H) ppm.

4-Chlorophenyl 3,4,6-tri-O-benzyl-1-thio-β-D-glucopyranoside (19). Following general process A, starting from **2** (100 mg, 0.24 mmol), after 30 min of ring-opening reaction for the crude anhydro sugar, purification by silica gel flash column chromatography afforded **19** as colorless syrup (96 mg, 70%). Rf = 0.33 (petroleum ether/ethyl acetate 8:1); ^{1}H NMR (400 MHz, chloroform-d) δ 7.56–7.49 (m, 2H), 7.42–7.18 (m, 17H), 4.94–4.79 (m, 3H), 4.64–4.54 (m, 3H), 4.49 (d, J = 9.6 Hz, 1H), 3.83–3.71 (m, 2H), 3.66–3.42 (m, 4H), 2.38 (d, J = 2.2 Hz, 1H) ppm. ^{13}C NMR (100 MHz, chloroform-d) δ 138.38, 138.18, 137.97, 134.41, 134.32, 130.15, 129.10, 129.05, 128.81, 128.58, 128.47, 128.42, 128.00, 127.99, 127.91, 127.88, 127.70, 127.66, 87.61, 85.89, 79.35, 75.38, 75.09, 73.43, 72.44, 68.93, 38.00 ppm. $[\alpha]^{20}_D$ = −115 (c 0.04, CH_2Cl_2); HRMS (ESI-TOF) (m/z): $[M + Na]^+$ calculated for $C_{33}H_{33}O_5ClSNa^+$, 599.1635; found, 599.1611.

4-Aminophenyl 3,4,6-tri-O-benzyl-1-thio-β-D-glucopyranoside (20). Following general process A, starting from **2** (50 mg, 0.12 mmol), after 30 min of ring-opening reaction for the crude anhydro sugar, purification by silica gel flash column chromatography afforded **20** as colorless syrup (45 mg, 68%). Rf = 0.35 (petroleum ether/ethyl acetate 2:1); ^{1}H NMR (600 MHz, $CDCl_3$) δ 7.41–7.25 (m, 15H), 7.24–7.20 (m, 2H), 6.59–6.54 (m, 2H), 4.94 (d, J = 11.2 Hz, 1H), 4.88–4.81 (m, 2H), 4.66–4.54 (m, 3H), 4.33 (d, J = 9.6 Hz, 1H), 4.15 (q, J = 7.1 Hz, 1H), 3.82–3.74 (m, 2H), 3.64–3.54 (m, 2H), 3.42 (t, J = 9.0 Hz, 1H), 2.43 (s, 1H) ppm. ^{13}C NMR (100 MHz, chloroform-d) δ 147.25, 138.57, 136.38, 128.48, 128.40, 128.33, 127.98, 127.76, 127.64, 127.49, 115.33, 88.26, 85.91, 79.46, 75.26, 75.06, 73.43, 72.21, 69.04, 29.71, 29.33 ppm. $[\alpha]^{20}_D$ = +61.5 (c 0.026, CH_2Cl_2); HRMS (ESI-TOF) (m/z): $[M + Na]^+$ calculated for $C_{31}H_{32}O_5S_2Na^+$, 580.2130; found, 580.2120.

Thiophen-2-ylthio 3,4,6-tri-O-benzyl-1-thio-β-D-glucopyranoside (21). Following general process A, starting from **2** (50 mg, 0.12 mol), after 30 min of ring-opening reaction for the crude anhydro sugar, purification by silica gel flash column chromatography afforded **21** as a red solid (41 mg, 63%): mp 90.9–92.1 °C; Rf = 0.52 (petroleum ether/ethyl acetate 5:1); ^{1}H NMR (400 MHz, chloroform-d) δ 7.47–7.09 (m, 17H), 6.98 (dd, J = 5.5, 3.5 Hz, 1H), 4.92–4.76 (m, 3H), 4.70–4.51 (m, 3H), 4.30 (d, J = 9.4 Hz, 1H), 3.82–3.70 (m, 2H), 3.62–3.46 (m, 3H), 3.42 (ddd, J = 9.0, 6.2, 2.7 Hz, 1H), 2.36 (d, J = 2.4 Hz, 1H) ppm. ^{13}C NMR (100 MHz, chloroform-d) δ 138.39, 138.04, 136.17, 131.08, 128.56, 128.43, 128.34, 128.00, 127.96, 127.88, 127.81, 127.63, 127.53, 87.53, 85.80, 79.69, 75.39, 75.06, 73.50, 71.86, 68.87, 29.39 ppm. $[\alpha]^{20}_D$ = −46.6 (c 0.058, CH_2Cl_2); HRMS (ESI-TOF) (m/z): $[M + Na]^+$ calculated for $C_{31}H_{32}O_5S_2Na^+$, 571.1589; found, 571.1602.

Benzothiazol-2-yl 3,4,6-tri-O-benzyl-1-thio-β-D-glucopyranoside (22). Following general process A, starting from **2** (50 mg, 0.12 mmol), after 30 min of ring-opening reaction for the crude anhydro sugar, purification by silica gel flash column chromatography afforded **22** as a white solid (51.3 mg, 72%): mp 119.9–123.0 °C; Rf = 0.36 (petroleum ether/ethyl acetate 6:1); ^{1}H NMR (400 MHz, chloroform-*d*) δ 7.94 (d, *J* = 8.1 Hz, 1H), 7.72 (d, *J* = 8.0 Hz, 1H), 7.45 (t, *J* = 7.7 Hz, 1H), 7.39–7.23 (m, 14H), 7.23–7.17 (m, 2H), 5.07 (d, *J* = 9.5 Hz, 1H), 4.99–4.80 (m, 3H), 4.67–4.47 (m, 3H), 3.86–3.63 (m, 6H), 3.13 (d, *J* = 3.1 Hz, 1H) ppm. ^{13}C NMR (100 MHz, chloroform-*d*) δ 211.54, 152.68, 138.40, 138.11, 138.00, 128.56, 128.45, 128.36, 128.01, 127.95, 127.89, 127.85, 127.76, 127.61, 126.31, 125.06, 122.43, 121.02, 86.49, 85.95, 79.94, 75.49, 75.10, 73.50, 68.65, 29.34, 14.14 ppm. [α]$^{20}_D$ = −64.3 (c 0.07, CH$_2$Cl$_2$); HRMS (ESI-TOF) (*m/z*): [M + Na]$^+$ calculated for C$_{34}$H$_{33}$NO$_5$S$_2$Na$^+$, 622.1698; found, 622.1707.

Phenyl 3,4,6-tri-O-benzyl-1-seleno-β-D-glucopyranoside (**26**) [34]. Following general process A, starting from **2** (100 mg, 0.24 mmol), after 12 h of ring-opening reaction for the crude anhydro sugar, purification by silica gel flash column chromatography afforded **26** as a white solid (85.5 mg, 61%). Rf = 0.43 (petroleum ether/ethyl acetate 8:1); ^{1}H NMR (400 MHz, chloroform-*d*) δ 7.72–7.66 (m, 2H), 7.41–7.19 (m, 18H), 4.93 (d, *J* = 11.2 Hz, 1H), 4.89–4.82 (m, 2H), 4.75 (d, *J* = 9.7 Hz, 1H), 4.67–4.54 (m, 3H), 3.85–3.74 (m, 2H), 3.67–3.57 (m, 2H), 3.56–3.47 (m, 2H) ppm.

Phenyl 3,4,6-tri-O-benzyl-2-O-picolyl-β-D-1-thio-glucopyranoside (**28**). Following the general process B, starting from **2** (100 mg, 0.24 mmol), after 2 h of ring-opening reaction for the crude anhydro sugar and then reaction with PicBr•HBr (2.0 equiv) for 1 h in the presence of sodium hydride (6 equiv), the residue was purified by column chromatography on silica gel (ethyl acetate-hexane gradient elution) to afford **28** as a colorless syrup (101 mg, 66%). Rf = 0.33 (petroleum ether/ethyl acetate 4:1); ^{1}H NMR (400 MHz, chloroform-*d*) δ 8.58–8.50 (m, 1H), 7.65 (td, *J* = 7.7, 1.9 Hz, 1H), 7.58–7.49 (m, 3H), 7.40–7.09 (m, 19H), 5.04 (d, *J* = 12.4 Hz, 1H), 4.96–4.75 (m, 4H), 4.69 (d, *J* = 9.6 Hz, 1H), 4.64–4.49 (m, 3H), 3.84–3.69 (m, 3H), 3.65 (t, *J* = 9.3 Hz, 1H), 3.55 (t, *J* = 9.1 Hz, 2H) ppm. ^{13}C NMR (100 MHz, chloroform-*d*) δ 158.30, 149.00, 138.29, 138.20, 138.06, 136.52, 133.54, 132.04, 128.90, 128.44, 128.41, 128.35, 128.00, 127.92, 127.82, 127.70, 127.65, 127.56, 127.50, 122.35, 121.65, 87.19, 86.55, 81.32, 79.13, 75.93, 75.81, 75.06, 73.43, 69.04 ppm. [α]$^{20}_D$ = −45 (c 0.04, CH$_2$Cl$_2$); HRMS (ESI-TOF) (*m/z*): [M + Na]$^+$ calculated for C$_{39}$H$_{39}$O$_5$NSNa$^+$, 656.2447; found, 656.2416.

Phenyl 3,4,6-tri-O-benzyl-2-O-picolyl-β-D-1-thio-galactopyranoside (**29**). Following the general process B, starting from **12a** (100 mg, 0.24 mmol), after 2 h of ring-opening reaction for the crude anhydro sugar, and then reaction with PicBr•HBr (2.0 equiv) for 1 h in the presence of sodium hydride (6 equiv), the residue was purified by column chromatography on silica gel (ethyl acetate-hexane gradient elution) to afford **29** as a colorless syrup (73 mg, 48%). Rf = 0.35 (petroleum ether/ethyl acetate 4:1); ^{1}H NMR (400 MHz, chloroform-*d*) δ 8.53 (d, *J* = 5.0 Hz, 1H), 7.62 (t, *J* = 7.6 Hz, 1H), 7.56–7.43 (m, 3H), 7.41–7.09 (m, 19H), 4.98–4.87 (m, 3H), 4.75–4.63 (m, 3H), 4.58 (d, *J* = 11.5 Hz, 1H), 4.51–4.37 (m, 2H), 4.01–3.90 (m, 2H), 3.72–3.57 (m, 4H) ppm. ^{13}C NMR (100 MHz, chloroform-*d*) δ 158.67, 148.91, 138.75, 138.14, 137.89, 136.40, 133.94, 131.56, 128.80, 128.46, 128.39, 128.22, 127.96, 127.86, 127.84, 127.64, 127.49, 127.11, 122.24, 121.79, 87.52, 83.91, 78.04, 77.40, 77.34, 77.08, 76.76, 76.28, 74.49, 73.62, 73.48, 72.57, 68.79 ppm. [α]$^{20}_D$ = +9.1 (c 0.33, CH$_2$Cl$_2$); HRMS (ESI-TOF) (*m/z*): [M + Na]$^+$ calculated for C$_{39}$H$_{39}$O$_5$NSNa$^+$, 656.2447; found, 656.2433.

Phenyl 3,4,6-Tri-O-benzyl-2-O-(phenylmethoxy)methyl-β-D-1-thio-glucopyranoside (**30**) [21]. Following the general process B, starting from **2** (100 mg, 0.24 mmol), after 2 h of ring-opening reaction for the crude anhydro sugar and then reaction with BOMCl (2.0 equiv) for 3 h in the presence of sodium hydride (6 equiv), the residue was purified by column chromatography on silica gel (ethyl acetate-hexane gradient elution) to afford **30** as a white solid (100.3 mg, 63%). Rf = 0.53 (petroleum ether/ethyl acetate 6:1); ^{1}H NMR (400 MHz, CDCl$_3$) δ 7.59–7.49 (m, 2H), 7.39–7.12 (m, 23H), 5.06 (d, *J* = 6.4 Hz, 1H), 4.94 (d, *J* = 6.4 Hz, 1H), 4.92–4.83 (m, 3H), 4.79 (d, *J* = 10.8 Hz, 1H), 4.69–4.58 (m, 3H), 4.58–4.49 (m, 2H), 3.77 (dd, *J* = 10.9, 2.0 Hz, 1H), 3.74–3.61 (m, 4H), 3.52 (m, 1H) ppm.

Phenyl 3,4,6-Tri-O-benzyl-2-O-(cyanomethyl)-β-D-1-thio-glucopyranoside (**31**) [18]. Following the general process B, starting from **2** (100 mg, 0.24 mmol), after 2 h of ring-opening reaction for the crude anhydro sugar and then reaction with bromoacetonitrile (2.5 equiv) for 2 h in the presence of sodium hydride (6 equiv), the residue was purified by column chromatography on silica gel (ethyl acetate-hexane gradient elution) to afford **31** as a white solid (76.8 mg, 55%). Rf = 0.53 (petroleum ether/ethyl acetate 6:1); ^{1}H NMR (400 MHz, CDCl$_3$) δ 7.61–7.52 (m, 2H), 7.41–7.22 (m, 16H), 7.19 (dd, J = 7.2, 2.3 Hz, 2H), 4.90–4.82 (m, 2H), 4.80 (d, J = 10.9 Hz, 1H), 4.64–4.56 (m, 2H), 4.55 (s, 1H), 4.52 (t, J = 1.8 Hz, 1H), 4.51–4.39 (m, 2H), 3.77 (dd, J = 10.9, 2.1 Hz, 1H), 3.72 (dd, J = 10.9, 4.3 Hz, 1H), 3.67–3.62 (m, 2H), 3.47 (m, 1H), 3.35 (m, 1H) ppm.

Phenyl 3,4,6-Tri-O-benzyl-2-O-(2-cyanobenzyl)-β-D-1-thio-glucopyranoside (32) [22]. Following the general process B, starting from **2** (100 mg, 0.24 mmol), after 2 h of ring-opening reaction for the crude anhydro sugar and then reaction with 2-cyanobenzyl bromide (2.0 equiv) for 1 h in the presence of sodium hydride (6 equiv), the residue was purified by column chromatography on silica gel (ethyl acetate-hexane gradient elution) to afford **32** as a colorless oil (107.4 mg, 68%). Rf = 0.31 (petroleum ether/ethyl acetate 6:1); ^{1}H NMR (400 MHz, CDCl$_3$) δ 7.73–7.53 (m, 6H), 7.43–7.19 (m, 18H), 5.11 (d, J = 12.5 Hz, 1H), 5.03 (d, J = 12.5 Hz, 1H), 4.91–4.79 (m, 3H), 4.73–4.56 (m, 4H), 3.83 (dd, J = 10.9, 2.0 Hz, 1H), 3.78 (d, J = 4.4 Hz, 1H), 3.77–3.66 (m, 2H), 3.60–3.51 (m, 2H) ppm.

Phenyl 3,4,6-Tri-O-benzyl-2-O-benzoyl-β-D-1-thio-glucopyranoside (33) [35]. Following the general process B, starting from **2** (100 mg, 0.24 mmol), after 2 h of ring-opening reaction for the crude anhydro sugar and then reaction with BzCl (3.0 equiv) for 2 h in the presence of sodium hydride (6 equiv), the residue was purified by column chromatography on silica gel (ethyl acetate-hexane gradient elution) to afford **33** as a white solid (97.8 mg, 63%). Rf = 0.61 (petroleum ether/ethyl acetate 6:1); ^{1}H NMR (400 MHz, CDCl$_3$) δ 8.11–8.03 (m, 2H), 7.60 (d, J = 7.4 Hz, 1H), 7.54–7.44 (m, 4H), 7.43–7.20 (m, 13H), 7.18–7.11 (m, 5H), 5.31 (dd, J = 10.0, 9.0 Hz, 1H), 4.87–4.78 (m, 2H), 4.75 (d, J = 11.0 Hz, 1H), 4.70–4.55 (m, 4H), 3.91–3.82 (m, 2H), 3.81–3.74 (m, 2H), 3.64 (m, 1H) ppm.

Phenyl 3,4,6-Tri-O-benzyl-2-O-pivaloyl-β-D-1-thio-glucopyranoside (34) [36]. Following the general process B, starting from **2** (100 mg, 0.24 mmol), after 2 h of ring-opening reaction for the crude anhydro sugar and then reaction with PivCl (3.0 equiv) for 2 h in the presence of sodium hydride (6 equiv), the residue was purified by column chromatography on silica gel (ethyl acetate-hexane gradient elution) to afford **34** as a white solid (78.3 mg, 52%). Rf = 0.73 (petroleum ether/ethyl acetate 8:1); ^{1}H NMR (400 MHz, CDCl$_3$) δ 7.58–7.49 (m, 2H), 7.41–7.22 (m, 16H), 7.19 (dd, J = 7.2, 2.4 Hz, 2H), 5.13 (t, 1H), 4.84–4.76 (m, 2H), 4.75–4.53 (m, 5H), 3.81 (dd, 1 H, J = 1.5, 11.0 Hz, 1H), 3.79–3.67 (m, 3H), 3.59 (m, 1H), 1.26 (s, 9H) ppm.

Methyl 2,3,4-tri-O-benzoyl-6-O-(3,4,6-tri-O-benzyl-2-O-picolyl-β-D-glucopyranosyl)-α-D-glucopyranoside (35). Following the general process C, the glycosylation between **28** (100 mg, 0.16 mmol, 1.3 equiv) and methyl 2,3,4-tri-O-benzoyl-α-D-glucopyranoside (1.0 equiv) led to **30**. Purification by silica gel flash column chromatography afforded **35** as a colorless syrup (87.5 mg, 70%, β-only). Rf = 0.26 (petroleum ether/ethyl acetate 2:1); ^{1}H NMR (400 MHz, chloroform-d) δ 8.53 (d, J = 4.9 Hz, 1H), 8.03–7.76 (m, 7H), 7.66–7.58 (m, 1H), 7.55–7.46 (m, 3H), 7.45–7.20 (m, 19H), 7.17–7.07 (m, 3H), 6.14 (t, J = 9.8 Hz, 1H), 5.42 (t, J = 9.9 Hz, 1H), 5.24–5.16 (m, 2H), 5.11 (d, J = 3.6 Hz, 1H), 4.95–4.83 (m, 2H), 4.82–4.73 (m, 2H), 4.56–4.46 (m, 3H), 4.44 (d, J = 12.3 Hz, 1H), 4.34 (td, J = 8.7, 8.2, 4.1 Hz, 1H), 4.10 (dd, J = 10.9, 2.1 Hz, 1H), 3.77 (dd, J = 11.1, 7.8 Hz, 1H), 3.73–3.57 (m, 4H), 3.53–3.41 (m, 2H), 3.34 (s, 3H) ppm. ^{13}C NMR (100 MHz, chloroform-d) δ 165.81, 165.74, 165.50, 158.78, 148.95, 138.50, 138.14, 138.09, 136.46, 133.42, 133.33, 133.05, 129.93, 129.90, 129.66, 129.28, 129.11, 128.90, 128.41, 128.34, 128.25, 127.98, 127.94, 127.75, 127.58, 122.18, 121.45, 103.94, 96.68, 84.47, 82.78, 77.65, 75.68, 75.40, 75.00, 74.95, 73.45, 72.11, 70.49, 69.91, 68.97, 68.61, 55.52 ppm. [α]$^{20}_D$ = +40.9 (c 0.22, CH$_2$Cl$_2$); HRMS (ESI-TOF) (*m/z*): [M + Na]$^+$ calculated for C$_{61}$H$_{59}$O$_{14}$NNa$^+$, 1052.3833 ; found, 1052.3814.

Methyl 2,3,4-tri-O-benzyl-6-O-(3,4,6-tri-O-benzyl-2-O-picolyl-β-D-glucopyranosyl)-α-D-glucopyranoside (36) [37]. Following the general process C, the glycosylation between **28** (100 mg, 0.16 mmol, 1.3 equiv) and methyl 2,3,4-tri-O-benzyl-α-D-glucopyranoside (1.0 equiv) led to **36**. Purification by silica gel flash column chromatography afforded **36** as a colorless syrup (97.7 mg, 82%, β-only). Rf = 0.46 (petroleum ether/ethyl acetate 2:1); ^{1}H NMR (400 MHz, chloroform-d) δ 8.44 (d, J = 4.8 Hz, 1H), 7.48–7.20 (m, 28H), 7.19–7.11 (m,

4H), 7.02 (t, *J* = 6.1 Hz, 1H), 5.13 (d, *J* = 12.9 Hz, 1H), 4.96–4.85 (m, 3H), 4.84–4.70 (m, 4H), 4.68–4.59 (m, 2H), 4.58–4.51 (m, 4H), 4.47 (d, *J* = 11.0 Hz, 1H), 4.39 (d, *J* = 7.8 Hz, 1H), 4.17 (d, *J* = 10.7 Hz, 1H), 3.95 (t, *J* = 9.3 Hz, 1H), 3.78 (dt, *J* = 14.3, 7.1 Hz, 1H), 3.72–3.61 (m, 4H), 3.60–3.40 (m, 5H), 3.31 (s, 3H) ppm.

Methyl 2,3,6-tri-O-benzyl-4-O-(3,4,6-tri-O-benzyl-2-O-picolyl-β-D-glucopyranosyl)-α-D-glucopyranoside (37) [37]. Following the general process C, the glycosylation between **28** (100 mg, 0.16 mmol, 1.3 equiv) and methyl 2,3,6-tri-*O*-benzyl-α-D-glucopyranoside (1.0 equiv) led to **37**. Purification by silica gel flash column chromatography afforded **37** as a colorless syrup (59.6 mg, 50%, β-only). Rf = 0.48 (petroleum ether/ethyl acetate 2:1); ^{1}H NMR (400 MHz, chloroform-*d*) δ 8.50 (d, *J* = 4.8 Hz, 1H), 7.54 (t, *J* = 7.7 Hz, 1H), 7.45–7.06 (m, 32H), 5.07 (d, *J* = 11.3 Hz, 1H), 4.96 (d, *J* = 13.5 Hz, 1H), 4.90–4.71 (m, 6H), 4.64–4.52 (m, 4H), 4.48–4.34 (m, 4H), 3.96 (t, *J* = 9.5 Hz, 1H), 3.87–3.77 (m, 2H), 3.72 (d, *J* = 10.9 Hz, 1H), 3.65–3.57 (m, 2H), 3.48 (m, 5H), 3.34 (s, 3H), 3.31–3.27 (m, 1H) ppm.

Methyl 2,4,6-tri-O-benzyl-3-O-(3,4,6-tri-O-benzyl-2-O-picolyl-β-D-glucopyranosyl)-α-D-glucopyranoside (38) [37]. Following the general process C, the glycosylation between **28** (100 mg, 0.16 mmol, 1.3 equiv) and methyl 2,4,6-tri-*O*-benzyl-α-D-glucopyranoside (1.0 equiv) led to **38**. Purification by silica gel flash column chromatography afforded **38** as a colorless syrup (95.4 mg, 80%, β-only). Rf = 0.65 (petroleum ether/ethyl acetate 2:1); ^{1}H NMR (400 MHz, chloroform-*d*) δ 8.56 (d, *J* = 4.9 Hz, 1H), 7.56–7.47 (m, 2H), 7.40–7.04 (m, 31H), 5.29 (d, *J* = 13.5 Hz, 1H), 5.14–4.95 (m, 4H), 4.89 (d, *J* = 10.9 Hz, 1H), 4.83 (d, *J* = 10.6 Hz, 1H), 4.70–4.53 (m, 3H), 4.53–4.33 (m, 7H), 3.81–3.64 (m, 6H), 3.63–3.48 (m, 4H), 3.45 (d, *J* = 9.4 Hz, 1H), 3.30 (s, 3H) ppm.

Methyl 3,4,6-tri-O-benzyl-2-O-(3,4,6-tri-O-benzyl-2-O-picolyl-β-D-glucopyranosyl)-α-D-glucopyranoside (39) [37]. Following the general process C, the glycosylation between **28** (100 mg, 0.16 mmol, 1.3 equiv) and methyl 3,4,6-tri-*O*-benzyl-α-D-glucopyranoside led to **39**. Purification by silica gel flash column chromatography afforded **39** as a colorless syrup (91.8 mg, 77%, β-only). Rf = 0.55 (petroleum ether/ethyl acetate 2:1); ^{1}H NMR (400 MHz, chloroform-*d*) δ 8.42 (d, *J* = 5.0 Hz, 1H), 7.45–7.10 (m, 30H), 7.11–6.96 (m, 3H), 5.18 (d, *J* = 13.4 Hz, 1H), 5.03–4.85 (m, 3H), 4.85–4.70 (m, 4H), 4.67–4.56 (m, 3H), 4.55–4.40 (m, 4H), 4.01 (t, *J* = 9.2 Hz, 1H), 3.89–3.61 (m, 10H), 3.58 (t, *J* = 7.4 Hz, 1H), 3.43 (m, 1H), 3.39 (s, 3H) ppm.

1,2:5,6-Di-O-isopropylidine-3-O-(3,4,6-tri-O-benzyl-2-O-picolyl-β-D-glucopyranosyl)-α-D-glucofuranose (40) [37]. Following the general process C, the glycosylation between **28** (100 mg, 0.16 mmol, 1.3 equiv) and 1,2:5,6-bis-*O*-(1-methylethylidene)-α-D-glucofuranose (1.0 equiv) led to **40**. Purification by silica gel flash column chromatography afforded **40** as colorless syrup (49.2 mg, 52%, β-only). Rf = 0.35 (petroleum ether/ethyl acetate 2:1); ^{1}H NMR (400 MHz, chloroform-*d*) δ 8.56 (d, *J* = 4.9 Hz, 1H), 7.67–7.56 (m, 1H), 7.43–7.22 (m, 14H), 7.21–7.08 (m, 3H), 5.81 (d, *J* = 3.7 Hz, 1H), 4.93–4.86 (m, 2H), 4.85–4.75 (m, 3H), 4.66–4.47 (m, 5H), 4.43 (q, *J* = 5.9 Hz, 1H), 4.38–4.28 (m, 2H), 4.12–4.01 (m, 2H), 3.78–3.58 (m, 4H), 3.47–3.38 (m, 2H), 1.48 (s, 3H), 1.41 (s, 3H), 1.31 (s, 3H), 1.25 (s, 3H) ppm.

Methyl 2,3,4-tri-O-benzyl-6-O-(3,4,6-tri-O-benzyl-2-O-picolyl-β-D-galactopyranosyl)-α-D-glucopyranoside (41) [37]. Following the general process C, the glycosylation between **29** (100 mg, 0.16 mmol, 1.3 equiv) and methyl 2,3,4-tri-*O*-benzyl-α-D-glucopyranoside (1.0 equiv) led to **41**. Purification by silica gel flash column chromatography afforded **41** as a colorless syrup (102.5 mg, 86%, β-only). Rf = 0.36 (petroleum ether/ethyl acetate 2:1); ^{1}H NMR (400 MHz, chloroform-*d*) δ 8.43 (d, *J* = 4.8 Hz, 1H), 7.48–7.39 (m, 2H), 7.39–7.08 (m, 30H), 7.04 (q, *J* = 4.7 Hz, 1H), 5.09 (d, *J* = 13.2 Hz, 1H), 4.98–4.87 (m, 3H), 4.80–4.38 (m, 11H), 4.35 (d, *J* = 7.7 Hz, 1H), 4.13 (d, *J* = 10.8 Hz, 1H), 3.97–3.84 (m, 3H), 3.79 (dd, *J* = 10.5, 5.0 Hz, 1H), 3.52 (m, 7H), 3.28 (s, 3H) ppm.

Supplementary Materials: The following supporting information can be downloaded at: https://www.mdpi.com/article/10.3390/molecules27185980/s1, Experimental methods, synthesis of thio-containing glycosides, Characterization of unknown compounds, ^{1}H NMR and ^{13}C NMR spectra, references [38–46].

Author Contributions: Conceptualization, Y.-F.G., T.L. and H.D.; methodology, Y.-F.G. and T.L.; validation, Y.-F.G. and T.L.; formal analysis, G.-J.F. and C.-Y.L.; data curation, G.-J.F. and C.-Y.L.; writing—original draft preparation, Y.-F.G., T.L. and H.D.; writing—review and editing, Y.-F.G. and H.D.; supervision, H.D. project administration, H.D.; funding acquisition, H.D. All authors have read and agreed to the published version of the manuscript.

Funding: This research was founded by the National Nature Science Foundation of China (Nos. 21772049).

Institutional Review Board Statement: Not applicable.

Informed Consent Statement: Not applicable.

Data Availability Statement: Not applicable.

Acknowledgments: The authors aregrateful to the staffs in the Analytical and Test Center of HUST for support with the NMR instruments.

Conflicts of Interest: The authors declare no conflict of interest.

Sample Availability: Samples of the compounds are available from the authors.

References

1. Ge, J.-T.; Zhou, L.; Luo, T.; Lv, J.; Dong, H. A One-Pot Method for Removal of Thioacetyl Group via Desulfurization under Ultraviolet Light To Synthesize Deoxyglycosides. *Org. Lett.* **2019**, *21*, 5903–5906. [CrossRef] [PubMed]
2. Zhang, Y.; Zhao, F.-L.; Luo, T.; Pei, Z.; Dong, H. Regio/Stereoselective Glycosylation of Diol and Polyol Acceptors in Efficient Synthesis of Neu5Ac-α-2,3-LacNPhth Trisaccharide. *Chem. Asian J.* **2019**, *14*, 223–234. [CrossRef]
3. Varki, A. Biological roles of glycans. *Glycobiology* **2017**, *27*, 3–49. [CrossRef]
4. Nigudkar, S.S.; Demchenko, A.V. Stereocontrolled 1,2-cis glycosylation as the driving force of progress in synthetic carbohydrate chemistry. *Chem. Sci.* **2015**, *6*, 2687–2704. [CrossRef]
5. Crich, D. Mechanism of a Chemical Glycosylation Reaction. *Acc. Chem. Res.* **2010**, *43*, 1144–1153. [CrossRef]
6. Lv, J.; Liu, C.-Y.; Guo, Y.-F.; Feng, G.-J.; Dong, H. $SnCl_2$-Catalyzed Acetalation/Selective Benzoylation Sequence for the Synthesis of Orthogonally Protected Glycosyl Acceptors. *Eur. J. Org. Chem.* **2022**, *33*, e202101565.
7. Carthy, C.M.; Zhu, X.-M. Chemoselective activation of ethyl vs. phenyl thioglycosides: One-pot synthesis of oligosaccharides. *Org. Biomol. Chem.* **2020**, *18*, 9029–9034. [CrossRef]
8. Lacey, K.D.; Quarels, R.D.; Du, S.; Fulton, A.; Reid, N.J.; Firesheets, A.; Ragains, J.R. Acid-Catalyzed O-Glycosylation with Stable Thioglycoside Donors. *Org. Lett.* **2018**, *20*, 5181–5185. [CrossRef]
9. Saliba, R.C.; Wooke, Z.J.; Nieves, G.A.; Chu, A.A.; Bennett, C.S.; Pohl, N.L.B. Challenges in the Conversion of Manual Processes to Machine-Assisted Syntheses: Activation of Thioglycoside Donors with Aryl(trifluoroethyl) iodonium Triflimide. *Org. Lett.* **2018**, *20*, 800–803. [CrossRef]
10. Vibhute, A.M.; Dhaka, A.; Athiyarath, V.; Sureshan, K.M. A versatile glycosylation strategy via Au(III) catalyzed activation of thioglycoside donors. *Chem. Sci.* **2016**, *7*, 4259–4263. [CrossRef]
11. Goswami, M.; Ashley, D.C.; Baik, M.H.; Pohl, N.L.B. Mechanistic Studies of Bismuth(V)-Mediated Thioglycoside Activation Reveal Differential Reactivity of Anomers. *J. Org. Chem.* **2016**, *81*, 5949–5962. [CrossRef] [PubMed]
12. Lian, G.; Zhang, X.; Yu, B. Thioglycosides in Carbohydrate Research. *Carbohydr. Res.* **2015**, *403*, 13–22. [CrossRef] [PubMed]
13. Luo, T.; Zhang, Q.; Guo, Y.-F.; Pei, Z.-C.; Dong, H. Efficient Preparation of 2-SAc-Glycosyl Donors and Investigation of Their Application in the Synthesis of 2-Deoxyglycosides. *Eur. J. Org. Chem.* **2022**, *2022*, e202200533. [CrossRef]
14. Poulsen, L.T.; Heuckendorff, M.; Jensen, H.H. Effect of 2-O-Benzoyl para-Substituents on Glycosylation Rates. *ACS Omega.* **2018**, *3*, 7117–7123. [CrossRef] [PubMed]
15. Speciale, G.; Farren-Dai, M.; Shidmoossavee, F.S.; Williams, S.J.; Bennet, A.J. C2-Oxyanion Neighboring Group Participation: Transition State Structure for the Hydroxide-Promoted Hydrolysis of 4-Nitrophenyl α-D-Mannopyranoside. *J. Am. Chem. Soc.* **2016**, *138*, 14012–14019. [CrossRef]
16. Elferink, H.; Mensink, R.A.; White, P.B.; Boltje, T.J. Stereoselective β-Mannosylation by Neighboring-Group Participation. *Angew. Chem. Int. Ed.* **2016**, *55*, 11217–11220. [CrossRef]
17. Buda, S.; Nawój, M.; Gołębiowska, P.; Dyduch, K.; Michalak, A.; Mlynarski, J. Application of 2-Substituted Benzyl Groups in Stereoselective Glycosylation. *J. Org. Chem.* **2015**, *80*, 770–780. [CrossRef]
18. Molla, M.R.; Das, P.; Guleria, K.; Subramanian, R.; Kumar, A.; Thakur, R. Cyanomethyl Ether as an Orthogonal Participating Group for Stereoselective Synthesis of 1,2-trans-β-O-Glycosides. *J. Org. Chem.* **2020**, *85*, 9955–9968. [CrossRef]
19. de Kleijne, F.F.J.; Moons, S.J.; White, P.B.; Boltje, T.J. C-2 auxiliaries for stereoselective glycosylation based on common additive functional groups. *Org. Biomol. Chem.* **2020**, *18*, 1165–1184. [CrossRef]
20. Moons, S.J.; Mensink, R.A.; Bruekers, J.P.J.; Vercammen, M.L.A.; Jansen, L.M.; Boltje, T.J. α-Selective Glycosylation with β-Glycosyl Sulfonium Ions Prepared via Intramolecular Alkylation. *J. Org. Chem.* **2019**, *84*, 4486–4500. [CrossRef]

21. Karak, M.; Joh, Y.; Suenaga, M.; Oishi, T.; Torikai, K. 1,2-trans Glycosylation via Neighboring Group Participation of 2-O-Alkoxymethyl Groups: Application to One-Pot Oligosaccharide Synthesis. *Org. Lett.* **2019**, *21*, 1221–1225. [CrossRef] [PubMed]
22. Hoang, K.L.M.; Liu, X.-W. The Intriguing Dual-Directing Effect of 2-Cyanobenzyl Ether for A Highly Stereospecific Glycosylation Reaction. *Nat. Commun.* **2014**, *5*, 5051–5060. [CrossRef] [PubMed]
23. Pozsgay, V.; Kubler-Kielb, J.; Coxon, B.; Santacroce, P.; Robbins, J.B.; Schneerson, R. Synthetic Oligosaccharides as Tools to Demonstrate Cross-Reactivity between Polysaccharide Antigens. *J. Org. Chem.* **2012**, *77*, 5922–5941. [CrossRef] [PubMed]
24. Tiwari, V.K.; Kumar, A.; Schmidt, R.R. Disaccharide-Containing Macrocycles by Click Chemistry and Intramolecular Glycosylation. *Eur. J. Org. Chem.* **2012**, *15*, 2945–2956. [CrossRef]
25. Cox, D.J.; Fairbanks, A.J. Stereoselective synthesis of α-glucosides by neighbouring group participation via an intermediate thiophenium ion. *Tetrahedron: Asymmetry.* **2009**, *20*, 773–780. [CrossRef]
26. Gordon, D.M.; Danishefsky, S.J. Displacement reactions of a 1,2-anhydro-α-D-hexopyranose: Installation of useful functionality at the anomeric carbon. *Carbohydr. Res.* **1990**, *206*, 361–366. [CrossRef]
27. Marzabadi, C.H.; Spilling, C.D. Stereoselective Glucal Epoxide Formation1. *J. Org. Chem.* **1993**, *58*, 3761–3766. [CrossRef]
28. Ge, J.-T.; Zhang, L.-F.; Pu, L.; Zhang, Y.; Pei, Z.-C.; Dong, H. The Oxidation of S-Acetyl by Nitrite: Mechanism and Application. *ChemistrySelect.* **2020**, *5*, 14549–14553. [CrossRef]
29. Feng, G.-J.; Wang, S.-S.; Lv, J.; Luo, T.; Wu, Y.; Dong, H. Improved Synthesis of 1-Glycosyl Thioacetates and Its Application in the Synthesis of Thioglucoside Gliflozin Analogues. *Eur. J. Org. Chem.* **2021**, *2021*, 2940–2949. [CrossRef]
30. Feng, G.-J.; Luo, T.; Guo, Y.-F.; Liu, C.-Y.; Dong, H. Concise Synthesis of 1-Thioalkyl Glycoside Donors by Reaction of Per-O-acetylated Sugars with Sodium Alkanethiolates under Solvent-Free Conditions. *J. Org. Chem.* **2022**, *87*, 3638–3646. [CrossRef]
31. Valerio, S.; Iadonisi, A.; Adinolfi, M.; Ravida, A. Novel Approaches for the Synthesis and Activation of Thio- and Selenoglycoside Donors. *J. Org. Chem.* **2007**, *72*, 6097–6106. [CrossRef] [PubMed]
32. Smoot, J.T.; Pornsuriyasak, P.; Demchenko, A.V. Development of an Arming Participating Group for Stereoselective Glycosylation and Chemoselective Oligosaccharide Synthesis. *Angew. Chem. Int. Ed.* **2005**, *44*, 7123–7126. [CrossRef] [PubMed]
33. Chayajarus, K.; Chambers, D.J.; Chughtai, M.J.; Fairbanks, A.J. Stereospecific Synthesis of 1,2-cis Glycosides by Vinyl-Mediated IAD. *Org. Lett.* **2004**, *6*, 3797–3800. [CrossRef] [PubMed]
34. Bussolo, V.D.; Fialella, A.; Balzano, F.; Barretta, G.U.; Crotti, P. Stereoselective Synthesis of β-Phenylselenoglycosides from Glycals and Rationalization of the Selenoglycosylation Processes. *J. Org. Chem.* **2010**, *75*, 4284–4287. [CrossRef]
35. Behera, A.; Rai, D.; Kulkarni, S.S. Total Syntheses of Conjugation-Ready Trisaccharide Repeating Units of Pseudomonas aeruginosa O11 and Staphylococcus aureus Type 5 Capsular Polysaccharide for Vaccine Development. *J. Am. Chem. Soc.* **2020**, *142*, 456–467. [CrossRef]
36. Zhang, Y.-L.; Knapp, S. Glycosylation of Nucleosides. *J. Org. Chem.* **2016**, *81*, 2228–2242. [CrossRef]
37. Smoot, J.T.; Demchenko, A.V. How the Arming Participating Moieties can Broaden the Scope of Chemoselective Oligosaccharide Synthesis by Allowing the Inverse Armed−Disarmed Approach. *J. Org. Chem.* **2008**, *73*, 8838–8850. [CrossRef]
38. Chennaiah, A.; Verma, A.K.; Vankar, Y.D. TEMPO-Catalyzed Oxidation of 3-O-Benzylated/Silylated Glycals to the Corresponding Enones Using a PIFA–Water Reagent System. *J. Org. Chem.* **2018**, *83*, 10535–10540. [CrossRef]
39. Lellouche, J.-P.; Koeller, S. The Particular Sensitivity of Silyl Ethers of d-Glucal toward Two Vilsmeier−Haack Reagents $POCl_3 \cdot DMF$ and $(CF_3SO_2)_2O \cdot DMF$. Their Unique and Selective Conversion to the Corresponding C(6)-O-Formates. *J. Org. Chem.* **2001**, *66*, 693–696. [CrossRef]
40. Balijepalli, A.S.; McNeely, J.H.; Hamoud, A.; Grinstaff, M.W. Guidelines for β-lactam synthesis: Glycal protecting groups dictate stereoelectronics and [2 + 2] cycloaddition kinetics. *J. Org. Chem.* **2020**, *85*, 12044–12057. [CrossRef]
41. Kim, Y.; Oh, K.; Song, H.; Lee, D.-S.; Park, S.B. Synthesis and Biological Evaluation of α-Galactosylceramide Analogues with Heteroaromatic Rings and Varying Positions of a Phenyl Group in the Sphingosine Backbone. *J. Med. Chem.* **2013**, *56*, 7100–7109. [CrossRef] [PubMed]
42. Bi, J.-J.; Tan, Q.; Wu, H.; Liu, Q.-F.; Zhang, G.-S. Rhodium-Catalyzed Denitrogenative Transannulation of N-Sulfonyl-1,2,3-triazoles with Glycals Giving Pyrroline-Fused N-Glycosides. *Org. Lett.* **2021**, *23*, 6357–6361. [CrossRef]
43. Chen, H.; Xian, T.; Zhang, W.; Si, W.; Luo, X.; Zhang, B.; Zhang, M.; Wang, Z.; Zhang, J.-B. An efficient method for the synthesis of pyranoid glycals. *Carbohydr. Res.* **2016**, *431*, 42–46. [CrossRef] [PubMed]
44. Aurrecoechea, J.M.; Arrate, M.; Gil, J.H.; Lopez, B. Direct carbohydrate to carbocycle conversions via intramolecular allylation with Et2Zn/Pd (0). *Tetrahedron.* **2003**, *59*, 5515–5522. [CrossRef]
45. Bieg, T.; Kral, K.; Paszkowska, J.; Szeja, W.; Wandzik, I. Microwave-assisted regioselective benzylation: An access to glycal derivatives with a free hydroxyl group at C4. *J. Carbohydr. Chem.* **2012**, *31*, 593–601. [CrossRef]
46. Liu, M.; Luo, Z.-X.; Li, T.; Xiong, D.-C.; Ye, X.-S. Electrochemical Trifluoromethylation of Glycals. *J. Org. Chem.* **2021**, *86*, 16187–16194. [CrossRef]

Article

Preparation of Chiral Enantioenriched Densely Substituted Cyclopropyl Azoles, Amines, and Ethers via Formal S_N2' Substitution of Bromocylopropanes

Hillary Straub [1], Pavel Ryabchuk [1,†], Marina Rubina [1,2] and Michael Rubin [1,2,*]

1. Department of Chemistry, University of Kansas, Lawrence, KS 66045, USA
2. Department of Chemistry, North Caucasus Federal University, 355009 Stavropol, Russia
* Correspondence: mrubin@ku.edu
† Current Address: Discovery Chemistry, Janssen Research & Development, Janssen Pharmaceutica N.V., Turnhoutseweg 30, B-2340 Beerse, Belgium.

Abstract: Enantiomerically enriched cyclopropyl ethers, amines, and cyclopropylazole derivatives possessing three stereogenic carbon atoms in a small cycle are obtained via the diastereoselective, formal nucleophilic substitution of chiral, non-racemic bromocyclopropanes. The key feature of this methodology is the utilization of the chiral center of the cyclopropene intermediate, which governs the configuration of the two adjacent stereocenters that are successively installed via 1,4-addition/epimerization sequence.

Keywords: cyclopropenes; cyclopropanes; nucleophilic addition; metal-templated reactions

1. Introduction

Enantiomerically pure cyclopropane derivatives are ubiquitous, nature inspired [1–8] building blocks abundantly employed in organic synthesis [9–13], asymmetric catalysis [14–17], and medicinal chemistry [18–25]. These advanced synthons are typically accessed via diastereoselective 1,3-ring closure reactions [26–30] or asymmetric cyclopropanation [31–41]. A less established, complementary approach relies on chemo- and diastereoselective installation of additional substituents into pre-formed chiral or prochiral cyclopropanes [42–45]. Strain-release-driven additions of different entities to cyclopropenes proved useful for the assembly of enantiomerically enriched cyclopropane derivatives that are not easily accessible via other methods [46–53]. Synthetic methodologies exploiting stereoselective ring-retentive, metal-catalyzed [13,54,55], and organocatalytic [56,57] additions to cyclopropenes were developed by several research groups and have eventually evolved into a rapidly growing area. Our group recently disclosed an efficient diastereoselective route to cyclopropanes **3** via a formal substitution of bromocyclopropanes **1** with oxygen, nitrogen, or sulfur-based nucleophiles (Scheme 1) [51,58]. The reaction proceeds via a base-assisted dehydrohalogenation, affording a highly reactive cyclopropene intermediate **2** and the subsequent nucleophilic addition across the double bond of cyclopropene. Herein, we report our progress on extending this methodology for the preparation of enantiomerically enriched cyclopropanes.

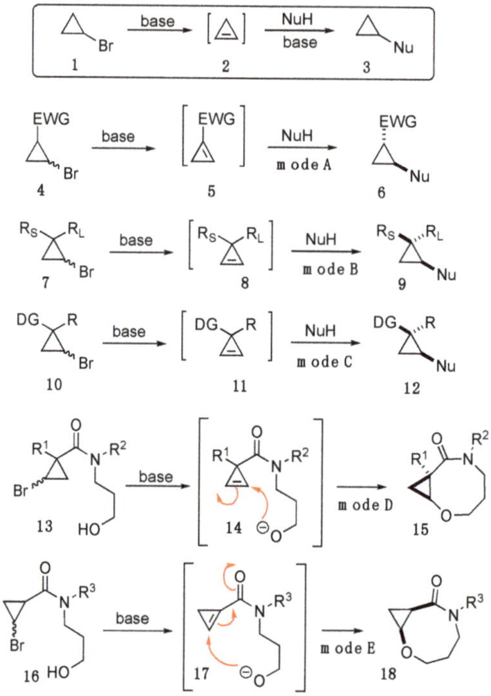

Scheme 1. Different modes of formal nucleophilic substitution of bromocyclopropanes (notations used: EWG—electron-withdrawing group; DG—directing group; R_S—small substituent; R_L—large substituent).

2. Results and Discussion

This section may be divided by subheadings. It should provide a concise and precise description of the experimental results, their interpretation, and the experimental conclusions that can be drawn. In our earlier studies of the formal nucleophilic substitution of bromocyclopropanes, we have demonstrated several reaction modes that allow for efficient control of the diastereoselectivity of this transformation (Scheme 1). Thus, it was shown that the derivatives of 2-bromocyclopropylcarboxylic acid **4** produced achiral cyclopropene **5** upon treatment with base. The latter underwent in situ addition of nucleophiles to afford *trans*-cyclopropane **6**. The high diastereoselectivity of the addition was attributed to a base-assisted, thermodynamically driven epimerization of the tertiary carbon atom (C-1, mode **A**, Scheme 1) [59–63]. Alternative approaches to control the diastereoselectivity of the intermolecular nucleophilic substitution were also developed by utilizing 1,2,2-trisubstituted cyclopropanes as the starting materials. The first approach employs substrates bearing two substituents with significantly different steric demands (**7**, small R_S, and large R_L). The in situ generated achiral cyclopropene **8** undergoes nucleophilic attack at the least hindered face, resulting in selective formation of product **9** (Scheme 1, mode **B**) [62,63]. The second approach takes advantage of bromocyclopropane **10**, bearing a directing functionality (DG, typically carboxamide or carboxylic acid group) capable of efficient coordination to the potassium cation, which serves as a delivery vehicle for the nucleophilic counter-anion. Overall, the addition to the double bond of cyclopropene **11** proceeds in cis fashion, with respect to the directing functional group furnishing **12** with high diastereoselectivity (Scheme 1, mode **C**) [62,63]. The addition of a tethered alkoxide entity was also investigated; both *exo-trig* (Scheme 1, mode **D**) [64,65] and *endo-trig* (mode **E**) [66] modes efficiently provided the corresponding medium-size heterocycles **15** and **18**.

Attempts to extend this methodology beyond the trisubstituted cyclopropane substrates greatly amplify the challenge of controlling the stereoselectivity of the addition. Indeed, all the modes discussed above require the control of a single center only, since the two forming chiral centers are linked to each other. In 2013, we communicated on the realization of a more advanced strategy, involving two modes of diastereoselctivity control providing tetrasubstituted cyclopropyl ethers **23** (mode **F**, Scheme 2) [67]. The proof of concept of such a strategy was showcased on racemic bromocyclorporpanes **20**. We also demonstrated employing racemic substrates, i.e., that the relative configuration of the center at C-3 can be efficiently controlled by steric environment employing appropriate substituents R_S, R_L; thus, control of this step is related to mode **B**. Finally, relative configuration at C-1 was installed via the base-assisted epimerization of this center, in a process identical to the one, previously used in mode **A** (Scheme 2) [66,67]. We reasoned that the absolute configuration of the quaternary stereogenic center at C-2 in chiral non-racemic amide **20** would be preserved during the dehydrohalogenation/nucleophilic addition sequence, which can be used to access to enantiomerically enriched compounds **23**.

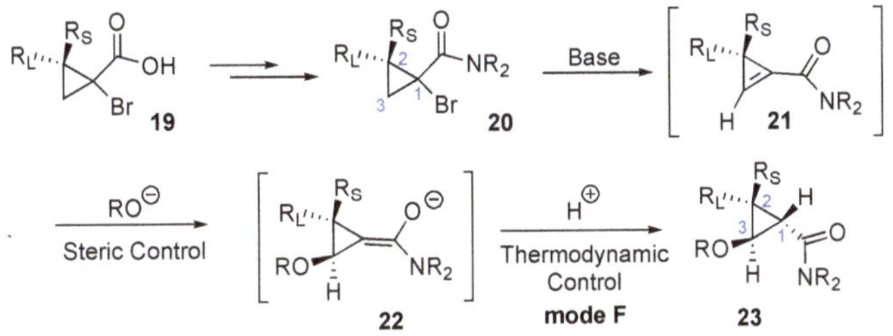
19
20
Base
21
Steric Control
22
Thermodynamic Control
mode F
23

Scheme 2. "Dual" control of diastereoselectivity—a new mode of formal nucleophilic substitution of bromocyclopropanes.

In order to access the densely substituted enantiopure cyclopropanes, we have developed a very facile protocol for the chiral resolution of carboxylic acids **19**, utilizing the re-crystallization of racemic acids with cinchona alkaloids [68]. It was shown that a variety of enantiomerically enriched acids **19** with ee > 95% were available in multi-gram scale, in both enantiomeric forms after single crystallization of either cinchonine or cinchonidine salts. Enantiopure acids can easily be converted into amides **20**, as a precursor for enantioenriched cyclopropenes (Scheme 3).

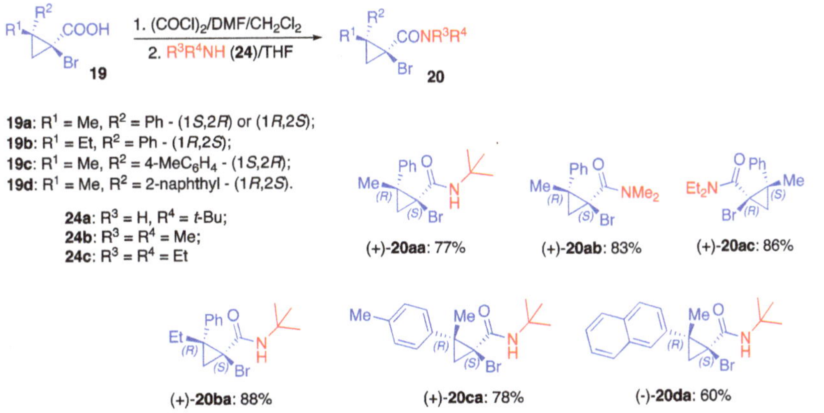
1. (COCl)2/DMF/CH2Cl2
2. R3R4NH (24)/THF
19
20
19a: R1 = Me, R2 = Ph - (1S,2R) or (1R,2S);
19b: R1 = Et, R2 = Ph - (1R,2S);
19c: R1 = Me, R2 = 4-MeC6H4 - (1S,2R);
19d: R1 = Me, R2 = 2-naphthyl - (1R,2S).
24a: R3 = H, R4 = t-Bu;
24b: R3 = R4 = Me;
24c: R3 = R4 = Et
(+)-20aa: 77%
(+)-20ab: 83%
(+)-20ac: 86%
(+)-20ba: 88%
(+)-20ca: 78%
(-)-20da: 60%

Scheme 3. Preparation of homochiral 1-bromocyclopropylcarboxamides employed as starting materials in these studies.

2.1. Alcohol Nucleophiles

With enantiomerically pure amides in hand, we have utilized the "dual control" mode of the formal nucleophilic substitution of bromide with various alkoxides (Scheme 4). At 40 °C in DMSO, bromocyclopropanes **20** were converted to the corresponding cyclopropanes **23**. Primary alcohols served as excellent nucleophiles for the title reaction, with diastereoselectivity greater than 25:1 in all cases. We demonstrated that this methodology is complementary to our previous report, providing an easy access to the enantiopure cyclopropyl ethers.

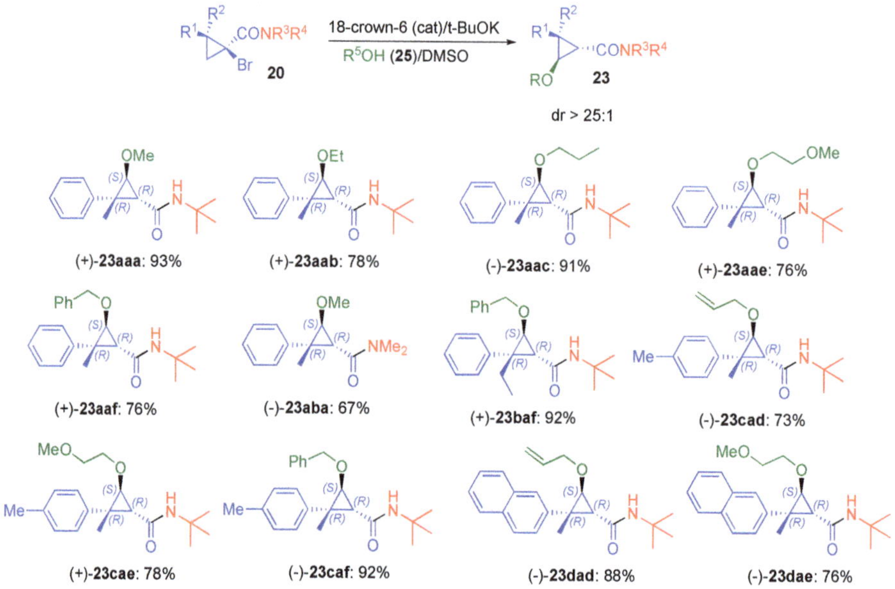

Scheme 4. Preparation of homochiral cyclopropyl ethers via diastereoselective formal substitution of bromocyclopropanes **20** with alkoxides.

2.2. Azole Nucleophiles

The nitrogen-based nucleophiles in our original report have been explored to a lesser extent. Therefore, we became interested in utilizing homochiral amides **20** for the generation of the corresponding cyclopropyl amines. We tested a series of different amines as N-pronucleophiles; however, our initial attempts to induce the addition primary and secondary alkyl amines, as well as carboxamides and sulfonamides, were unsuccessful. We were pleased to find that the azoles underwent a facile addition to cyclopropenes to provide substituted hetarylcyclopropanes in an optically pure form (Scheme 5). The reaction in the presence of pyrrole afforded the corresponding tetrasubstituted cyclopropanes (+)-**23cag** and (+)-**23dag** in high yields and with excellent diastereoselectivities. We were glad to find that such problematic nucleophiles, such as indoles, known for their susceptibility to Friedel–Crafts alkylation, dimerization, and polymerization, afforded good, isolated yields of the corresponding adducts. Substituted indoles and 7-azaindole proceeded cleanly to afford the corresponding cyclopropanes. Similarly, pyrazole was engaged in a very efficient transformation with enantiomerically pure cyclopropyl bromide, providing (+)-**23cah** in good, isolated yield, although longer reaction times were reacquired, and the diastereoselectivity was slightly lower. More acidic azoles, including imidazoles, benzimidazoles, and triazoles, did not participate in the title reaction, due to deactivation of the base in the reaction media, thus preventing the generation of the cyclopropane intermediate. The sensitivity of the reaction to sterics can be seen by comparing the reactivity of bromocyclopropanes possessing a methyl and an ethyl group, respectively, at the β-quaternary center. Compared to methyl-tolyl cyclopropane (+)-**23cag**, its ethyl/phenyl isomer reacted very sluggishly at 40 °C and required higher temperature to achieve full conversion, which led to a lower, although still respectable, diastereoselectivity of 15:1 for **23aag**. To our delight, the carboxamide, possessing a larger naphthyl substituent, also participated in the substitution reaction with pyrrole, giving (+)-**23dag** as a single enantiomer.

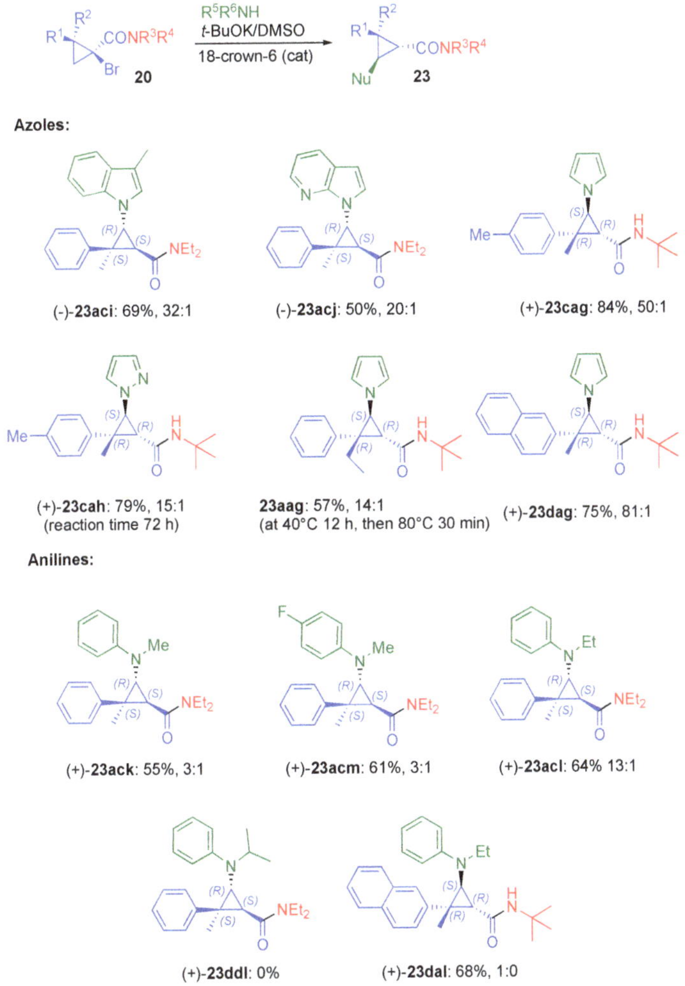

Scheme 5. Preparation of homochiral cyclopropyl amines via diastereoselective formal substitution of bromocyclopropanes **20** with azoles and anilines.

2.3. Aniline Nucleophiles

Anilines were also tested in this reaction, and, to our delight, *N*-methylaniline gave a cyclopropyl amine (+)-**23ack** in 55% and dr 3:1. *p*-Flouro-N-methylaniline can be utilized in the described reaction, providing a tetrasubstituted cyclopropane (+)-**23acm** with similar diastereoselectivity and yield. It was found that increased steric hindrance at the N-termini of the pronucleophile had a significant effect on the reaction course. Thus, aniline bearing an ethyl substituent greatly increased the reaction's efficacy; the diastereoselectivity increased to 13:1 for (+)-**23acl**. The utilization of naphthyl-substituted bromocyclopropane precursor gave exclusive formation of (+)-**23dal** (Scheme 5). Unfortunately, anilines with the secondary alkyl group at the nitrogen atom did not participate in this reaction, most likely due to the excessive steric demands (compound **23ddl** in Scheme 5).

3. Materials and Methods

3.1. General

NMR spectra (See Supporting Information, Figures S1–S32) were recorded on a Bruker Avance DRX-500 (500 MHz) with a dual carbon/proton cryoprobe (CPDUL). ^{13}C NMR spectra were registered with broadband decoupling. The (+) and (−) designations represent positive and negative intensities of signals in ^{13}C DEPT-135 experiments. Numbers of magnetically equivalent carbons for each signal in ^{13}C NMR spectra (unless it is one) are also reported. IR spectra were recorded on a ThermoFisher Nicolet iS 5 FT-IR Spectrometer. HRMS was carried out on LCT Premier (Micromass Technologies) instrument, employing ESI TOF detection techniques. Glassware used in moisture-free syntheses was flame-dried in vacuum prior to use. Column chromatography was carried out on silica gel (Sorbent Technologies, 40–63 mm). Precoated silica gel plates (Sorbent Technologies Silica XG 200 mm) were used for TLC analyses. Anhydrous dichloromethane was obtained by passing degassed commercially available HPLC-grade inhibitor-free solvent consecutively through two columns filled with activated alumina and stored over molecular sieves under nitrogen. Water was purified by dual stage deionization, followed by dual stage reverse osmosis. Anhydrous THF was obtained by refluxing commercially available solvent over calcium hydride, followed by distillation in a stream of dry nitrogen. All other reagents and solvents were purchased from commercial vendors and used as received. Diastereomeric ratios of products were measured by GC and NMR analyses of crude reaction mixtures. In the event where minor diastereomer was not detected by either of these methods, ratio >100:1 was reported.

3.2. Preparation of Starting Materials

(+)-(1S,2R)-1-Bromo-N-(tert-butyl)-2-methyl-2-phenylcyclopropane-1-carboxamide (20aa). Typical procedure A. A flame dried 100 mL round-bottom flask, equipped with drying tube and magnetic stir bar, was charged with (1S,2R)-1-bromo-2-methyl-2-phenylcyclopropane carboxylic acid (**19a**) (1.10 g, 4.33 mmol, 1.00 equiv.), DMF (10 mL), and anhydrous dichloromethane (40 mL). The mixture was treated with oxalyl chloride (563 µL, 6.50 mmol, 1.50 equiv.) at 0 °C, stirred for 15 min, warmed to room temperature, and additionally stirred for 2 h. The solvent was removed in vacuum, and the crude acyl chloride was dissolved in dry THF (20 mL), followed by addition of a solution of *tert*-butyl amine (**24a**) (1.35 mL, 12.8 mmol, 2.97 equiv.) in THF (20 mL). The reaction mixture was stirred overnight. After the reaction was complete, the solvent was removed in vacuum, and the residue was partitioned between EtOAc (25 mL) and water (25 mL). The organic phase was separated, and the aqueous layer was extracted with EtOAc (2 × 25 mL). The combined organic phases were dried (MgSO$_4$), filtered, and concentrated. The residual crude oil was purified by column chromatography on silica gel. The titled compound obtained a colorless solid, mp: 83.2–86.0 °C, R$_f$ 0.55 (hexanes/EtOAc 6:1), [α]$_D$ = +14.0° (c 0.172, CH$_2$Cl$_2$). Yield 1.03 g (3.32 mmol, 77%). Spectral properties of this material were identical to those reported for the racemic amide [67].

(+)-(1R,2S)-1-Bromo-N,N,2-trimethyl-2-phenylcyclopropane-1-carboxamide (20ab). Compound was obtained via typical procedure A, employing (1R,2S)-1-bromo-2-methyl-2-phenylcyclopropane-1-carboxylic acid (**19a**) (510 mg, 2.01 mmol, 1.00 equiv.), oxalyl chloride (260 µL, 3.03 mmol, 1.51 equiv), and 40 wt.% aq solution of dimethyl amine (**24b**) (753 µL, 8.91 mmol, 4.43 equiv.). Chromatographic purification afforded title compound as a colorless solid, mp: 81.7–83.3 °C, R$_f$ 0.34 (hexanes/EtOAc 10:1), [α]$_D$ = +13.3° (c 0.098, CH$_2$Cl$_2$). Yield 466 mg (1.66 mmol, 83%). ^{1}H NMR (500 MHz, CDCl$_3$) δ$_H$ 7.43–7.10 (m, 5H), 2.65 (s, 3H), 2.57 (d, $^2J_{H,H}$ = 7.4 Hz, 1H), 2.56 (s, 3H) 1.85 (s, 3H), 1.37 (d, $^2J_{H,H}$ = 7.4 Hz, 1H); ^{13}C NMR (126 MHz, CDCl$_3$) δ$_C$ 166.5, 138.2, 128.1 (+, 2C), 127.1 (+, 2C), 126.3, 42.6,

38.5 (+), 30.8, 27.0 (-), 24.1 (+); FT IR (KBr, cm^{-1}): 2927, 1647, 1558, 1496, 1396, 1272, 1176, 1082, 1058, 1029, 954, 763, 696, 680, 669, 650; HRMS (TOF ES): found 281.0415, calculated for $C_{13}H_{16}BrNO$ (M$^+$) 281.0415 (0.0 ppm).

(+)-(1S,2R)-1-Bromo-N,N-diethyl-2-methyl-2-phenylcyclopropane-1-carboxamide (**20ac**). Compound was obtained via typical procedure A, employing (1S,2R)-1-bromo-2-methyl-2-phenylcyclopropane-1-carboxylic acid (**19a**) (515 mg, 2.02 mmol, 1.00 equiv.), oxalyl chloride (260 µL, 3.03 mmol, 1.50 equiv.), and diethyl amine (**24c**) (823 µL, 7.96 mmol, 3.96 equiv.). Chromatographic purification afforded title compound as a light yellow solid, mp: 74.3–76.2 °C, R_f 0.34 (hexanes/EtOAc 10:1), $[α]^{25}_D$ = +17.0° (c 0.194, CH_2Cl_2). Yield 537 mg (1.74 mmol, 86%). ^{1}H NMR (500 MHz, $CDCl_3$) $δ_H$ 7.43–7.04 (m, 5H), 3.48 (dq, $^2J_{H,H}$ = 14.2, $^3J_{H,H}$ = 7.1 Hz, 1H), 3.37 (dq, $^2J_{H,H}$ = 14.2, $^3J_{H,H}$ = 7.1 Hz, 1H), 2.71 (dq, $^2J_{H,H}$ = 14.0, $^3J_{H,H}$ = 7.1 Hz, 1H) 2.65 (d, $^2J_{H,H}$ = 7.3 Hz, 1H), 2.58 (dq, $^2J_{H,H}$ = 14.0, $^3J_{H,H}$ = 7.0 Hz, 1H), 1.85 (s, 3H), 1.33 (d, $^2J_{H,H}$ = 7.3 Hz, 1H), 1.0 (t, $^3J_{H,H}$ = 7.1 Hz, 3H) 0.47 (t, $^3J_{H,H}$ = 7.1 Hz, 3H); ^{13}C NMR (126 MHz, $CDCl_3$) $δ_C$ 165.8, 138.1, 128.1 (+), 127.0 (+), 126.7 (+), 42.4, 42.0 (-), 38.3 (-), 31.0, 26.9 (-), 24.3 (+), 12.6 (+), 11.1 (+); FT IR (KBr, cm^{-1}): 2977, 2933, 1643, 1639, 1498, 1456, 1433, 1380, 1282, 1219, 1064, 719, 696, 582; HRMS (TOF ES): found 309.0724, calculated for $C_{15}H_{20}BrNO$ (M$^+$) 309.0728 (1.3 ppm).

(-)-(1S,2R)-1-Bromo-N-benzyl-1-bromo-2-methyl-2-phenylcyclopropane-1-carboxamide (**20ad**). Compound was obtained via typical procedure A, employing (1S,2R)-1-bromo-2-methyl-2-phenylcyclopropane carboxylic acid (**19a**) (255 mg, 1.00 mmol, 1.00 equiv.), oxalyl chloride (130 µL, 1.50 mmol, 1.50 equiv.), and benzyl amine (**24d**) (327 µL, 3.00 mmol, 3.0 equiv). Chromatographic purification afforded a colorless solid, mp: 88.2-91.3 °C, R_f 0.36 (hexanes/EtOAc 6:1), $[α]^{25}_D$ = −115.2° (c 0.046, CH_2Cl_2). Yield 240 mg (0.700 mmol, 70%). Spectral properties of this material were identical to those reported for the racemic amide [67].

(+)-(1S,2R)-1-Bromo-2-methyl-2-phenylcyclopropyl)(pyrrolidin-1-yl)methanone (**20ae**). Compound was obtained via typical procedure A, employing (1S,2R)-1-bromo-2-methyl-2-phenylcyclopropane carboxylic acid (**19a**) (255 mg, 1.00 mmol, 1.00 equiv.), oxalyl chloride (130 µL, 1.50 mmol, 1.50 equiv.) and pyrrolidine (**24e**) (246 µL, 3.00 mmol, 3.00 equiv). Chromatographic purification afforded a colorless oil, R_f 0.39 (hexanes/EtOAc, 3:1), $[α]^{25}_D$ = +12.5° (c 0.172, CH_2Cl_2). Yield 289 mg (0.941 mmol, 94%). Spectral properties of this material were identical to those reported for the racemic amide [67].

(+)-(1R,2S)-1-Bromo-N-(tert-butyl)-2-ethyl-2-phenylcyclopropane-1-carboxamide (**20ba**). Compound was obtained via typical procedure A, employing (1R,2S)-1-bromo-2-ethyl-2-phenylcyclopropane carboxylic acid (**19b**) (1.00 g, 3.73 mmol, 1.00 equiv.), oxalyl chloride (711 µL, 5.60 mmol, 1.50 equiv.) and *tert*-butyl amine (**24a**) (1.18 mL, 11.2 mmol, 3.00 equiv.). Chromatographic purification afforded a colorless solid, mp: 63.8–65.7 °C, R_f 0.52 (hexanes/EtOAc 9:1), $[α]^{25}_D$ = +5.8° (c 0.052, CH_2Cl_2). Yield 1.06 g (3.27 mmol, 88%). Spectral properties of this material were identical to those reported for the racemic amide [67].

(+)-(1S,2R)-1-Bromo-N-(tert-butyl)-2-methyl-2-(p-tolyl)cyclopropane-1-carboxamide (**20ca**). Compound was obtained via typical procedure A (1S,2R)-1-bromo-2-methyl-2-(p-tolyl)cyclopropane-1-carboxylic acid (**19c**) (240 mg, 0.89 mmol, 1.00 equiv.), oxalyl chloride (116 µL, 1.35 mmol, 1.52 equiv.) and *tert*-butyl amine (**24a**) (280 µL, 2.67 mmol, 3.00 equiv.). Chromatographic purification afforded a colorless solid, mp: 78.4–81.2 °C, R_f 0.35 (hexanes/EtOAc 20:1), $[α]^{25}_D$ = +14.3° (c 0.071, CH_2Cl_2). Yield 225 mg (0.694 mmol, 78%). Spectral properties of this material were identical to those reported previously [67].

(-)-(1R,2S)-1-Bromo-N-(tert-butyl)-2-methyl-2-(naphthalen-2-yl)cyclopropane-1-carboxamide (**20da**). Compound was obtained via typical procedure A, employing (1R,2S)-1-bromo-2-methyl-2-naphthalen-2-yl)cyclopropane-1-carboxylic acid (**19d**) (608 mg, 1.99 mmol, 1.00 equiv.), oxalyl chloride (260 µL, 3.00 mmol, 1.51 equiv.), and *tert*-butyl amine (**24a**) (630 µL, 6.00 mmol, 3.02 equiv). Chromatographic purification afforded title compound as a colorless solid, mp: 88.1–89.6 °C, R_f 0.32 (hexanes/EtOAc 20:1), $[α]^{25}_D$ = −41.9° (c 0.418, CH_2Cl_2). Yield 427 mg (1.19 mmol, 60%). ^{1}H NMR (400 MHz, $CDCl_3$)

δ_H 7.82–7.68 (m, 3H), 7.65 (d, $^3J_{H,H}$ = 1.2 Hz, 1H), 7.46–7.39 (m, 2H) 7.35 (dd, $^3J_{H,H}$ = 8.5, $^4J_{H,H}$ = 1.8 Hz, 1H), 6.32 (br. s, 1H), 2.69 (d, $^2J_{H,H}$ = 6.3 Hz, 1H), 1.78 (s, 3H), 1.34 (d, $^2J_{H,H}$ = 6.3 Hz, 1H), 1.03 (s, 9H); ^{13}C NMR (126 MHz, CDCl$_3$) δ_C 165.3, 138.2, 133.3, 132,5, 128.0 (+), 127.8 (+), 127.7 (+), 126.8 (+) 126.3 (+), 126.1 (+), 125.7 (+), 51.6, 45.2, 35.1, 28.4 (+, 3C), 28.1 (+), 26.6(-); FT IR (KBr, cm^{-1}): 3421, 3053, 2964, 2925, 1678,1599, 1512, 1454, 1392, 1363, 1290, 1221, 1134, 1063, 958, 893, 856, 815, 750; HRMS (TOF ES): found 359.0883, calculated for C$_{19}$H$_{22}$BrNO (M$^+$) 359.0885 (0.6 ppm).

3.3. Nucleophilic Addition Reactions

(+)-(1R,2R,3S)-3-(Benzyloxy)-N-(*tert*-butyl)-2-methyl-2-phenylcyclopropane-1-carboxamide (23aaf). Typical procedure B. An oven-dried 10 mL Weaton vial was charged with 18-crown-6 ether (5.3 mg, 20 µmol, 10 mol%), t-BuOK (134 mg, 1.20 mmol, 6.00 equiv.), benzyl alcohol (**25f**) (62.2 µL, 0.598 mmol, 2.92 equiv.), and anhydrous DMSO (10.0 mL). The mixture was stirred at room temperature for 1 min, and (1*S*,2*R*)-1-bromo-*N*-(*tert*-butyl)-2-methyl-2-phenylcyclopropane-1-carboxamide (**20aa**) 63.5 mg (0.205 mmol, 1.00 equiv.) was added in single portion. The reaction mixture was stirred overnight at 80 °C, then solvent was removed in vacuum, and the residue was partitioned between water (15 mL) and EtOAc (15 mL). The organic layer was separated, and the aqueous phase was extracted with EtOAc (3 × 15 mL). Combined organic extracts were washed with brine, dried over MgSO$_4$, filtered, and evaporated. Flash column chromatography on silica gel afforded the titled compound as a colorless solid, mp: 139.9–142.3 °C; R_f 0.32 (hexanes/EtOAc 4:1), $[\alpha]^{25}_D$ = +14.5° (c 0.076, CH$_2$Cl$_2$). dr 60:1. Yield 52.0 mg (0.154 mmol, 76%). Spectral properties of this material were identical to those reported earlier for the racemic compound [67].

(+)-(1R,2R,3S)-N-(*tert*-Butyl)-3-methoxy-2-methyl-2-phenylcyclopropane-1-carboxamide (23aaa). Compound was obtained according to typical procedure B from 65.6 mg (0.211 mmol, 1.00 equiv.) of (1*S*,2*R*)-1-bromo-*N*-(*tert*-butyl)-2-methyl-2-phenylcyclopropane-1-carboxamide (**20aa**) employing methanol (**25a**) (24.2 µL, 0.598 mmol, 2.84 equiv.) as pronucleophile. Chromatographic purification afforded 53.1 mg (0.197 mmol, 93%) of the title compound as a colorless solid, mp: 120.6-122.9 °C; R_f 0.26 (hexanes/EtOAc 3:1), $[\alpha]^{25}_D$ = +17.6° (c 0.068, CH$_2$Cl$_2$). dr 42:1. Spectral properties of this material were identical to those reported for the racemic compound [67].

(+)-(1R,2R,3S)-N-(*tert*-Butyl)-3-ethoxy-2-methyl-2-phenylcyclopropane-1-carboxamide (23aab). Compound was obtained according to typical procedure B from 63.2 mg (0.204 mmol, 1.00 equiv.) of (1*S*,2*R*)-1-bromo-*N*-(*tert*-butyl)-2-methyl-2-phenylcyclopropane-1-carboxamide (**20aa**), employing ethanol (**25b**) (35.0 µL, 0.600 mmol, 2.95 equiv.) as pronucleophile. Chromatographic purification afforded 42.3 mg (0.155 mmol, 78%) of the title compound as a white solid, mp: 130.4–131.6 °C; R_f 0.33 (hexanes/EtOAc 3:1), $[\alpha]^{25}_D$ = +10.5° (c 0.048, CH$_2$Cl$_2$). dr 30:1. Spectral properties of this material were identical to those reported for the racemic compound [67].

(-)-(1R,2R,3S)-N-(*tert*-Butyl)-2-methyl-2-phenyl-3-propoxycyclopropane-1-carboxamide (23aac). Compound was obtained according to typical procedure B from 62.0 mg (0.200 mmol, 1.00 equiv.) of (1*S*,2*R*)-1-bromo-*N*-(*tert*-butyl)-2-methyl-2-phenylcyclopropane-1-carboxamide (**20aa**), employing *n*-propanol (**25c**) (44.9 µL, 0.600 mmol, 3.00 equiv.) as pronucleophile. Chromatographic purification afforded 52.9 mg (0.182 mmol, 91%) of the title compound as a colorless solid, mp: 122.1–124.9 °C; R_f 0.37 (hexanes/EtOAc 3:1), $[\alpha]^{25}_D$ = −25.0° (c 0.044, CH$_2$Cl$_2$). dr 44:1. Spectral properties of this material were identical to those reported for the racemic compound [67].

(+)-(1R,2R,3S)-N-(*tert*-Butyl)-3-(2-methoxyethoxy)-2-methyl-2-phenylcyclopropane-1-carboxamide (23aae). Compound was obtained according to typical procedure B from 63.5 mg (0.205 mmol, 1.00 equi v.) of (1*S*,2*R*)-1-bromo-*N*-(*tert*-butyl)-2-methyl-2-phenylcyclopropane-1-carboxamide (**20aa**), employing 2-methoxyethanol (**25e**) (47.3 µL, 0.601 mmol, 2.93 equiv.) as pronucleophile. Chromatographic purification afforded 46.1 mg (0.152 mmol, 76%) of the title compound as a colorless solid, mp: 101.9–104.2 °C; R_f 0.34

(hexanes/EtOAc 1:1), [α]^{25}$_D$ = +70.0° (c 0.056, CH$_2$Cl$_2$). dr 58:1. Spectral properties of this material were identical to those reported for the racemic compound [67].

(-)-(1R,2R,3S)-3-Methoxy-N,N,2-trimethyl-2-phenylcyclopropane-1-carboxamide (23aba). Compound was obtained according to typical procedure B from 56.4 mg (0.201 mmol, 1.00 equiv.) of (1S,2R)-1-bromo-N,N,2-trimethyl-2-phenylcyclopropane-1-carboxamide (**20ab**), employing methanol (25.8 µL, 0.633 mmol, 3.15 equiv.) as pronucleophile. Chromatographic purification afforded 31.2 mg (0.134 mmol, 67%) as a colorless solid, mp: 108.9–111.3 °C, R$_f$ 0.28 (hexanes/EtOAc 1:1), [α]^{25}$_D$ = −96.6° (c 0.089, CH$_2$Cl$_2$). dr >100:1. ^{1}H NMR (500 MHz, CDCl$_3$) δ$_H$ 7.34–7.12 (m, 5H), 4.25 (d, ^{3}J$_{H,H}$ = 5.8 Hz, 1H), 4.07 (s, 3H), 3.55 (s, 3H), 2.83 (s, 3H), 2.00 (d, ^{3}J$_{H,H}$ = 5.8 Hz, 1H), 1.65 (s, 3H); ^{13}C NMR (126 MHz, CDCl$_3$) δ$_C$ 167.5, 140.2, 127,4 (+, 2C), 126,9 (+, 2C), 125.7 (+), 66.8 (+), 57.3 (+), 36.3 (+), 36.0, 34.2 (+), 33.3 (+), 19.6 (+); FT IR (KBr, cm^{-1}): 2972, 2929, 1734, 1637, 1479, 1460, 1430, 1377, 1265, 1230, 1143, 1070, 952, 847, 796, 763, 759, 740, 698, 624; HRMS (TOF ES): found 234.1498, calculated for C$_{14}$H$_{20}$NO$_2$ (M+H)$^+$ 234.1494 (1.7 ppm).

(-)-(1S,2S,3R)-N,N-Diethyl-2-methyl-3-(3-methyl-1H-indol-1-yl)-2-phenylcyclopropane-1-carboxamide (23aci). Compound was obtained according to typical procedure B from 62.0 mg (0.201 mmol, 1.00 equiv.) of (1R,2S)-1-bromo-N,N-diethyl-2-methyl-2-phenylcyclopropane-1-carboxamide (**20ac**), employing skatole (**25i**) (79.0 mg, 0.602 mmol, 3.00 equiv.) as pronucleophile. Chromatographic purification afforded 49.6 mg (0.138 mmol, 69%) as a colorless solid, mp: 116.2–117.5 °C, R$_f$ 0.21 (hexanes/EtOAc 5:1), [α]^{25}$_D$ = −64.0° (c 0.050, CH$_2$Cl$_2$). dr 32:1. ^{1}H NMR (500 MHz, CDCl$_3$) δ$_H$ 7.58 (d, ^{3}J$_{H,H}$ = 7.8 Hz, 1H), 7.48 (d, ^{3}J$_{H,H}$ = 8.1 Hz, 1H), 7.41 (d, ^{3}J$_{H,H}$ = 7.8 Hz, 2H), 7.27 (d, ^{3}J$_{H,H}$ = 6.0 Hz, 1H), 7.24–7.20 (m, 1H), 7.14 (dd, ^{3}J$_{H,H}$ = 11.0, 3.9 Hz, 1H), 6.92 (s, 1H), 4.63 (d, ^{3}J$_{H,H}$ = 4.2 Hz, 1H), 3.76 (td, ^{2}J$_{H,H}$ = 14.3, ^{3}J$_{H,H}$ = 7.0 Hz, 1H), 3.66 (td, ^{2}J$_{H,H}$ = 13.9, ^{3}J$_{H,H}$ = 7.0 Hz, 1H), 3.34 (dq, ^{2}J$_{H,H}$ = 14.3, ^{3}J$_{H,H}$ = 7.0 Hz, 1H), 2.93 (dq, ^{2}J$_{H,H}$ = 13.9, ^{3}J$_{H,H}$ = 7.0 Hz, 1H), 2.55 (d, ^{3}J$_{H,H}$ = 4.2 Hz, 1H), 2.34 (s, 3H), 1.42 (s, 3H), 1.32 (t, ^{3}J$_{H,H}$ = 7.1 Hz, 3H), 0.86 (t, ^{3}J$_{H,H}$ = 7.1 Hz, 3H); ^{13}C NMR (126 MHz, CDCl$_3$) δ$_C$ 166.5, 140.3, 137.9, 129.4, 128.7, (+, 2C), 128.1 (+, 2C), 127.2 (+), 125.7 (+), 122.0 (+), 119.3 (+), 119.2 (+), 111.1, 110.5 (+), 42.3 (+), 41.9 (-), 40.2 (-), 37.0, 35.9 (+), 22.1 (+), 15.9 (+), 12.7 (+), 9.8 (+); FT IR (KBr, cm^{-1}): 2972, 2927, 1639, 1465, 1379, 1309, 1263, 1230, 1143, 759, 740, 698; HRMS (TOF ES): found 360.2200, calculated for C$_{24}$H$_{28}$N$_2$O (M$^+$) 360.2202 (0.6 ppm).

(-)-(1S,2S,3R)-N,N-Diethyl-2-methyl-2-phenyl-3-(1H-pyrrolo[2,3-b]pyridin-1-yl) cyclopropane-1-carboxamide (23acj). Compound was obtained according to typical procedure B from 62.1 mg (0.201 mmol, 1.00 equiv.) of (1R,2S)-1-bromo-N,N-diethyl-2-methyl-2-phenylcyclopropane-1-carboxamide (**20ac**), employing 7-azaindole (**25j**) (71.0 mg, 0.600 mmol, 3.00 equiv.) as pronucleophile. Chromatographic purification afforded 35.4 mg (0.102 mmol, 50%) as a colorless solid, mp: 113.2–114.0 °C, R$_f$ 0.42 (hexanes/EtOAc 2:1), [α]^{25}$_D$ = −5.2° (c 0.669, CH$_2$Cl$_2$). dr 20:1. ^{1}H NMR (500 MHz, CDCl$_3$) δ$_H$ 8.38 (dd, ^{3}J$_{H,H}$ = 4.7, ^{4}J$_{H,H}$ = 1.5 Hz, 1H), 7.90 (dd, ^{3}J$_{H,H}$ = 7.8 Hz, 1H), 7.65-7.51 (m, 2H), 7.34 (t, ^{3}J$_{H,H}$ = 7.7 Hz, 2H), 7.25 (d, ^{3}J$_{H,H}$ = 7.5 Hz, 1H), 7.22 (d, ^{3}J$_{H,H}$ = 3.6 Hz, 1H), 7.08 (dd, ^{3}J$_{H,H}$ = 7.8, 4.7 Hz, 1H), 6.47 (d, ^{3}J$_{H,H}$ = 3.5 Hz, 1H), 4.72 (d, ^{3}J$_{H,H}$ = 4.3 Hz, 1H), 3.85 (dq, ^{2}J$_{H,H}$ = 14.4, ^{3}J$_{H,H}$ = 7.1 Hz, 1H), 3.78-3.62 (m, 1H), 3.43 (dq, ^{2}J$_{H,H}$ = 14.3, ^{3}J$_{H,H}$ = 7.1 Hz, 1H), 2.97 (dq, ^{2}J$_{H,H}$ = 14.0, ^{3}J$_{H,H}$ = 7.0 Hz, 1H), 2.73 (d, ^{3}J$_{H,H}$ = 4.3 Hz, 1H), 1.40 (t, ^{3}J$_{H,H}$ = 7.1 Hz, 3H), 1.26 (s, 3H), 0.97 (t, ^{3}J$_{H,H}$ = 7.1Hz, 3H); ^{13}C NMR (126 MHz, CDCl$_3$) δ$_C$ 166.9, 149.4, 143.6 (+), 141.1, 128.9 (+, 2C), 128.6 (+), 128.6 (+, 2C), 128.3 (+), 127.1 (+), 120.9 (+), 116.3 (+), 100.0 (+), 42.7 (+), 42.1 (+), 40.4 (-), 37.5, 33.3 (+), 22.6 (+), 14.9 (+), 12.8 (+); FT IR (KBr, cm^{-1}): 2972, 2929, 1733, 1637, 1479, 1460, 1448, 1433, 1377, 1265, 1220, 1143, 1097, 1070, 952, 846, 796, 775, 763, 723, 702, 624, 598; HRMS (TOF ES): found 348.2076, calculated for C$_{22}$H$_{26}$N$_3$O (M+H)$^+$ 348.2070 (1.7 ppm).

(+)-(1S,2S,3R)-N,N-Diethyl-2-methyl-3-(methyl(phenyl)amino)-2-phenylcyclopropane-1-carboxamide (23ack). Compound was obtained according to typical procedure B from 62.0 mg (.201 mmol, 1.00 equiv.) of (1R,2S)-1-bromo-N,N-diethyl-2-methyl-2-phenylcyclopropane-1-carboxamide (**20ac**), employing N-methylaniline (**25k**) (65.0 µL, 0.600 mmol, 3.00 equiv.) as pronucleophile. Chromatographic purification afforded

36.4 mg (0.110 mmol, 55.0%) as yellow oil, R_f 0.22 (hexanes/EtOAc 5:1), dr 3:1 ^{1}H NMR (500 MHz, CDCl$_3$) δ_H 7.43–7.12 (m, 7H), 6.95 (d, $^3J_{H,H}$ = 7.9 Hz, 2H), 6.79 (d, $^3J_{H,H}$ = 7.3 Hz, 3H), 3.78 (d, $^3J_{H,H}$ = 4.4 Hz, 1H), 3.68–3.55 (m, 2H), 3.19 (dq, $^2J_{H,H}$ = 14.7, $^3J_{H,H}$ = 7.2 Hz, 1H), 3.11 (s, 3H), 2.79 (dq, $^2J_{H,H}$ = 14.7, $^3J_{H,H}$ = 7.2 Hz, 1H), 1.97 (d, $^3J_{H,H}$ = 4.4 Hz, 1H), 1.62 (s, 3H), 1.17 (t, $^3J_{H,H}$ = 7.1 Hz, 3H), 0.79 (t, $^3J_{H,H}$ = 7.1 Hz, 3H); ^{13}C NMR (126 MHz, CDCl$_3$) δ_C 167.2, 151.0, 141.0, 129.2 (+, 2C), 128.5 (+, 2C), 127.8 (+, 2C), 126.8 (+), 118.1, 114.8 (+, 2C), 48.5 (+), 41.6 (-), 40.4 (+), 39.8 (-), 38.3, 36.9 (+), 21.0 (+), 14.6 (+), 12.6 (+); FT IR (KBr, cm^{-1}): 2972, 2929, 1639, 1598, 1500, 1479, 1444, 1433, 1379, 1305, 1220, 1143, 1116, 1029, 950, 904, 698, 611; HRMS (TOF ES): found 337.2281, calculated for C$_{22}$H$_{29}$N$_2$O (M+H)$^+$ 3370.2280 (0.3 ppm). $[\alpha]^{25}_D$ = +33.1° (c 0.366, CH$_2$Cl$_2$).

(+)-(1S,2S,3R)-N,N-Diethyl-3-(ethyl(phenyl)amino)-2-methyl-2-phenylcyclopropane-1-carboxamide (23acl). Compound was obtained according to typical procedure B from 62.0 mg (0.201 mmol, 1.00 equiv.) of (1R,2S)-1-bromo-N,N-diethyl-2-methyl-2-phenyl-cyclopropane-1-carboxamide (20ac), employing N-ethylaniline (25l) (75.0 µL, 0.600 mmol, 3.00 equiv.) as pronucleophile. Chromatographic purification afforded 44.8 mg (0.128 mmol, 64%) as a yellow oil, R_f 0.31 (hexanes/EtOAc 5:1), $[\alpha]^{25}_D$ = +11.6° (c 0.160, CH$_2$Cl$_2$). dr 13:1. ^{1}H NMR (500 MHz, CDCl$_3$) δ_H 7.51–7.14 (m, 7H), 7.01–6.89 (m, 2H), 6.76 (t, $^3J_{H,H}$ = 7.3 Hz, 1H), 3.84 (d, $^3J_{H,H}$ = 4.6 Hz, 1H), 3.70–3.42 (m, 4H), 3.14 (dq, $^2J_{H,H}$ = 14.3, $^3J_{H,H}$ = 7.1 Hz, 1H), 2.84 (dq, $^2J_{H,H}$ = 13.9, $^3J_{H,H}$ = 7.0 Hz, 1H), 1.95 (d, $^3J_{H,H}$ = 4.6 Hz, 1H), 1.60 (s, 3H), 1.21 (t, $^3J_{H,H}$ = 7.0 Hz, 3H), 1.14 (t, $^3J_{H,H}$ = 7.1 Hz, 3H), 0.76 (t, $^3J_{H,H}$ = 7.1 Hz, 3H); ^{13}C NMR (126 MHz, CDCl$_3$) δ_C 167.3, 149.2, 141.0, 129.2 (+, 2C), 128.4 (+, 2C), 127.6 (+, 2C), 126.7 (+, 2C), 118.0 (+), 115.7 (+), 46.3 (+), 46.2 (-), 41.5 (-), 39.7 (-), 36.9 (+), 14.5 (+), 12.5 (+), 11.2 (+); FT IR (KBr, cm^{-1}): 3085, 2972, 2358, 1637, 1598, 1498, 1458, 1444, 1434, 1377, 1259, 1143, 1080, 831, 752, 696, 613; HRMS (TOF ES): found 350.2358, calculated for C$_{23}$H$_{30}$N$_2$O (M$^+$) 350.2358 (0.0 ppm).

(+)-(1S,2S,3R)-N,N-Diethyl-3-((4-fluorophenyl)(methyl)amino)-2-methyl-2-phenyl-cyclopropane-1-carboxamide (23acm). Compound was obtained according to typical procedure B from 59.8 mg (0.193 mmol, 1.00 equiv.) of (1R,2S)-1-bromo-N,N-diethyl-2-methyl-2-phenylcyclopropane-1-carboxamide (20ac), employing 4-fluoro-N-methylaniline (25m) (72.2 µL, 0.600 mmol, 3.11 equiv.) as pronucleophile. Chromatographic purification afforded 41.6 mg (0.118 mmol, 61%) of the title compound as a yellow oil, R_f 0.25 (hexanes/EtOAc 4:1), $[\alpha]^{25}_D$ = +35.4° (c 0.362, CH$_2$Cl$_2$). dr 3:1. ^{1}H NMR (500 MHz, CDCl$_3$) δ_H 7.35–7.27 (m, 4H), 7.21 (ddd, $^3J_{H,F}$ = 5.0 Hz, $^3J_{H,H}$ = 4.5, 1.9 Hz, 1H), 7.00–6.84 (m, 4H), 3.71 (d, $^3J_{H,H}$ = 4.4 Hz, 1H), 3.67–3.55 (m, 2H), 3.18 (dd, $^2J_{H,H}$ = 14.7, $^3J_{H,H}$ = 7.2 Hz, 1H), 3.08 (s, 3H), 2.85 (dd, $^2J_{H,H}$ = 13.6, $^3J_{H,H}$ = 7.0 Hz, 1H), 1.91 (d, $^3J_{H,H}$ = 4.4 Hz, 1H), 1.62 (s, 3H), 1.16 (t, $^3J_{H,H}$ = 7.2 Hz, 3H), 0.79 (t, $^3J_{H,H}$ = 7.1 Hz, 3H); ^{13}C NMR (126 MHz, CDCl$_3$) δ_C 167.1, 156.3 (d, $^1J_{C,F}$ = 236.5 Hz), 147.5 (d, $^4J_{C,F}$ = 1.9 Hz), 140.8, 128.4 (+, 2C), 127.6 (+, 2C), 126.7 (+), 116.0 (d, $^2J_{C,F}$ = 7.4 Hz, +, 2C), 115.4 (d, $^3J_{C,F}$ = 22.0 Hz, +, 2C), 48.8 (+), 41.5 (-), 41.1 (+), 39.7 (-), 38.1 (+), 36.8(+), 20.8 (+), 14.4 (+), 12.5 (+); FT IR (KBr, cm^{-1}): 2974, 2873, 1635, 1510, 1479, 1458, 1446, 1379, 1263, 1224, 1143, 1022, 825, 763, 698; HRMS (TOF ES): found 353.2028, calculated for C$_{22}$H$_{26}$FN$_2$O (M-H)$^+$ 353.2029 (0.3 ppm).

(+)-(1R,2R,3S)-3-(Benzyloxy)-N-tert-butyl-2-ethyl-2-phenylcyclopropane-1-carboxamide (23baf). Compound was obtained according to typical procedure B from 72.2 mg (0.223 mmol, 1.00 equiv) of (1S,2R)-1-bromo-N-(tert-butyl)-2-ethyl-2-phenylcyclopropane-1-carboxamide (20ba), employing benzyl alcohol (25f) (64.9 µL, 0.624 mmol, 2.80 equiv.) as pronucleophile. Chromatographic purification afforded 49.0 mg (0.139 mmol, 70%) of the title compound as a colorless solid, mp: 136.2–137.1 °C, R_f 0.23 (hexanes/EtOAc 9:1), $[\alpha]^{25}_D$ = +50.0° (c 0.11, CH$_2$Cl$_2$). dr 39:1. Spectral properties of this material were identical to those reported for the racemic compound [67].

(−)-(1R,2R,3S)-3-(Allyloxy)-N-tert-butyl-2-methyl-2-(p-tolyl)cyclopropane-1-carboxamide (23cad). Compound was obtained according to typical procedure B from 65.0 mg (0.200 mmol, 1.00 equiv.) of (1S,2R)-1-bromo-N-(tert-butyl)-2-methyl-2-(p-tolyl)cyclopropane-1-carboxamide (20ca), employing allyl alcohol (25e) (41.0 µL, 0.600 mmol, 3.00 equiv) as pronucleophile. Chromatographic purification afforded 42.0 mg (0.146 mmol,

73%) of the title compound as a colorless solid, mp: 120.6–122.7 °C, R_f 0.38 (hexanes/EtOAc 5:1), $[\alpha]^{25}_D$ = −21.8° (c 0.16, CH_2Cl_2). dr 50:1. ^1H NMR (500 MHz, $CDCl_3$) δ_H 7.12 (d, $^3J_{H,H}$ = 8.1 Hz, 2H), 7.08 (d, $^3J_{H,H}$ = 7.9 Hz, 2H), 5.99 (ddt, $^3J_{H,H}$ = 16.2, 10.5, 5.7 Hz, 1H), 5.36 (dd, $^3J_{H,H}$ = 17.2, $^2J_{H,H}$ = 1.6 Hz, 1H), 5.24 (dd, $^3J_{H,H}$ = 10.4, $^2J_{H,H}$ = 1.4 Hz, 1H), 5.08 (br. s, 1H), 4.27–4.10 (m, 2H), 4.07 (d, ^3J = 3.3 Hz, 1H), 2.29 (s, 3H), 1.65 (d, ^3J = 3.3 Hz, 1H), 1.50 (s, 3H), 1.16 (s, 9H); ^{13}C NMR (126 MHz, $CDCl_3$) δ_C 168.1, 138.3, 136.4, 134.3 (+), 129.3 (+, 2C), 128.5 (+, 2C), 117.7 (-), 72.3 (-), 66.4 (+), 51.1, 37.0 (+), 37.0, 28.8 (+, 3C), 22.0 (+), 21.2 (+); FT IR (KBr, cm^{-1}): 3301, 2966, 2923, 1643, 1546, 1515, 1454, 1226, 1145, 985, 925, 817; HRMS (TOF ES): found 300.1967, calculated for $C_{19}H_{26}NO_2$ (M-H)$^+$ 300.1964 (1.0 ppm).

(+)-(1R,2R,3S)-N-(tert-Butyl)-3-(2-methoxyethoxy)-2-methyl-2-(p-tolyl)cyclopropane-1-carboxamide (**23cae**). Compound was obtained according to typical procedure B from 62.5 mg (0.193 mmol, 1.00 equiv.) of (1S,2R)-1-bromo-N-(tert-butyl)-2-methyl-2-(p-tolyl) cyclopropane-1-carboxamide (**20ca**), employing 2-methoxyethanol (**25e**) (47.3 µL, 0.645 mmol, 3.22 equiv.) as pronucleophile. Chromatographic purification afforded 49.5 mg (0.155 mmol, 78%) of the title compound as a colorless solid, mp: 94.9–97.1 °C, R_f 0.26 (hexanes/EtOAc 1:1), $[\alpha]^{25}_D$ = +32.0° (c 0.05, CH_2Cl_2). dr > 100:1. Spectral properties of this material were identical to those reported for the racemic compound [67].

(-)-(1R,2R,3S)-3-(Benzyloxy)-N-(tert-butyl)-2-methyl-2-(p-tolyl)cyclopropanecarboxamide (**23caf**) [67]. Compound was obtained according to typical procedure B from 130 mg (0.401 mmol, 1.00 equiv.) of (1S,2R)-1-bromo-N-(tert-butyl)-2-methyl-2-(p-tolyl)cyclopropane-1-carboxamide (**20ca**), employing benzyl alcohol (**25f**) (124 µL, 1.20 mmol, 3.00 equiv.) as pronucleophiles. The subsequent chromatographic purification afforded 129 mg (0.367 mmol, 92%) of the title compound as a colorless solid, mp: 139.8–140.6 °C, R_f 0.33 (hexanes/EtOAc 6:1), $[\alpha]^{25}_D$ = −24.5° (c 1.10, CH_2Cl_2). dr 44:1. ^1H NMR (500 MHz, $CDCl_3$) δ_H 7.54–7.17 (m, 5H), 7.10 (q, $^3J_{H,H}$ = 8.1 Hz, 4H), 5.03 (br. s, 1H), 4.87–4.53 (m, 2H), 4.12 (d, $^3J_{H,H}$ = 3.3 Hz, 1H), 2.30 (s, 3H), 1.66 (d, $^3J_{H,H}$ = 3.3 Hz, 1H), 1.54 (s, 3H), 1.16 (s, 9H); ^{13}C NMR (126 MHz, $CDCl_3$) δ_C 168.0, 138.3, 137.7, 136.4, 129.3 (+, 2C), 128.6 (+, 2C), 128.5 (+, 2C), 128.3 (+, 2C), 128.0 (+), 73.5 (-), 66.6 (+), 51.1, 37.1 (+), 28.8 (+, 3C), 22.1 (+), 21.2 (+); FT IR (KBr, cm^{-1}): 3308, 3063, 3030, 2966, 2926, 2864, 1643, 1543, 1516, 1454, 1431, 1375, 1364, 1346, 1277, 1226, 1204, 1144, 1099, 987, 817, 750, 734, 698; HRMS (TOF ES): found 374.2098, calculated for $C_{23}H_{29}NO_2$ (M+Na)$^+$ 374.2096 (0.5 ppm).

(+)-(1R,2R,3S)-N-(tert-Butyl)-2-methyl-3-(1H-pyrrol-1-yl)-2-(p-tolyl)cyclopropane-1-carboxamide (**23cag**). Compound was obtained according to typical procedure B from 64.6 mg (0.199 mmol, 1.00 equiv.) of (1S,2R)-1-bromo-N-(tert-butyl)-2-methyl-2-(p-tolyl) cyclopropane-1-carboxamide (**20ca**), employing pyrrole (**25g**) (42.0 µL, 0.600 mmol, 3.00 equiv.) as pronucleophile. Chromatographic purification afforded 52.0 mg (0.168 mmol, 84%) of the title compound as a colorless solid, mp: 208.0–210.1 °C, R_f 0.31 (hexanes/EtOAc 6:1), $[\alpha]^{25}_D$ = +46.5° (c 0.142, CH_2Cl_2). dr > 99:1. ^1H NMR (500 MHz, $CDCl_3$) δ_H 7.22 (d, $^3J_{H,H}$ = 8.0 Hz, 2H), 7.14 (d, $^3J_{H,H}$ = 7.9 Hz, 2H), 6.77 (t, $^3J_{H,H}$ = 2.1 Hz, 2H), 6.19 (t, $^3J_{H,H}$ = 2.1 Hz, 2H), 5.35 (br. s, 1H), 4.31 (d, $^3J_{H,H}$ = 4.2 Hz, 1H), 2.32 (s, 3H), 2.17 (d, $^3J_{H,H}$ = 4.2 Hz, 1H), 1.26 (s, 3H), 1.22 (s, 9H); ^{13}C NMR (126 MHz, $CDCl_3$) δ_C 166.9, 137.7, 137.0, 129.5 (+, 2C), 128.4 (+, 2C), 121.6 (+, 2C), 108.6 (+, 2C), 51.6, 45.0 (+), 37.1, 36.7 (+), 28.8 (+, 3C), 23.1 (+), 21.3 (+); FT IR (KBr, cm^{-1}): 3319, 2966, 2925, 1645, 1546, 1539, 1492, 1454, 1361, 1265, 1224, 1093, 1066, 981, 817, 721, 700; HRMS (TOF ES): found 309.1974, calculated for $C_{20}H_{25}N_2O$ (M-H)$^+$ 309.1967 (2.3 ppm).

(+)-(1R,2R,3S)-N-(tert-Butyl)-2-methyl-3-(1H-pyrazol-1-yl)-2-(p-tolyl)cyclopropane-1-carboxamide (**23cah**): Compound was obtained according to typical procedure B from 65.0 mg (0.200 mmol, 1.00 equiv.) of (1S,2R)-1-bromo-N-(tert-butyl)-2-methyl-2-(p-tolyl) cyclopropane-1-carboxamide (**20ca**), employing pyrazole (**25h**) (41.0 mg, 0.600 mmol, 3.00 equiv) as pronucleophiles. The subsequent chromatographic purification afforded 47.0 mg (0.158 mmol, 79%) of the title compound as a colorless solid, mp: 147.9–148.5 °C, R_f 0.35 (CH_2Cl_2/MeOH 20:1), $[\alpha]^{25}_D$ = +40.6° (c 0.35, CH_2Cl_2). dr 15:1. ^1H NMR (500 MHz, $CDCl_3$) δ_H 7.51 (d, $^3J_{H,H}$ = 2.1 Hz, 1H), 7.47 (d, $^3J_{H,H}$ = 1.2 Hz, 1H), 7.17 (d, $^3J_{H,H}$ = 7.9 Hz, 2H), 7.05 (d, $^3J_{H,H}$ = 7.9 Hz, 2H), 6.25 (t, $^3J_{H,H}$ = 1.9 Hz, 1H), 5.69 (br. s, 1H), 4.46 (d,

$^3J_{H,H}$ = 4.0 Hz, 1H), 2.61–2.45 (m, 1H), 2.24 (s, 3H), 1.16 (s, 9H), 1.12 (s, 3H); ^{13}C NMR (126 MHz, CDCl$_3$) δ_C 166.5, 139.4 (+), 137.6, 136.8, 130.6 (+), 129.4 (+, 2C), 128.4 (+, 2C), 106.2 (+), 51.6, 47.1 (+), 37.4, 35.8 (+), 28.8 (+, 3C), 22.7 (+), 21.3 (+); FT IR (KBr, cm^{-1}): 3306, 2964, 2925, 1649, 1544, 1452, 1392, 1274, 1224, 1089, 1047, 820, 752, 615; HRMS (TOF ES): found 312.2081, calculated for C$_{19}$H$_{26}$N$_3$O (M+H)$^+$ 312.2076 (1.6 ppm).

(−)-(1R,2R,3S)-3-(Allyloxy)-N-(tert-butyl)-2-methyl-2-(naphthalen-2-yl)cyclopropane-1-carboxamide (**23dad**). Compound was obtained according to typical procedure B from 69.9 mg (0.194 mmol, 1.00 equiv.) of (1S,2R)-1-bromo-N-(*tert*-butyl)-2-methyl-2-(naphthalen-2-yl)cyclopropane-1-carboxamide (**20da**), employing allyl alcohol (**25d**) (40.8 µL, 0.600 mmol, 3.09 equiv.) as pronucleophile. Chromatographic purification afforded 57.9 mg (0.172 mmol, 88%) of the title compound as a colorless solid, mp: 122.2–125.1 °C, R$_f$ 0.26 (hexanes/EtOAc 4:1), [α]^{25}$_D$ = −7.87° (c 0.178, CH$_2$Cl$_2$). dr > 99:1. ^{1}H NMR (500 MHz, CDCl$_3$) δ_H 7.85–7.72 (m, 3H), 7.69 (s, 1H), 7.50–7.36 (m, 2H), 7.35 (dd, $^3J_{H,H}$ = 8.4, $^4J_{H,H}$ = 1.7 Hz, 1H), 6.04 (ddt, $^3J_{H,H}$ = 17.2, 10.5, 5.7 Hz, 1H), 5.41 (dd, $^3J_{H,H}$ = 17.2, $^2J_{H,H}$ = 1.6 Hz, 1H), 5.27 (dd, $^3J_{H,H}$ = 10.4, $^2J_{H,H}$ = 1.4 Hz, 1H), 5.17 (br. s, 1H), 4.34–4.05 (m, 2H), 4.22 (d, $^3J_{H,H}$ = 3.3 Hz, 1H), 1.75 (d, $^3J_{H,H}$ = 3.3 Hz, 1H), 1.59 (s, 3H), 1.13 (s, 9H); ^{13}C NMR (126 MHz, CDCl$_3$) δ_C 167.9, 139.0, 134.3 (+), 133.6, 132.6, 128.2 (+), 127.8 (+), 127.8 (+), 127.3 (+), 127.0 (+), 126.1 (+), 125.7 (+), 117.8 (-), 72.4 (-), 66.5 (+), 51.2, 37.5, 37.1 (+), 28.8 (+, 3C), 22.0 (+); FT IR (KBr, cm^{-1}): 3319, 2966, 2925, 1643, 1542, 1454, 1361, 1269, 1226, 1147, 1128, 1087, 1062, 1041, 985, 923, 856, 817, 744, 667; HRMS (TOF ES): found 338.2119, calculated for C$_{22}$H$_{28}$NO$_2$ (M+H)$^+$ 338.2120 (0.3 ppm).

(−)-(1R,2R,3S)-N-(tert-Butyl)-3-(2-methoxyethoxy)-2-methyl-2-(naphthalen-2-yl)cyclopropane-1-carboxamide (**23dae**). Compound was obtained according to typical procedure B from 71.8 mg (0.199 mmol, 1.00 equiv.) of (1S,2R)-1-bromo-N-(*tert*-butyl)-2-methyl-2-(naphthalen-2-yl)cyclopropane-1-carboxamide (**20da**), employing 2-methoxyethanol (**25e**) (47.3 µL, 0.600 mmol, 3.02 equiv.) as pronucleophile. Chromatographic purification afforded 53.8 mg (0.151 mmol, 76%) of the title compound as a colorless solid, mp: 143.2–144.2 °C, R$_f$ 0.25 (hexanes/EtOAc 2:1), [α]^{25}$_D$ = −6.1° (c 0.214, CH$_2$Cl$_2$). dr > 99:1. ^{1}H NMR (500 MHz, CDCl$_3$) δ_H 7.86–7.65 (m, 4H), 7.54–7.32 (m, 3H), 5.19 (br. s, 1H), 4.23 (d, $^3J_{H,H}$ = 3.3 Hz, 1H), 3.96–3.77 (m, 2H), 3.73–3.60 (m, 2H), 3.44 (s, 3H), 2.17 (s, 3H), 1.76 (d, $^3J_{H,H}$ = 3.3 Hz, 1H), 1.12 (s, 9H); ^{13}C NMR (126 MHz, CDCl$_3$) δ_C 167.9, 139.0, 133.6, 132.6, 128.2 (+), 127.8 (+), 127.3 (+), 127.0 (+), 126.0 (+), 125.7 (+), 71.9 (-), 70.5 (-), 67.0 (+), 59.3 (+), 51.2, 37.6, 37.0 (+), 31.1 (+), 28.8 (+, 3C), 21.9 (+); FT IR (KBr, cm^{-1}): 3323, 2966, 2871, 1645, 1541, 1454, 1390, 1363, 1269, 1226, 1151, 1124, 956, 856, 817, 742, 667; HRMS (TOF ES): found 356.2225, calculated for C$_{22}$H$_{30}$NO$_3$ (M+H) 356.2226 (0.3 ppm).

(+)-(1R,2R,3S)-N-(tert-Butyl)-2-methyl-2-(naphthalen-2-yl)-3-(1H-pyrrol-1-yl)cyclopropane-1-carboxamide (**23dag**). Compound was obtained according to typical procedure B from 71.4 mg (0.198 mmol, 1.00 equiv.) of (1S,2R)-1-bromo-N-(*tert*-butyl)-2-methyl-2-(naphthalen-2-yl)cyclopropane-1-carboxamide (**20da**), employing pyrrole (**25g**) (42.0 µL, 0.606 mmol, 3.06 equiv.) as pronucleophile. Chromatographic purification afforded 51.9 mg (0.149 mmol, 75%) of the title compound as a colorless solid, mp: 181.4–183.2 °C, R$_f$ 0.23 (hexanes/EtOAc 6:1), [α]^{25}$_D$ = +15.6° (c 0.096 CH$_2$Cl$_2$). dr 81:1. ^{1}H NMR (500 MHz, CDCl$_3$) δ_H 8.01–7.67 (m, 4H), 7.57–7.35 (m, 3H), 6.85 (t, $^3J_{H,H}$ = 2.1 Hz, 2H), 6.23 (t, $^3J_{H,H}$ = 2.1 Hz, 2H), 5.44 (br., s, 1H), 4.46 (d, $^3J_{H,H}$ = 4.1 Hz, 1H), 2.28 (d, $^3J_{H,H}$ = 4.2 Hz, 1H), 1.36 (s, H), 1.19 (s, 9H); ^{13}C NMR (126 MHz, CDCl$_3$) δ_C 166.7, 138.3, 133.6, 132.7, 128.5 (+), 127.8 (+), 127.8 (+), 127.3 (+), 126.7 (+), 126.3 (+), 126.0 (+), 121.7 (+, 2C), 108.7 (+, 2C), 51.7, 45.2 (+), 37.7, 36.8 (+), 28.8 (+, 3C), 23.0 (+); FT IR (KBr, cm^{-1}): 3305, 2968, 1650, 1548, 1492, 1454, 1392, 1265, 1224, 1132, 1091, 1064, 981, 854, 817, 721, 680, 657; HRMS (TOF ES): found 346.2047, calculated for C$_{23}$H$_{26}$N$_2$O (M$^+$) 346.2045 (0.6 ppm).

(1R*,2R*,3S*)-N-(tert-butyl)-2-ethyl-2-phenyl-3-(1H-pyrrol-1-yl)cyclopropane-1-carboxamide (**23aag**). This compound was obtained according to procedure B, employing pyrrole (42 µL, 0.60 mmol, 3.0 equiv.) as pronucleophile. The reaction mixture was stirred overnight at 40 °C, and GC analysis showed incomplete conversion (75% based on starting material) and dr 17:1; after heating at 80 °C for 30 min, the reaction was complete. The

subsequent chromatographic purification afforded 35 mg (0.114 mmol, 57%) of the title compound as a white solid, mp: 177.5–120.0 °C, R_f 0.29 (hexanes/EtOAc 6:1), dr 14:1. ^{1}H NMR (500 MHz, CDCl$_3$) δ_H 7.44–7.07 (m, 5H), 6.72 (t, $^3J_{H,H}$ = 2.1 Hz, 2H), 6.12 (t, $^3J_{H,H}$ = 2.1 Hz, 2H), 5.28 (br. s, 1H), 4.25 (d, $^3J_{H,H}$ = 4.3 Hz, 1H), 2.12 (d, $^3J_{H,H}$ = 4.3 Hz, 1H), 1.45–1.32 (m, 2H), 1.16 (s, 9H), 0.71 (t, $J_{H,H}$ = 7.4 Hz, 3H); ^{13}C NMR (126 MHz, CDCl$_3$) δ_C 166.9, 138.6, 129.7 (+, 2C), 128.4 (+, 2C), 127.4 (+), 121.7 (+, 2C), 108.6 (+, 2C), 51.6, 46.3 (+), 43.0, 35.1 (+), 28.8 (+, 3C), 28.5 (-), 11.3 (+); FT IR (KBr, cm^{-1}): 3317, 3060, 2968, 2931, 1647, 1545, 1492, 1446, 1263, 1224, 721, 698; HRMS (TOF ES): found 309.1968, calculated for C$_{20}$H$_{25}$N$_2$ONa (M-H) 342.2045 (0.6 ppm).

(+)-(1R,2R,3S)-N-(tert-Butyl)-3-(ethyl(phenyl)amino)-2-methyl-2-(naphthalen-2-yl) cyclopropane-1-carboxamide (23dal). Compound was obtained according to typical procedure B from 39.1 mg (0.109 mmol, 1.00 equiv.) of (1S,2R)-1-bromo-N-(tert-butyl)-2-methyl-2-(naphthalen-2-yl)cyclopropane-1-carboxamide (**20da**), employing N-ethylaniline (**25l**) (40 µL, 0.318 mmol, 2.92 equiv.) as pronucleophile. Chromatographic purification afforded 29.8 mg (0.074 mmol, 68%) of the title compound as a colorless solid, m.p. 132.3–135.1 °C, R_f 0.31 (hexanes/EtOAc 6:1), $[\alpha]^{25}_D$ = +10.8° (c 0.074, CH$_2$Cl$_2$). dr > 99:1. ^{1}H NMR (500 MHz, CDCl$_3$) δ_H 7.97–7.66 (m, 4H), 7.55–7.39 (m, 3H), 7.27–7.19 (m, 2H), 7.06–6.93 (m, 2H), 6.80 (t, $^3J_{H,H}$ = 7.3 Hz, 1H), 5.18 (br. s, 1H), 3.71 (d, $^3J_{H,H}$ = 4.4 Hz, 1H), 3.77–3.64 (m, 1H), 3.61–3.48 (m, 1H), 1.73 (d, $^3J_{H,H}$ = 4.4 Hz, 1H), 1.61 (s, 3H), 1.25 (t, $^3J_{H,H}$ = 7.0 Hz, 3H), 1.10 (s, 9H); ^{13}C NMR (126 MHz, CDCl$_3$) δ_C 167.8, 149.2, 139.2, 133.6, 132.5, 129.2 (+, 2C), 128.3 (+), 127.9 (+), 127.8 (+), 126.9 (+), 126.8 (+), 126.2 (+), 125.8 (+), 118.3 (+), 116.0 (+), 51.3, 46.8 (+), 46.4 (-), 39.2 (+), 39.0, 28.8 (+, 3C), 21.9 (+), 11.2 (+); FT IR (KBr, cm^{-1}): 2968, 2358, 1645, 1595, 1531, 1498, 1454, 1366, 1255, 1188, 817, 742, 692; HRMS (TOF ES): found 399.2438, calculated for C$_{27}$H$_{31}$N$_2$O (M-H)$^+$ 399.2436 (0.5 ppm).

4. Conclusions

In conclusion, a highly efficient method for the assembly of tetrasubstituted chiral non-racemic cyclopropanes with all three asymmetric carbons in the strained ring was demonstrated. This method utilizes a "dual-control" strategy, which was successfully employed for the highly diastereoselective addition of the nucleophilic species to in situ generated enantiomerically enriched cyclopropenes. The chiral integrity of the starting material was translated to the product via the sequential installation of two stereogenic centers that were efficiently controlled by steric and thermodynamic effects. Alkoxides, as well as nitrogen-based nucleophiles (azoles and anilines), were used to access the homochiral derivatives of cyclopropyl ethers and cyclopropylamines. These reactions proceeded smoothly, affording unusually conformationally constrained amide derivatives of densely substituted enantiomerically enriched β-amino acids possessing three contiguous stereogenic carbon atoms. It should be also pointed out that one of these centers is an all-carbon-substituted quaternary stereocenter, the installation of which, by traditional methods, represents a long-standing challenge.

Supplementary Materials: The following supporting information can be downloaded at: https://www.mdpi.com/article/10.3390/molecules27207069/s1, ^{1}H and ^{13}C NMR spectral charts (Figures S1–S32). Figure S1. 1H NMR spectrum of compound **20ab**. Figure S2. 13C NMR spectrum of compound **20ab**. Figure S3. 1H NMR spectrum of compound **20ac**. Figure S4. 13C NMR spectrum of compound **20ac**. Figure S5. 1H NMR spectrum of compound **20da**. Figure S6. 13C NMR spectrum of compound **20da**. Figure S7. 1H NMR spectrum of compound **23aaa**. Figure S8. 13C NMR spectrum of compound **23aaa**. Figure S9. 1H NMR spectrum of compound **23aci**. Figure S10. 13C NMR spectrum of compound **23aci**. Figure S11. 1H NMR spectrum of compound **23acj**. Figure S12. 13C NMR spectrum of compound **23acj**. Figure S13. 1H NMR spectrum of compound **23acl**. Figure S14. 13C NMR spectrum of compound **23acl**. Figure S15. 1H NMR spectrum of compound **23ack**. Figure S16. 13C NMR spectrum of compound **23ack**. Figure S17. 1H NMR spectrum of compound **23acm**. Figure S18. 13C NMR spectrum of compound **23acm**. Figure S19. 1H NMR spectrum of compound **23aag**. Figure S20. 13C NMR spectrum of compound **23aag**. Figure S21. 1H NMR spectrum of compound **23dad**. Figure S22. 13C NMR spectrum of compound **23dad**. Figure S23. 1H NMR spectrum of compound **23dae**. Figure S24. 13C NMR spectrum of compound **23dae**. Figure S25. 1H NMR spectrum of compound **23dal**. Figure S26. 13C NMR spectrum of compound **23dal**. Figure S27. 1H NMR spectrum of compound **23cag**. Figure S28. 13C NMR spectrum of compound **23cag**. Figure S29. 1H NMR spectrum of compound **23cah**. Figure S30. 13C NMR spectrum of compound **23cah**. Figure S31. 1H NMR spectrum of compound **23cad**. Figure S32. 13C NMR spectrum of compound **23cad**.

Author Contributions: H.S.—investigation, formal data analysis; P.R.—investigation, formal data analysis, review, and editing; M.R. (Marina Rubina)—formal data analysis, review, and editing; M.R. (Michael Rubin)—conceptualization, supervision, data analysis, writing (original draft, review, and editing). All authors have read and agreed to the published version of the manuscript.

Funding: This work was financed by a grant from the Ministry of Education and Science of the Russian Federation (grant #0795-2020-0031).

Institutional Review Board Statement: Not applicable.

Informed Consent Statement: Not applicable.

Data Availability Statement: Supporting Information data include NMR spectral charts.

Conflicts of Interest: The authors declare no conflict of interest.

References

1. Chen, D.Y.K.; Pouwer, R.H.; Richard, J.-A. Recent advances in the total synthesis of cyclopropane-containing natural products. *Chem. Soc. Rev.* **2012**, *41*, 4631–4642. [CrossRef] [PubMed]
2. Nam, D.; Steck, V.; Potenzino, R.J.; Fasan, R. A diverse library of chiral cyclopropane scaffolds via chemoenzymic assembly and diversification of cyclopropyl ketones. *J. Am. Chem. Soc.* **2021**, *143*, 2221–2231. [CrossRef] [PubMed]
3. Donaldson, W.A. Synthesis of cyclopropane containing natural products. *Tetrahedron* **2001**, *57*, 8589–8627. [CrossRef]
4. Keglevich, P.; Keglevich, A.; Hazai, L.; Kalaus, G.; Szantay, C. Natural Compounds Containing a Condensed Cyclopropane Ring. Natural and Synthetic Aspects. *Curr. Org. Chem.* **2014**, *18*, 2037–2042. [CrossRef]
5. Shim, S.Y.; Kim, J.Y.; Nam, M.; Hwang, G.-S.; Ryu, D.H. Enantioselective Cyclopropanation with α-Alkyl-α-diazoesters Catalyzed by Chiral Oxazaborolidinium Ion: Total Synthesis of (+)-Hamavellone B. *Org. Lett.* **2016**, *18*, 160–163. [CrossRef]
6. Williams, J.D.; Yazarians, J.A.; Almeyda, C.C.; Anderson, K.; Boyce, G.R. Detection of the Previously Unobserved Stereoisomers of Thujone in the Essential Oil and Consumable Products of Sage (*Salvia officinalis* L.) Using Headspace Solid-Phase Microextraction-Gas Chromatography-Mass Spectrometry. *J. Agric. Food Chem.* **2016**, *64*, 4319–4326. [CrossRef]
7. Green, R.; Cheeseman, M.; Duffill, S.; Merritt, A.; Bull, S.D. An efficient asymmetric synthesis of grenadamide. *Tetrahedron Lett.* **2005**, *46*, 7931–7934. [CrossRef]
8. Ren, X.; Chandgude, A.L.; Fasan, R. Highly Stereoselective Synthesis of Fused Cyclopropane-γ-Lactams via Biocatalytic Iron-Catalyzed Intramolecular Cyclopropanation. *ACS Catal.* **2020**, *10*, 2308–2313. [CrossRef]
9. Wang, L.; Tang, Y. Asymmetric Ring-Opening Reactions of Donor-Acceptor Cyclopropanes and Cyclobutanes. *Isr. J. Chem.* **2016**, *56*, 463–475. [CrossRef]
10. Sanchez-Diez, E.; Vesga, D.L.; Reyes, E.; Uria, U.; Carrillo, L.; Vicario, J.L. Organocatalytically Generated Donor-Acceptor Cyclopropanes in Domino Reactions. One-Step Enantioselective Synthesis of Pyrrolo[1,2-a]quinolines. *Org. Lett.* **2016**, *18*, 1270–1273. [CrossRef]
11. Cao, B.; Mei, L.-Y.; Li, X.-G.; Shi, M. Palladium-catalyzed asymmetric [3+2] cycloaddition to construct 1,3-indandione and oxindole-fused spiropyrazolidine scaffolds. *RSC Adv.* **2015**, *5*, 92545–92548. [CrossRef]
12. Gharpure, S.J.; Nanda, L.N.; Shukla, M.K. Donor-Acceptor Substituted Cyclopropane to Butanolide and Butenolide Natural Products: Enantiospecific First Total Synthesis of (+)-Hydroxyancepsenolide. *Org. Lett.* **2014**, *16*, 6424–6427. [CrossRef] [PubMed]
13. Cohen, Y.; Cohen, A.; Marek, I. Creating Stereocenters Within Acyclic Systems by C-C Bond Cleavage of Cyclopropanes. *Chem. Rev.* **2021**, *121*, 140–161. [CrossRef] [PubMed]
14. Rubina, M.; Sherrill, W.M.; Rubin, M. Dramatic Stereo- and Enantiodivergency in the Intermolecular Asymmetric Heck Reaction Catalyzed by Palladium Complexes with Cyclopropane-Based PHOX Ligands. *Organometallics* **2008**, *27*, 6393–6395. [CrossRef]
15. Rubina, M.; Sherrill, W.M.; Barkov, A.Y.; Rubin, M. Rational design of cyclopropane-based chiral PHOX ligands for intermolecular asymmetric Heck reaction. *Beilstein J. Org. Chem.* **2014**, *10*, 1536–1548. [CrossRef]
16. Khlebnikov, A.F.; Kozhushkov, S.I.; Yufit, D.S.; Schill, H.; Reggelin, M.; Spohr, V.; de Meijere, A. A Novel Type of Chiral Triangulane-Based Diphosphane Ligands for Transition Metals. *Eur. J. Org. Chem.* **2012**, *2012*, 1530–1545. [CrossRef]
17. Molander, G.A.; Burke, J.P.; Carroll, P.J. Synthesis and Application of Chiral Cyclopropane-Based Ligands in Palladium-Catalyzed Allylic Alkylation. *J. Org. Chem.* **2004**, *69*, 8062–8069. [CrossRef]
18. Onajole, O.K.; Vallerini, G.P.; Eaton, J.B.; Lukas, R.J.; Brunner, D.; Caldarone, B.J.; Kozikowski, A.P. Synthesis and Behavioral Studies of Chiral Cyclopropanes as Selective α4β2-Nicotinic Acetylcholine Receptor Partial Agonists Exhibiting an Antidepressant Profile. Part III. *ACS Chem. Neurosci.* **2016**, *7*, 811–822. [CrossRef]

19. Reddy, C.N.; Nayak, V.L.; Mani, G.S.; Kapure, J.S.; Adiyala, P.R.; Maurya, R.A.; Kamal, A. Synthesis and biological evaluation of spiro[cyclopropane-1,3′-indolin]-2′-ones as potential anticancer agents. *Bioorg. Med. Chem. Lett.* **2015**, *25*, 4580–4586. [CrossRef]
20. Mei, H.; Pan, G.; Zhang, X.; Lin, L.; Liu, X.; Feng, X. Catalytic Asymmetric Ring-Opening/Cyclopropanation of Cyclic Sulfur Ylides: Construction of Sulfur-Containing Spirocyclopropyloxindoles with Three Vicinal Stereocenters. *Org. Lett.* **2018**, *20*, 7794–7797. [CrossRef]
21. Rodriguez, K.X.; Howe, E.N.; Bacher, E.P.; Burnette, M.; Meloche, J.L.; Meisel, J.; Schnepp, P.; Tan, X.; Chang, M.; Zartman, J.; et al. Combined Scaffold Evaluation and Systems-Level Transcriptome-Based Analysis for Accelerated Lead Optimization Reveals Ribosomal Targeting Spirooxindole Cyclopropanes. *ChemMedChem* **2019**, *14*, 1653–1661. [CrossRef] [PubMed]
22. Riss, P.J.; Roesch, F. A convenient chemo-enzymatic synthesis and 18F-labelling of both enantiomers of trans-1-toluenesulfonyloxymethyl-2-fluoromethyl-cyclopropane. *Org. Biomol. Chem.* **2008**, *6*, 4567–4574. [CrossRef] [PubMed]
23. Unzner, T.A.; Grossmann, A.S.; Magauer, T. Rapid Access to Orthogonally Functionalized Naphthalenes: Application to the Total Synthesis of the Anticancer Agent Chartarin. *Angew. Chem. Int. Ed.* **2016**, *55*, 9763–9767. [CrossRef] [PubMed]
24. Hopkins, C.D.; Schmitz, J.C.; Chu, E.; Wipf, P. Total Synthesis of (-)-CP2-Disorazole C1. *Org. Lett.* **2011**, *13*, 4088–4091. [CrossRef]
25. Casar, Z. Synthetic Approaches to Contemporary Drugs that Contain the Cyclopropyl Moiety. *Synthesis* **2020**, *52*, 1315–1345. [CrossRef]
26. Bender, T.A.; Dabrowski, J.A.; Zhong, H.; Gagne, M.R. Diastereoselective B(C6F5)3-Catalyzed Reductive Carbocyclization of Unsaturated Carbohydrates. *Org. Lett.* **2016**, *18*, 4120–4123. [CrossRef]
27. Wu, W.; Lin, Z.; Jiang, H. Recent advances in the synthesis of cyclopropanes. *Org. Biomol. Chem.* **2018**, *16*, 7315–7329. [CrossRef]
28. Li, J.-H.; Feng, T.-F.; Du, D.-M. Construction of Spirocyclopropane-Linked Heterocycles Containing Both Pyrazolones and Oxindoles through Michael/Alkylation Cascade Reactions. *J. Org. Chem.* **2015**, *80*, 11369–11377. [CrossRef]
29. Tollefson, E.J.; Erickson, L.W.; Jarvo, E.R. Stereospecific Intramolecular Reductive Cross-Electrophile Coupling Reactions for Cyclopropane Synthesis. *J. Am. Chem. Soc.* **2015**, *137*, 9760–9763. [CrossRef]
30. Munnuri, S.; Falck, J.R. Directed, Remote Dirhodium C(sp3)-H Functionalization, Desaturative Annulation, and Desaturation. *J. Am. Chem. Soc.* **2022**, *144*, 17989–17998. [CrossRef]
31. Calo, F.P.; Fuerstner, A. A Heteroleptic Dirhodium Catalyst for Asymmetric Cyclopropanation with α-Stannyl α-Diazoacetate. "Stereoretentive" Stille Coupling with Formation of Chiral Quarternary Carbon Centers. *Angew. Chem.* **2020**, *59*, 13900–13907. [CrossRef] [PubMed]
32. Chanthamath, S.; Nguyen, D.T.; Shibatomi, K.; Iwasa, S. Highly Enantioselective Synthesis of Cyclopropylamine Derivatives via Ru(II)-Pheox-Catalyzed Direct Asymmetric Cyclopropanation of Vinylcarbamates. *Org. Lett.* **2013**, *15*, 772–775. [CrossRef] [PubMed]
33. Huang, J.-Q.; Liu, W.; Zheng, B.-H.; Liu, X.Y.; Yang, Z.; Ding, C.-H.; Li, H.; Peng, Q.; Hou, X.-L. Pd-Catalyzed Asymmetric Cyclopropanation Reaction of Acyclic Amides with Allyl and Polyenyl Carbonates. Experimental and Computational Studies for the Origin of Cyclopropane Formation. *ACS Catal.* **2018**, *8*, 1964–1972. [CrossRef]
34. Inoue, H.; Nga, P.T.T.; Fujisawa, I.; Iwasa, S. Synthesis of Forms of a Chiral Ruthenium Complex Containing a Ru-Colefin(sp2) Bond and Their Application to Catalytic Asymmetric Cyclopropanation Reactions. *Org. Lett.* **2020**, *22*, 1475–1479. [CrossRef] [PubMed]
35. Li, Z.; Roesler, L.; Wissel, T.; Breitzke, H.; Hofmann, K.; Limbach, H.-H.; Gutmann, T.; Buntkowsky, G. Design and characterization of novel dirhodium coordination polymers-the impact of ligand size on selectivity in asymmetric cyclopropanation. *Catal. Sci. Technol.* **2021**, *11*, 3481–3492. [CrossRef]
36. Purins, M.; Waser, J. Asymmetric Cyclopropanation and Epoxidation via a Catalytically Formed Chiral Auxiliary. *Angew. Chem. Int. Ed.* **2022**, *61*, e202113925. [CrossRef] [PubMed]
37. Shi, T.; Luo, Y.; Wang, X.-L.; Lu, S.; Zhao, Y.-L.; Zhang, J. Theoretical Studies on the Mechanism, Enantioselectivity, and Axial Ligand Effect of a Ru(salen)-Catalyzed Asymmetric Cyclopropanation Reaction. *Organometallics* **2014**, *33*, 3673–3682. [CrossRef]
38. Wang, H.-X.; Li, W.-P.; Zhang, M.-M.; Xie, M.-S.; Qu, G.-R.; Guo, H.-M. Synthesis of chiral pyrimidine-substituted diester D-A cyclopropanes via asymmetric cyclopropanation of ylides. *Chem. Commun.* **2020**, *56*, 11649–11652. [CrossRef]
39. Wei, B.; Sharland, J.C.; Lin, P.; Wilkerson-Hill, S.M.; Fullilove, F.A.; McKinnon, S.; Blackmond, D.G.; Davies, H.M.L. In Situ Kinetic Studies of Rh(II)-Catalyzed Asymmetric Cyclopropanation with Low Catalyst Loadings. *ACS Catal.* **2020**, *10*, 1161–1170. [CrossRef]
40. White, J.D.; Shaw, S. A New Cobalt-Salen Catalyst for Asymmetric Cyclopropanation. Synthesis of the Serotonin-Norepinephrine Reuptake Inhibitor (+)-Synosutine. *Org. Lett.* **2014**, *16*, 3880–3883. [CrossRef]
41. Yuan, W.-C.; Lei, C.-W.; Zhao, J.-Q.; Wang, Z.-H.; You, Y. Organocatalytic Asymmetric Cyclopropanation of 3-Acylcoumarins with 3-Halooxindoles: Access to Spirooxindole-cyclopropa[c]coumarin Compounds. *J. Org. Chem.* **2021**, *86*, 2534–2544. [CrossRef] [PubMed]
42. Hoshiya, N.; Takenaka, K.; Shuto, S.; Uenishi, J.-I. Pd(II)-Catalyzed Alkylation of Tertiary Carbon via Directing-Group-Mediated C(sp3)-H Activation: Synthesis of Chiral 1,1,2-Trialkyl Substituted Cyclopropanes. *Org. Lett.* **2016**, *18*, 48–51. [CrossRef] [PubMed]
43. Hoshiya, N.; Kondo, M.; Fukuda, H.; Arisawa, M.; Uenishi, J.I.; Shuto, S. Entry to Chiral 1,1,2,3-Tetrasubstituted Arylcyclopropanes by Pd(II)-Catalyzed Arylation via Directing Group-Mediated C(sp3)-H Activation. *J. Org. Chem.* **2017**, *82*, 2535–2544. [CrossRef]

44. Jerhaoui, S.; Poutrel, P.; Djukic, J.P.; Wencel-Delord, J.; Colobert, F. Stereospecific C-H activation as a key step for the asymmetric synthesis of various biologically active cyclopropanes. *Org. Chem. Front.* **2018**, *5*, 409–414. [CrossRef]
45. Minami, T.; Fukuda, K.; Hoshiya, N.; Fukuda, H.; Watanabe, M.; Shuto, S. Synthesis of Enantiomerically Pure 1,2,3-Trisubstituted Cyclopropane Nucleosides Using Pd-Catalyzed Substitution via Directing Group-Mediated C(sp3)-H Activation as a Key Step. *Org. Lett.* **2019**, *21*, 656–659. [CrossRef]
46. Vicente, R. Recent Progresses towards the Strengthening of Cyclopropene Chemistry. *Synthesis* **2016**, *48*, 2343–2360. [CrossRef]
47. Li, P.; Zhang, X.; Shi, M. Recent developments in cyclopropene chemistry. *Chem. Commun.* **2020**, *56*, 5457–5471. [CrossRef]
48. Muller, D.S.; Marek, I. Copper mediated carbometalation reactions. *Chem. Soc. Rev.* **2016**, *45*, 4552–4566. [CrossRef]
49. Zhu, Z.-B.; Wei, Y.; Shi, M. Recent developments of cyclopropene chemistry. *Chem. Soc. Rev.* **2011**, *40*, 5534–5563. [CrossRef]
50. Dian, L.; Marek, I. Asymmetric Preparation of Polysubstituted Cyclopropanes Based on Direct Functionalization of Achiral Three-Membered Carbocycles. *Chem. Rev.* **2018**, *118*, 8415–8434. [CrossRef]
51. Edwards, A.; Rubina, M.; Rubin, M. Nucleophilic Addition of Cyclopropenes. *Curr. Org. Chem.* **2016**, *20*, 1862–1877. [CrossRef]
52. Fumagalli, G.; Stanton, S.; Bower, J.F. Recent Methodologies That Exploit C-C Single-Bond Cleavage of Strained Ring Systems by Transition Metal Complexes. *Chem. Rev.* **2017**, *117*, 9404–9432. [CrossRef] [PubMed]
53. Rubin, M.; Rubina, M.; Gevorgyan, V. Transition Metal Chemistry of Cyclopropenes and Cyclopropanes. *Chem. Rev.* **2007**, *107*, 3117–3179. [CrossRef] [PubMed]
54. Nie, S.; Lu, A.; Kuker, E.L.; Dong, V.M. Enantioselective Hydrothiolation: Diverging Cyclopropenes through Ligand Control. *J. Am. Chem. Soc.* **2021**, *143*, 6176–6184. [CrossRef]
55. Yamanushkin, P.M.; Rubina, M.; Rubin, M. Amide-Directed Reactions of Small Carbocycles. *Curr. Org. Chem.* **2021**, *25*, 1686–1703. [CrossRef]
56. Liu, F.; Bugaut, X.; Schedler, M.; Froehlich, R.; Glorius, F. Designing N-Heterocyclic Carbenes: Simultaneous Enhancement of Reactivity and Enantioselectivity in the Asymmetric Hydroacylation of Cyclopropenes. *Angew. Chem., Int. Ed.* **2011**, *50*, 12626–12630. [CrossRef]
57. Liu, X.; Fox, J.M. Enantioselective, Facially Selective Carbomagnesation of Cyclopropenes. *J. Am. Chem. Soc.* **2006**, *128*, 5600–5601. [CrossRef]
58. Alnasleh, B.K.; Sherrill, W.M.; Rubina, M.; Banning, J.; Rubin, M. Highly Diastereoselective Formal Nucleophilic Substitution of Bromocyclopropanes. *J. Am. Chem. Soc.* **2009**, *131*, 6906–6907. [CrossRef]
59. Banning, J.E.; Prosser, A.R.; Rubin, M. Thermodynamic Control of Diastereoselectivity in the Formal Nucleophilic Substitution of Bromocyclopropanes. *Org. Lett.* **2010**, *12*, 1488–1491. [CrossRef]
60. Prosser, A.R.; Banning, J.E.; Rubina, M.; Rubin, M. Formal Nucleophilic Substitution of Bromocyclopropanes with Amides en route to Conformationally Constrained β-Amino Acid Derivatives. *Org. Lett.* **2010**, *12*, 3968–3971. [CrossRef]
61. Ryabchuk, P.; Rubina, M.; Xu, J.; Rubin, M. Formal Nucleophilic Substitution of Bromocyclopropanes with Azoles. *Org. Lett.* **2012**, *14*, 1752–1755. [CrossRef] [PubMed]
62. Banning, J.E.; Gentillon, J.; Ryabchuk, P.G.; Prosser, A.R.; Rogers, A.; Edwards, A.; Holtzen, A.; Babkov, I.A.; Rubina, M.; Rubin, M. Formal Substitution of Bromocyclopropanes with Nitrogen Nucleophiles. *J. Org. Chem.* **2013**, *78*, 7601–7616. [CrossRef] [PubMed]
63. Banning, J.E.; Prosser, A.R.; Alnasleh, B.K.; Smarker, J.; Rubina, M.; Rubin, M. Diastereoselectivity Control in Formal Nucleophilic Substitution of Bromocyclopropanes with Oxygen- and Sulfur-Based Nucleophiles. *J. Org. Chem.* **2011**, *76*, 3968–3986. [CrossRef]
64. Alnasleh, B.K.; Rubina, M.; Rubin, M. Templated assembly of medium cyclic ethers via exo-trig nucleophilic cyclization of cyclopropenes. *Chem. Commun.* **2016**, *52*, 7494–7496. [CrossRef] [PubMed]
65. Edwards, A.; Bennin, T.; Rubina, M.; Rubin, M. Synthesis of 1,5-dioxocanes via the two-fold C-O bond forming nucleophilic 4 + 4-cyclodimerization of cycloprop-2-en-1-ylmethanols. *RSC Adv.* **2015**, *5*, 71849–71853. [CrossRef]
66. Ryabchuk, P.; Matheny, J.P.; Rubina, M.; Rubin, M. Templated Assembly of Chiral Medium-Sized Cyclic Ethers via 8-endo-trig Nucleophilic Cyclization of Cyclopropenes. *Org. Lett.* **2016**, *18*, 6272–6275. [CrossRef]
67. Ryabchuk, P.; Edwards, A.; Gerasimchuk, N.; Rubina, M.; Rubin, M. Dual Control of the Selectivity in the Formal Nucleophilic Substitution of Bromocyclopropanes en Route to Densely Functionalized, Chirally Rich Cyclopropyl Derivatives. *Org. Lett.* **2013**, *15*, 6010–6013. [CrossRef]
68. Edwards, A.; Ryabchuk, P.; Barkov, A.; Rubina, M.; Rubin, M. Preparative resolution of bromocyclopropylcarboxylic acids. *Tetrahedron Asymmetry* **2014**, *25*, 1537–1549. [CrossRef]

Article

An Improved Synthetic Method for Sensitive Iodine Containing Tricyclic Flavonoids

Mihail Lucian Birsa * and Laura G. Sarbu *

Department of Chemistry, Alexandru Ioan Cuza University of Iasi, 11 Carol I Blvd., 700506 Iasi, Romania
* Correspondence: lbirsa@uaic.ro (M.L.B.); laura.sarbu@uaic.ro (L.G.S.)

Abstract: The synthesis of new iodine containing synthetic tricyclic flavonoids is reported. Due to the sensitivity of the precursors to the heat and acidic conditions required for the ring closure of the 1,3-dithiolium core, a new cyclization method has been developed. It consists in the treatment of the corresponding iodine-substituted 3-dithiocarbamic flavonoids with a 1:1 (v/v) mixture of glacial acetic acid–concentrated sulfuric acid at 40 °C. The synthesis of the iodine-substituted 3-dithiocarbamic flavonoids has also been tuned in terms of reaction conditions.

Keywords: flavonoids; 1,3-dithiolium salts; dithiocarbamates

1. Introduction

Flavonoids are a diverse group of polyphenolic plant secondary metabolites. Associated with the multitude of substitution patterns on the C-6–C-3–C-6 backbone, more than 9000 flavonoids are known [1]. The attention that they receive is a direct consequence of the many biological activities that this class of compounds displays. Studies performed on flavonoids found that they possess antioxidant, anti-inflammatory, antimicrobial, antitumoral, antiviral or cardioprotective properties [2–4]. The antimicrobial properties that some flavonoids display could be exploited for this purpose. In principle, flavonoids can act directly against the infectious microorganisms, they can be used in combination with other antibiotics (synergistic relationship), or they can act against bacterial virulence factors, such as the cell-binding ability or toxins released by the pathogens. Many flavonoids, such as quercetin and naringenin [5], apigenin [6] or epigallocatechin gallate [7], to name but a few, are known to possess antibacterial activity. More than that, epigallocatechin gallate was also shown to enhance the activity of other antibiotics against drug-resistant pathogens [8]. In the past few years, the subject of antibacterial research has often been related to semisynthetic and synthetic flavonoids, some of these compounds being more active than natural flavonoids [9]. Our recent review highlighted the synthetic flavonoids with antimicrobial activities known up to date in the literature [10].

The emergence of more and more nosocomial infections caused by multidrug-resistant organisms (MDROs) is one of the most worrying phenomena of recent years. The discovery of new and more efficient antimicrobial drugs is therefore a matter of high priority among scientists and clinicians worldwide. Ideally, antibacterial agents should belong to new classes, since the structural alteration of drugs to which resistance has already developed rarely provides a major solution [11]. Following the general interest for synthetic flavonoids, the synthesis of a new class of tricyclic flavonoids as a combination of a condensed benzopyran core and 1,3-dithiolium ring was reported [12]. Subsequently, this class of new synthetic flavonoids proved to exhibit good to excellent antibacterial activities against both Gram-positive and Gram-negative bacteria [13]. The tricyclic flavonoids developed by us inhibited and also killed bacterial cells at very low concentrations (up to 0.24 µg/mL MIC and MBC values) [14,15]. Moreover, some of these flavonoids exhibited a stronger inhibitory and bactericidal effect compared with some antibiotics and other natural or syn-

thetic flavonoids reported in the literature and inhibited to some degree the proliferation of cancer cells [16].

Recently, we reported a study on the influence of halogen substituents on the antibacterial properties of tricyclic flavonoids [17]. Upon going from fluorine to iodine, these compounds exhibited good to excellent antimicrobial properties against both Gram-positive and Gram-negative pathogens. The results suggested that halogen size was the main factor for the change in potency rather than polarity/electronics. Prompted by these findings, we decided to investigate the synthesis of sulphur-containing tricyclic flavonoids bearing two iodine substituents on the benzopyran moiety.

2. Results and Discussion

The synthetic route used to obtain 1,3-dithiolium flavonoids **5a–e** is described in Scheme 1 and follows the protocol used for the model compound **5a**. 2-Bromo-1-(2-hydroxy-3,5-diiodophenyl)ethan-1-one (**1**) [18] readily underwent nucleophilic substitution in the presence of the N,N-diethyldithiocarbamate anion, in acetone, yielding the desired phenacyl carbodithioate **2**. The incorporation of the N,N-diethyldithiocarbamic unit was confirmed by NMR spectral data. Thus, the ^{1}H NMR spectrum indicated the presence of two triplets, at 1.30 ppm and 1.38 ppm, corresponding to the two methyl groups, and also two quartets, 3.83 ppm and 4.03 ppm, provided by the two methylene units directly bound to the nitrogen atom. The ^{13}C NMR spectrum confirmed the presence of the two methyl groups (11.5 ppm and 12.6 ppm), the two nitrogen-bounded methylene groups (47.3 ppm and 50.5 ppm) and the thiocarbonyl carbon atom (192.8 ppm).

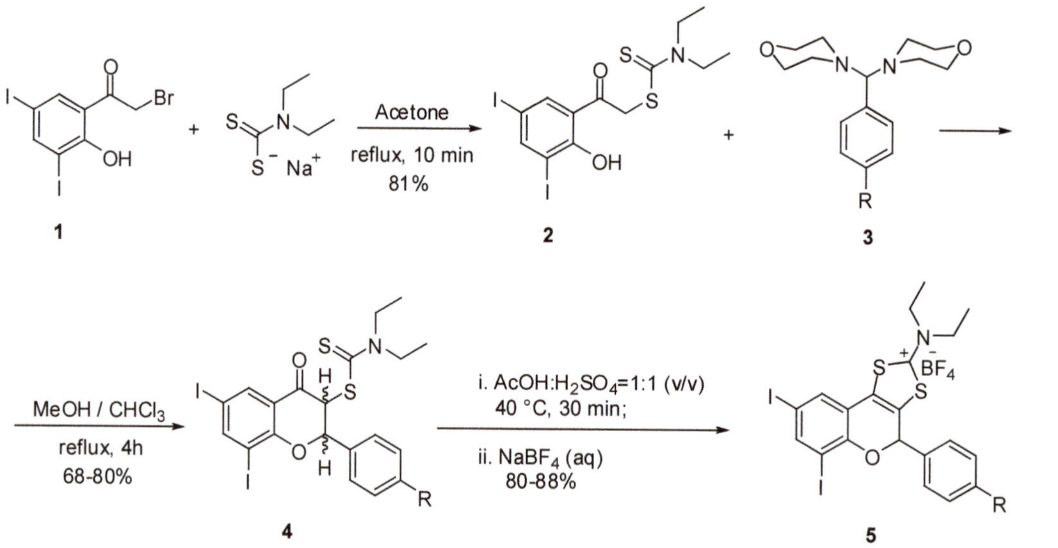

3, 4, 5	a	b	c	d	e
R	Me	Et	F	Br	OMe

Scheme 1. The synthesis of tricyclic flavonoids **5a–e** from flavanones **4a–e**. The later have been obtained starting from phenacyl bromide **1** through dithiocarbamate **2** and aminals **3a–e**.

The reaction of 1-(2-hydroxy-3,5-diiodophenyl)-1-oxa-ethan-2-yl N,N-diethylaminocarbodithioate (**2**) with aminals **3** provided 3-substituted dithiocarbamic flavanones **4a–e** as a mixture of diastereoisomers (Scheme 1). Aminals **3** were synthesized according to the literature procedures [19,20]. Due to the low solubility of dithiocarbamate **2** in ethanol, an improved experimental procedure using a mixture of chloroform and methanol (1:1) as solvent was developed. Thus, the homogeneous reaction mixture was heated at reflux for 4 h. After cooling, pale yellow precipitates were formed that were filtered, dried and recrystallized from ethanol to provide 3-dithiocarbamic flavanones **4a–e**, as an inseparable mixture of diastereoisomers, in 68–80% yields. NMR spectra supported the benzopyran ring closure. Thus, besides the NMR pattern of *para*-substituted aromatic ring originating from aminal **3**, we observed the disappearance of the signal of the methylene group from dithiocarbamate **2** (4.86 ppm) and the presence of the characteristic pattern of vicinal hydrogen atoms at the C-2 and C-3 positions of the benzopyran ring for both diastereoisomers between 5.7 and 6 ppm. Because these two protons can be located either on the same side or on opposite sides of the plane of the molecule, two stereoisomers, *anti*-**4'** and *syn*-**4"** can be obtained (Figure 1). The relative orientation of the two hydrogen atoms would, of course, be expected to have an influence on the magnitude of their coupling constants. The *anti* isomers always displayed a coupling constant between 6.2 and 7.3 Hz and the *syn* isomers around 4 Hz. The coupling constants and diastereoisomeric ratios of flavonoids **4a–e** are presented in Table 1. A ^{13}C NMR analysis confirmed the presence of the C-2 carbon atom, found around 80.0 ppm, while the C-3 carbon atom could be found around 60.0 ppm.

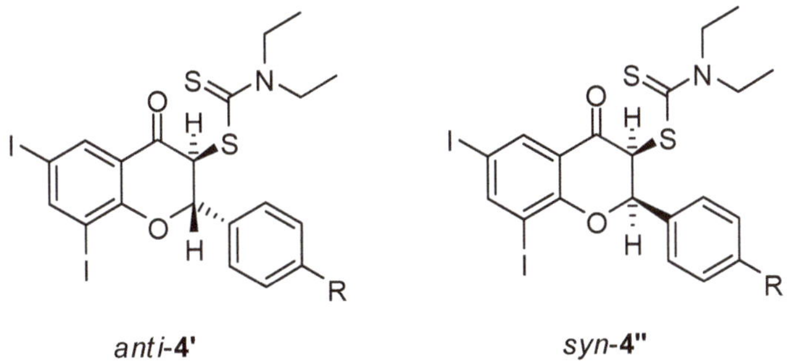

Figure 1. Diastereoisomers of flavonoids **4a–e**.

Table 1. Coupling constants H-2–H-3 and diastereoisomers ratio of flavanones **4a–e**.

4	a	b	c	d	e
^3J anti, Hz	6.2	6.3	7.3	7.1	6.7
^3J syn, Hz	4.1	4.3	3.8	3.6	3.7
anti/syn ratio	77:23	95:5	77:23	67:33	76:24

α-Ketodithiocarbamates are valuable precursors for 2-dialkylamino-1,3-dithiolium-2-yl cations [21–23]. Usually, the acid-catalysed cyclocondensation of these substrates is the method employed for the synthesis of the desired 1,3-dithiolium cations. This consisted in using a glacial acetic acid/sulfuric acid 3:1 (*v/v*) at 80 °C for 10 min [24]. Previously, we developed specific methods for the sensitive starting materials prone to decomposition under regular reaction conditions. In one such application, a mixture of phosphorus pentoxide–methanesulfonic acid 1:10 (*w/v*) was used for the synthesis of several 4-iodoaryl-1,3-dithiolium salts [25].

Despite our previous experience with the synthesis of tricyclic flavonoids of type **5** [13,14,17], attempts to close the 1,3-dithiolium ring on flavanones **4** led to a black intractable material. Even under mild reaction conditions described by us for iodine-substituted phenacyl dithiocarbamates [25,26], the cyclization reactions failed for all new reported flavanones **4**. Consequently, we tuned the reaction conditions in terms of reducing the reaction temperature and the composition of the cyclization mixture. The best results for our substrates were obtained using a mixture of glacial acetic acid/sulfuric acid 1:1 (*v/v*) at 40 °C for 30 min, followed by a treatment with an aqueous solution of sodium tetrafluoroborate. Thus, the tricyclic 1,3-dithiolium flavonoids **5** was obtained as white crystals in 80–88% yields.

The cyclization of dithiocarbamates **4** to tricyclic flavonoids **5** was accompanied by important spectral changes. Thus, IR spectroscopy showed the absence of the carbonyl absorption bands (1690–1700 cm^{-1}) and the presence of new strong and broad absorption bands (ca. 1070 cm^{-1}) from the tetrafluoroborate anion. In the ^{1}H NMR spectra, the doublets corresponding to the C-3 hydrogens disappeared; at the same time, the signals of the C-2 hydrogens were shifted to ca. 6.9 ppm and became singlets. The ^{13}C NMR spectra confirmed the absence of the carbonyl and thiocarbonyl atoms and showed a new signal at ca. 185 ppm corresponding to the 1,3-dithiol-2-ylium carbon atom.

3. Materials and Methods

3.1. Chemistry

Melting points were obtained on a *KSPI* melting-point meter (A. KRÜSS Optronic, Hamburg, Germany) and were uncorrected. IR spectra were recorded on a Bruker Tensor 27 instrument (Bruker Optik GmbH, Ettlingen, Germany). NMR spectra were recorded on a Bruker 500 MHz spectrometer (Bruker BioSpin, Rheinstetten, Germany). Chemical shifts are reported in ppm downfield from TMS. UV–vis spectra were recorded on a Varian BioChem 100 spectrophotometer. Mass spectra were recorded on a Thermo Scientific ISQ LT instrument (Thermo Fisher Scientific Inc., Waltham, MA, USA). All reagents were commercially available and used without further purification. Elemental analysis, nuclear magnetic resonance data and copies of ^{13}C-NMR spectra are included in the Supplementary Material.

3.1.1. 1-(2-Hydroxy-3,5-diiodophenyl)-1-oxoethan-2-yl N,N-diethylamino-1-carbodithioate (**2**)

To a solution of 2-bromo-1-(2-hydroxy-3,5-diiodophenyl)ethan-1-one (**1**, 1.4 g, 3 mmol) in acetone (10 mL), a solution of sodium N,N-diethyldithiocarbamate trihydrate (0.68 g, 3 mmol) in acetone/water (10 mL, 1:1 v/v) was added. The resulting mixture was refluxed for 10 min, cooled to room temperature and poured into water (100 mL) with vigorous stirring. The precipitate thus formed was vacuum-filtered and recrystallized from ethanol, yielding 1.3 g (81%) of yellow crystals; M.p. = 162–163 °C. IR (ATR, cm^{-1}) 1699, 1499, 1425, 1245, 1174, 821, 621. ^{1}H NMR (CDCl$_3$) δ 12.71 (s, 1H), 8.30 (d, J = 1.6 Hz, 1H), 8.25 (d, J = 1.7 Hz, 1H), 4.86 (s, 2H), 4.03 (q, J = 6.9 Hz, 2H), 3.83 (q, J = 6.9 Hz, 2H), 1.38 (t, J = 6.9 Hz, 3H), 1.30 (t, J = 6.9 Hz, 3H). ^{13}C NMR (CDCl$_3$) δ 198.1, 192.8, 160.6, 152.9, 138.7, 120.6, 88.2, 80.6, 50.5, 47.3, 43.5, 12.6, 11.5. UV–vis (λ_{max}, nm) 373. MS (EI) (m/z): 534.8 (M$^+$, 37%) for C$_{13}$H$_{15}$I$_2$NO$_2$S$_2$.

3.1.2. General Procedure for 6,8-Diiodo-2-(4-methylphenyl)-4-oxochroman-3-yl N,N-diethyldithiocarbamate (**4a**)

To a solution of 1-(3,5-diiodo-2-hydroxyphenyl)-1-oxoethan-2-yl N,N-diethyldithiocarbamate (**2**) (0.268 g, 0.5 mmol) in a mixture of CHCl$_3$/MeOH (12 mL, 1:1 v/v) aminal **3a** (0.13 g, 0.5 mmol) was added and the reaction mixture was heated under reflux for 4 h. After cooling, the solid material was filtered off and purified by recrystallization from ethanol to give **4a** (0.23 g, 72%) as colourless crystals. IR (ATR, cm^{-1}) 2738, 1698, 1419, 1255, 1203, 963, 811, 506, 485, 430. ^{1}H NMR (CDCl$_3$, selected data for the major isomer) δ 8.26 (d, J = 1.7 Hz, 1H), 8.13 (d, J = 1.7 Hz, 1H), 7.38 (d, J = 7.7 Hz, 2H), 7.16 (d, J = 7.7 Hz, 2H), 6.01 (d, J = 6.2 Hz, 1H), 5.75 (d, J = 6.2 Hz, 1H), 3.99 (m, 2H), 3.68 (m, 2H), 2.35 (s, 3H), 1.25 (t, J = 6.9 Hz, 6H). ^{13}C NMR (CDCl$_3$, selected data for the major isomer) δ 191.6, 186.2, 158.8, 152.6, 138.7, 136.3, 132.8, 129.3, 127.2, 122.9, 87.6, 84.7, 83.0, 57.8, 50.5, 47.3, 21.2, 12.6, 11.4. UV–vis (λ_{max}, nm) 314. MS (EI) m/z: 636.8 (M$^+$, 17%) for C$_{21}$H$_{21}$I$_2$NO$_2$S$_2$.

3.1.3. 6,8-Diiodo-2-(4-ethylphenyl)-4-oxochroman-3-yl N,N-diethyldithiocarbamate (**4b**)

Colourless crystals, 0.22 g, 68%. IR (ATR, cm^{-1}) 2965, 1697, 1417, 1256, 1202, 825, 641, 474, 438. ^{1}H NMR (CDCl$_3$, selected data for the major isomer) δ 8.26 (d, J = 1.8 Hz, 1H), 8.13

(d, *J* = 1.8 Hz, 1H), 7.4 (d, *J* = 7.9 Hz, 2H), 7.19 (d, *J* = 7.9 Hz, 2H), 6.01 (d, *J* = 6.3 Hz, 1H), 5.76 (d, *J* = 6.3 Hz, 1H), 3.98 (m, 2H), 3.69 (m, 2H), 2.65 (q, *J* = 7.5 Hz, 2H), 1.25 (t, *J* = 7.5 Hz, 3H), 1.23 (t, *J* = 7.6 Hz, 6H). ^{13}C NMR (CDCl$_3$, selected data for the major isomer) δ 191.2, 186.3, 158.8, 152.6, 144.9, 136.3, 133.0, 128.1, 127.2, 122.9, 87.6, 84.7, 83.0, 57.8, 50.5, 47.3, 28.5, 15.3, 12.6, 11.4. UV–vis (λ_{max}, nm) 313. MS (EI) *m/z*: 650.8 (M$^+$, 27%) for C$_{22}$H$_{23}$I$_2$NO$_2$S$_2$.

3.1.4. 6,8-Diiodo-2-(4-fluorophenyl)-4-oxochroman-3-yl *N,N*-diethyldithiocarbamate (**4c**)

Colourless crystals, 0.24 g, 75%. IR (ATR, cm^{-1}) 2980, 1685, 1419, 1226, 1201, 974, 825, 537, 474. ^{1}H NMR (CDCl$_3$, selected data for the major isomer) δ 8.27 (d, *J* = 1.9 Hz, 1H), 8.16 (d, *J* = 1.9 Hz, 1H), 7.49 (m, 2H), 7.06 (m, 2H), 5.96 (d, *J* = 7.3 Hz, 1H), 5.79 (d, *J* = 7.3 Hz, 1H), 3.95 (m, 2H), 3.71 (m, 2H), 1.24 (t, *J* = 6.8 Hz, 6H). ^{13}C NMR (CDCl$_3$, selected data for the major isomer) δ 190.9, 186.1, 162.9, 158.7, 152.7, 136.4, 131.7, 129.4, 122.4, 115.5, 87.4, 84.9, 82.6, 58.4, 50.7, 47.3, 12.6, 11.4. UV–vis (λ_{max}, nm) 308. MS (EI) *m/z*: 640.8 (M$^+$, 24%) for C$_{20}$H$_{18}$FI$_2$NO$_2$S$_2$.

3.1.5. 6,8-Diiodo-2-(4-bromophenyl)-4-oxochroman-3-yl *N,N*-diethyldithiocarbamate (**4d**)

Colourless crystals, 0.28 g, 80%. IR (ATR, cm^{-1}) 2977, 1700, 1421, 1252, 1201, 817, 507, 433, 417. ^{1}H NMR (CDCl$_3$, selected data for the major isomer) δ 8.28 (d, *J* = 1.9 Hz, 1H), 8.16 (d, *J* = 1.9 Hz, 1H), 7.52 (d, *J* = 8.2 Hz, 2H), 7.41 (d, *J* = 8.2 Hz, 2H), 5.95 (d, *J* = 7.1 Hz, 1H), 5.74 (d, *J* = 7.1 Hz, 1H), 3.92 (m, 2H), 3.66 (m, 2H), 1.24 (t, *J* = 6.7 Hz, 6H). ^{13}C NMR (CDCl$_3$, selected data for the major isomer) δ 190.8, 185.9, 158.7, 152.7, 136.5, 134.8, 131.7, 129.1, 123.1, 122.1, 87.4, 85.1, 82.6, 58.2, 50.7, 47.4, 12.6, 11.4. UV–vis (λ_{max}, nm) 311. MS (EI) *m/z*: 700.7 (M$^+$, 296%) for C$_{20}$H$_{18}$BrI$_2$NO$_2$S$_2$.

3.1.6. 6,8-Diiodo-2-(4-methoxyphenyl)-4-oxochroman-3-yl *N,N*-diethyldithiocarbamate (**4e**)

Colourless crystals, 0.25 g, 77%. IR (ATR, cm^{-1}) 1699, 1496, 1420, 1246, 1180, 1032, 828, 663. ^{1}H NMR (CDCl$_3$, selected data for the major isomer) δ 8.26 (m, 1H), 8.14 (m, 1H), 7.42 (d, *J* = 8.2 Hz, 2H), 6.89 (d, *J* = 8.2 Hz, 2H), 5.96 (d, *J* = 6.7 Hz, 1H), 5.78 (d, *J* = 6.7 Hz, 1H), 3.97 (m, 2H), 3.81 (s, 3H), 3.65 (m, 2H), 1.25 (t, *J* = 6.7 Hz, 6H). ^{13}C NMR (CDCl$_3$, selected data for the major isomer) δ 191.2, 186.4, 159.9, 158.8, 152.6, 136.4, 128.7, 127.9, 122.8, 113.9, 87.7, 87.6, 84.7, 58.1, 55.3, 50.5, 47.3, 12.6, 11.4. UV–vis (λ_{max}, nm) 315. MS (EI) *m/z*: 652.8 (M$^+$, 37%) for C$_{21}$H$_{21}$I$_2$NO$_3$S$_2$.

3.1.7. General Procedure for 2-*N,N*-diethylamino-6,8-diiodo-4-(4-methylphenyl)-4*H*-1,3-dithiol[4,5-c]chromen-2-ylium tetrafluoroborate (**5a**)

To a mixture of sulfuric acid (1 mL) and acetic acid (1 mL), flavanone **4a** (0.21 g, 0.33 mmol) was added and the resulting solution was heated to 40 °C for 30 min. The reaction mixture was then left to cool to room temperature and a solution of sodium tetrafluoroborate (0.2 g) in water (10 mL) was added dropwise, with vigorous stirring. The resulting precipitate was then filtered, washed thoroughly with water and recrystallized from ethanol, yielding the desired tetrafluoroborate **5a** in the form of colourless crystals (0.17 g, 85%). M.p. 260–261 °C. IR (ATR, cm^{-1}) 1554, 1438, 1224, 1045, 729, 494, 458. ^{1}H NMR (DMSO-*d*6) δ 8.04 (d, *J* = 1.8 Hz, 1H), 7.74 (d, *J* = 1.8 Hz, 1H), 7.37 (d, *J* = 8.0 Hz, 2H), 7.26 (d, *J* = 8.0 Hz, 2H), 6.89 (s, 1H), 3.89 (m, 4H), 2.31 (s, 3H), 1.40 (t, *J* = 7.1 Hz, 3H), 1.32 (t, *J* = 7.1 Hz, 3H). ^{13}C NMR (DMSO-*d*6) δ 184.9, 150.7, 147.8, 140.2, 133.8, 130.0, 129.7, 127.8, 126.8, 119.1, 88.7, 87.6, 76.2, 54.7, 54.6, 21.3, 10.8, 10.5. UV–vis (λ_{max}, nm) 343. MS (EI) *m/z*: 619.9 (M$^+$-BF$_4$, 7%) for C$_{21}$H$_{20}$I$_2$NOS$_2$]$^+$.

3.1.8. 2-*N,N*-Diethylamino-6,8-diiodo-4-(4-ethylphenyl)-4*H*-1,3-dithiol[4,5-c]chromen-2-ylium tetrafluoroborate (**5b**)

Colourless crystals, M.p. 201–202 °C, (0.17 g, 81%). IR (ATR, cm^{-1}) 1549, 1428, 1217, 1034, 719, 496, 448. ^{1}H NMR (DMSO-*d*6) δ 8.06 (d, *J* = 1.8 Hz, 1H), 7.73 (d, *J* = 1.8 Hz, 1H), 7.35 (d, *J* = 8.1 Hz, 2H), 7.24 (d, *J* = 8.1 Hz, 2H), 6.87 (s, 1H), 3.87 (m, 4H), 2.35 (q, *J* = 7.3 Hz, 2H), 1.40 (t, *J* = 7.1 Hz, 3H), 1.32 (t, *J* = 7.1 Hz, 3H), 1.26 (t, *J* = 7.3 Hz, 3H). ^{13}C NMR

(DMSO-d_6) δ 185, 150.5, 147.7, 140.1, 133.5, 129.9, 129.6, 127.5, 126.7, 119.0, 88.8, 87.5, 76.1, 54.8, 54.6, 25.4, 12.3, 10.7, 10.4. UV–vis (λ_{max}, nm) 344. MS (EI) m/z: 633.9 (M$^+$-BF$_4$, 5%) for $C_{22}H_{22}I_2NOS_2$]$^+$.

3.1.9. 2-N,N-diethylamino-6,8-diiodo-4-(4-fluorophenyl)-4H-1,3-dithiol[4,5-c]chromen-2-ylium tetrafluoroborate (5c)

Colourless crystals, M.p. 237–238 °C (0.17 g, 83%). IR (ATR, cm^{-1}) 1551, 1433, 1225, 1048, 685, 458, 409. ^{1}H NMR (DMSO-d_6) δ 8.09 (d, J = 1.8 Hz, 1H), 7.80 (d, J = 1.8 Hz, 1H), 7.55 (dd, $^3J_{H-H}$ = 8.7 Hz, $^4J_{H-F}$ = 5.3 Hz, 2H), 7.30 (dd, $^3J_{H-H}$ = 8.8 Hz, $^3J_{H-F}$ = 8.7 Hz, 2H), 6.96 (s, 1H), 3.90 (m, 4H), 1.40 (t, J = 7.1 Hz, 3H), 1.33 (t, J = 7.1 Hz, 3H). ^{13}C NMR (DMSO-d_6) δ 185.0, 164.2, 162.3, 150.4, 147.8, 133.0, 130.3, 129.2, 127.1, 119.0, 116.5, 88.7, 87.7, 75.5, 54.7, 54.6, 10.8, 10.5. UV–vis (λ_{max}, nm) 339. MS (EI) m/z: 623.8 (M$^+$-BF$_4$, 8%) for $C_{20}H_{17}FI_2NOS_2$]$^+$.

3.1.10. 2-N,N-diethylamino-6,8-diiodo-4-(4-bromophenyl)-4H-1,3-dithiol[4,5-c]chromen-2-ylium tetrafluoroborate (5d)

Colourless crystals, M.p. 219–220 °C (0.2 g, 88%). IR (ATR, cm^{-1}) 1546, 1429, 1225, 1049, 737, 441, 428. ^{1}H NMR (DMSO-d_6) δ 8.09 (d, J = 1.7 Hz, 1H), 7.79 (d, J = 1.7 Hz, 1H), 7.66 (d, J = 8.4 Hz, 2H), 7.43 (d, J = 8.4 Hz, 2H), 6.92 (s, 1H), 3.87 (m, 4H), 1.39 (t, J = 7.1 Hz, 3H), 1.32 (t, J = 7.1 Hz, 3H). ^{13}C NMR (DMSO-d_6) δ 185.0, 150.4, 147.9, 136.0, 133.1, 132.5, 130.0, 128.7, 127.2, 123.9, 119.0, 88.6, 87.8, 75.5, 54.7, 54.6, 10.7, 10.5. UV–vis (λ_{max}, nm) 341. MS (EI) m/z: 683.7 (M$^+$-BF$_4$, 5%) for $C_{20}H_{17}BrI_2NOS_2$]$^+$.

3.1.11. 2-N,N-diethylamino-6,8-diiodo-4-(4-methoxyphenyl)-4H-1,3-dithiol chromen-2-ylium tetrafluoroborate (5e)

Colourless crystals, M.p. 235–236 °C (0.17 g, 80%). IR (ATR, cm^{-1}) 1548, 1429, 1247, 1070, 851, 617. ^{1}H NMR (DMSO-d_6) δ 8.08 (d, J = 1.8 Hz, 1H), 7.78 (d, J = 1.8 Hz, 1H), 7.41 (d, J = 8.7 Hz, 2H), 6.99 (d, J = 8.7 Hz, 2H), 6.86 (s, 1H), 3.90 (m, 4H), 3.76 (s, 3H), 1.40 (t, J = 7.1 Hz, 3H), 1.32 (t, J = 7.1 Hz, 3H). ^{13}C NMR (DMSO-d_6) δ 184.9, 160.9, 150.7, 147.7, 132.9, 129.9, 129.6, 128.6, 126.7, 119.0, 114.8, 88.7, 87.4, 76.1, 55.7, 54.7, 54.6, 10.8, 10.5. UV–vis (λ_{max}, nm) 342. MS (EI) m/z: 635.8 (M$^+$-BF$_4$, 9%) for $C_{21}H_{20}I_2NO_2S_2$]$^+$.

4. Conclusions

In conclusion, we reported the synthesis of five iodine-containing tricyclic flavonoids, whose backbone is known to induce antimicrobial properties. This was performed through a new synthetic approach using a glacial acetic acid/sulfuric acid 1:1 (v/v) mixture at 40 °C as a cyclization agent. The synthesis of the precursors 3-dithiocarbamic flavanone was also tuned in terms of the reaction conditions.

Supplementary Materials: The following supporting information can be downloaded at: https://www.mdpi.com/article/10.3390/molecules27238430/s1. Elemental analysis, nuclear magnetic resonance data and copies of ^{13}C-NMR spectra.

Author Contributions: Both authors (M.L.B. and L.G.S.) contributed equally to the conceptualization, methodology, investigation, and writing. All authors have read and agreed to the published version of the manuscript.

Funding: This research received no external funding.

Institutional Review Board Statement: Not applicable.

Informed Consent Statement: Not applicable.

Data Availability Statement: Not applicable.

Acknowledgments: Thanks to the CERNESIM Center, within the Interdisciplinary Research Institute at the "Alexandru Ioan Cuza" University of Iasi, for recording the NMR experiments.

Conflicts of Interest: The authors declare no conflict of interest.

Sample Availability: Samples of the compounds are not available. They can be prepared according to the reported experimental procedures.

References

1. Williams, C.A.; Grayer, J. Anthocyanins and other flavonoids. *Nat. Prod. Rep.* **2004**, *21*, 539–573. [CrossRef] [PubMed]
2. Cushnie, T.P.T.; Lamb, A.J. Recent advances in understanding the antibacterial properties of flavonoids. *Int. J. Antimicrob. Agents* **2011**, *38*, 99–107. [CrossRef]
3. Jae, M.S.; Kwang, H.L.; Baik, L.S. Antiviral effect of catechins in green tea on influenza virus. *Antivir. Res.* **2005**, *68*, 66–74. [CrossRef]
4. Rana, A.C.; Gulliya, B. Chemistry and Pharmacology of Flavonoids- A Review. *Ind. J. Pharm. Ed. Res.* **2019**, *53*, 8–20. [CrossRef]
5. Rauha, J.P.; Remes, S.; Heinonen, M.; Hopia, A.; Kahkonen, M.; Kujala, T.; Pihlaja, K.; Vuorela, H.; Vuorela, P. Antimicrobial effects of Finnish plant extracts containing flavonoids and other phenolic compounds. *Int. J. Food Microbiol.* **2000**, *56*, 3–12. [CrossRef] [PubMed]
6. Sato, Y.; Suzaki, S.; Nishikawa, T.; Kihara, M.; Shibata, H.; Higuti, T. Phytochemical flavones isolated from *Scutellaria barbata* and antibacterial activity against methicillin-resistant *Staphylococcus aureus*. *J. Ethnopharmacol.* **2000**, *72*, 483–488. [CrossRef] [PubMed]
7. Ikigai, H.; Nakae, T.; Hara, Y.; Shimamura, T. Bactericidal catechins damage the lipid bilayer. *Biochim. Biophys. Acta* **1993**, *1147*, 132–136. [CrossRef]
8. Zhao, W.H.; Hu, Z.Q.; Okubo, S.; Hara, Y.; Shimamura, T. Mechanism of synergy between epigallocatechin gallate and beta-lactams against methicillin-resistant *Staphylococcus aureus*. *Antimicrob. Agents Chemother.* **2001**, *45*, 1737–1742. [CrossRef]
9. Manner, S.; Skogman, M.; Goeres, D.; Vuorela, P.; Fallarero, A. Systematic exploration of natural and synthetic flavonoids for the inhibition of *Staphylococcus aureus* biofilms. *Int. J. Mol. Sci.* **2013**, *14*, 19434–19451. [CrossRef]
10. Sarbu, L.G.; Bahrin, L.G.; Babii, C.; Stefan, M.; Birsa, M.L. Synthetic flavonoids with antimicrobial activity: A review. *J. Appl. Microbiol.* **2019**, *127*, 1282–1290. [CrossRef]
11. Aslam, B.; Wang, W.; Arshad, M.I.; Khurshid, M.; Muzammil, S.; Rasool, M.H.; Nisar, M.A.; Alvi, R.F.; Aslam, M.A.; Qamar, M.U.; et al. Antibiotic resistance: A rundown of a global crisis. *Infect. Drug Resist.* **2018**, *11*, 1645–1658. [CrossRef] [PubMed]
12. Bahrin, L.G.; Jones, P.G.; Hopf, H. Tricyclic flavonoids with 1,3-dithiolium substructure. *Beilstein J. Org. Chem.* **2012**, *8*, 1999–2003. [CrossRef] [PubMed]
13. Bahrin, L.G.; Apostu, M.O.; Birsa, M.L.; Stefan, M. The antibacterial properties of sulfur containing flavonoids. *Bioorg. Med. Chem. Lett.* **2014**, *24*, 2315–2318. [CrossRef] [PubMed]
14. Bahrin, L.G.; Hopf, H.; Jones, P.G.; Sarbu, L.G.; Babii, C.; Mihai, A.C.; Stefan, M.; Birsa, M.L. Antibacterial structure–activity relationship studies of several tricyclic sulfur-containing flavonoids. *Beilstein J. Org. Chem.* **2016**, *12*, 1065–1071. [CrossRef] [PubMed]
15. Babii, C.; Mihalache, G.; Bahrin, L.G.; Neagu, A.-N.; Gostin, I.; Mihai, C.T.; Sarbu, L.G.; Birsa, M.L.; Stefan, M. A novel synthetic flavonoid with potent antibacterial properties: In vitro activity and proposed mode of action. *PLoS ONE* **2018**, *13*, e0194898. [CrossRef]
16. Sarbu, L.G.; Shova, S.; Peptanariu, D.; Sandu, I.A.; Birsa, M.L.; Bahrin, L.G. The Cytotoxic Properties of Some Tricyclic 1,3-Dithiolium Flavonoids. *Molecules* **2019**, *24*, 154–162. [CrossRef]
17. Bahrin, L.G.; Sarbu, L.G.; Hopf, H.; Jones, P.G.; Babii, C.; Stefan, M.; Birsa, M.L. The influence of halogen substituents on the biological properties of sulfur-containing flavonoids. *Bioorg. Med. Chem.* **2016**, *24*, 3166–3173. [CrossRef]
18. Sandulache, A.; Cascaval, A.; Toniutti, N.; Giumanini, A.G. New flavones by a novel synthetic route. *Tetrahedron* **1997**, *53*, 9813–9822. [CrossRef]
19. Seliger, H.; Happ, E.; Cascaval, A.; Birsa, M.L.; Nicolaescu, T.; Poinescu, I.; Cojocariu, C. Synthesis and characterization of new photostabilizers from 2,4-dihydroxybenzophenone. *Eur. Polym. J.* **1999**, *35*, 827–833. [CrossRef]
20. Birsa, M.L. Synthesis of some new substituted flavanones and related 4-chromanones by a novel synthetic method. *Synth. Commun.* **2002**, *32*, 115–118. [CrossRef]
21. Birsa, M.L.; Ganju, D. Synthesis and UV/Vis spectroscopic properties of new [2-(*N,N*-dialkylamino)-1,3-dithiolium-4-yl]phenolates. *J. Phys. Org. Chem.* **2003**, *16*, 207–212. [CrossRef]
22. Liu, S.-Y.; Wang, D.-G.; Zhong, A.-G.; Wen, H.-R. One-step rapid synthesis of π-conjugated large oligomers via C–H activation coupling. *Org. Chem. Front.* **2018**, *5*, 653–661. [CrossRef]
23. Wang, Y.-F.; Wang, C.-J.; Feng, Q.-Z.; Zhai, J.-J.; Qi, S.-S.; Zhong, A.-G.; Chu, M.-M.; Xu, D.-Q. Copper-catalyzed asymmetric 1,6-conjugate addition of in situ generated *para*-quinone methides with β-ketoesters. *Chem. Commun.* **2022**, *58*, 6653–6656. [CrossRef] [PubMed]
24. Bahrin, L.G.; Asaftei, I.V.; Sandu, I.G.; Sarbu, L.G. Synthesis of (4-Methylpiperazin-1-yl)carbodithioates and of their 1,3-Dithiolium Derivatives. *Rev. Chim. Buchar.* **2014**, *65*, 1046–1048.
25. Birsa, M.L. A new approach to preparation of 1,3-dithiolium salts. *Synth. Commun.* **2001**, *31*, 1271–1275. [CrossRef]
26. Birsa, M.L.; Asaftei, I.V. Solvatochromism of mesoionic iodo(1,3-dithiol-2-ylium-4-yl)phenolates. *Monat. Chem.* **2008**, *139*, 1433–1438. [CrossRef]

Article

The Use of Aryl-Substituted Homophthalic Anhydrides in the Castagnoli–Cushman Reaction Provides Access to Novel Tetrahydroisoquinolone Carboxylic Acid Bearing an All-Carbon Quaternary Stereogenic Center

Nazar Moshnenko [1], Alexander Kazantsev [1], Olga Bakulina [1], Dmitry Dar'in [1] and Mikhail Krasavin [1,2,*]

1 Institute of Chemistry, Saint Petersburg State University, 26 Universitetskii Prospect, 198504 Peterhof, Russia
2 Institute of Living Systems, Immanuel Kant Baltic Federal University, 236041 Kaliningrad, Russia
* Correspondence: m.krasavin@spbu.ru

Abstract: Novel aryl-substituted homophthalic acids were cyclodehydrated to the respective homophthalic anhydrides for use in the Castagnoli–Cushman reaction. With a range of imines, this reaction proceeded smoothly and delivered hitherto undescribed 4-aryl-substituted tetrahydroisoquinolonic acids with remarkable diastereoselectivity, good yields and no need for chromatographic purification. These findings significantly extend the range of cyclic anhydrides employable in the Castagnoli–Cushman reaction and signify access to a novel substitution pattern around the medicinally relevant tetrahydroisoquinolonic acid scaffold.

Keywords: homophthalic anhydride; imine; Castagnoli–Cushman reaction; tetrahydroisoquinolone; lactam; all-carbon quaternary atom

1. Introduction

The Castagnoli–Cushman reaction (CCR) [1] is a remarkably versatile [4 + 2]-type cyclocondensation of a-C-H-acidic cyclic anhydrides **1** with imines **2** leading, depending on the specific anhydride employed [2], to skeletally diverse [3] lactams **3** bearing multiple substituents, which in many cases proceeds in diastereoselective fashion. This reaction is multicomponent in nature because the requisite imine can be generated in situ from the respective amine and aldehyde [4], which makes this reaction particularly suitable for generating compound libraries in array format for drug discovery (Figure 1).

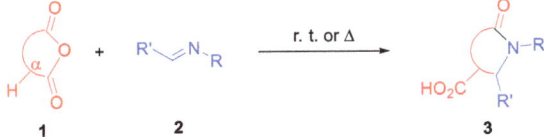

Figure 1. The Castagnoli–Cushman reaction.

Considering the fact that the cyclic anhydride (**1**) for the CCR input primarily controls the skeletal nature of the lactam product **3**, involvement of novel anhydrides in the reaction promises to deliver molecular frameworks which are either completely novel [5] or carry unprecedented substitution patterns around known cores.

Homophthalic anhydride (HPA) is one of the most popular and most reactive anhydrides used in the CCR. The reaction with HPA delivers tetrahydroisoquinolones (THIQs) with good control of diastereoselectivity [6–8]. The THIQ scaffold is of undisputable medicinal relevance, as evidenced by various molecular series possessing diverse biological activities reported in the literature. These can be exemplified by such compounds as adrenocorticotropic hormone receptor modulator **4** [9], apoptosis regulator **5** [10], trypanocidal cysteine protease inhibitor **6** [11], as well as antimalarial **7** [12] (Figure 2).

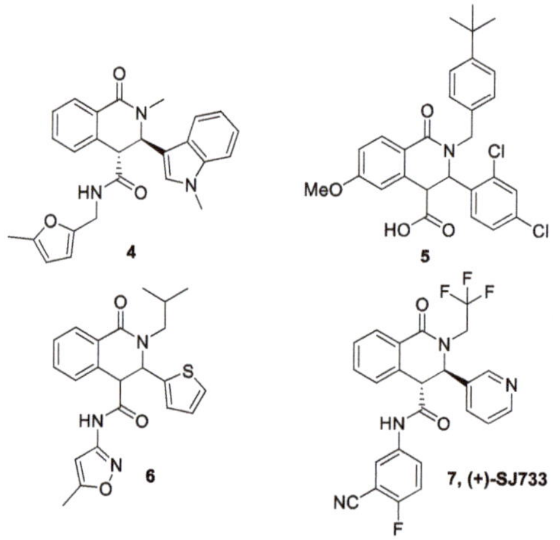

Figure 2. Examples of diversely biologically active tetrahydroisoquinolones.

The peripheral group diversity of HPA has been largely limited to the substitutions in the benzene ring [13], while substitutions at the methylene position remain almost completely unexplored except for methyl- [14,15] and benzyl- [15] substituted variants. We became interested in synthesizing novel HPA versions bearing an aryl group at the methylene linker (**8**) and exploring them as partners in the CCR. Our interest was fueled by the prospect of obtaining, possibly in diastereoselective manner, densely substituted THIQs **9** where the α-position (position 4 of the THIQ scaffold) of the hitherto undescribed carboxylic acid would be an all-carbon stereogenic center (Figure 3). Herein, we present the results obtained in the course of pursuing this goal.

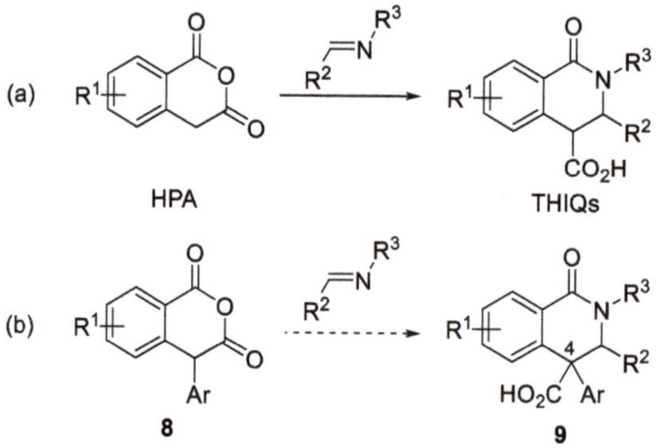

Figure 3. (**a**) Traditional CCR of HPA. (**b**) Synthetic goal pursued in this work.

2. Results

4-Aryl-substituted homophthalic acids **10** required for the preparation of anhydrides **8** were synthesized from indanones **11**. These, in turn, were prepared either by triflic acid-promoted arylation of cinnamic acids **12** [16] or by intramolecular Heck reaction of bromochalcone **13** [17]. The Heck reaction approach was used for the methoxy-substituted substrate because the respective TfOH-promoted arylation, when attempted, led to extensive tar formation. Indanones **11** were condensed with diethyl oxalate using either potassium or lithium *tert*-butoxide as the base, and the resulting condensation products **14** were oxidized with hydrogen peroxide in basic medium (as described previously [18]) to furnish novel homophthalic acids **10a–f** in modest to excellent yields over two steps from indanones **11** (Scheme 1).

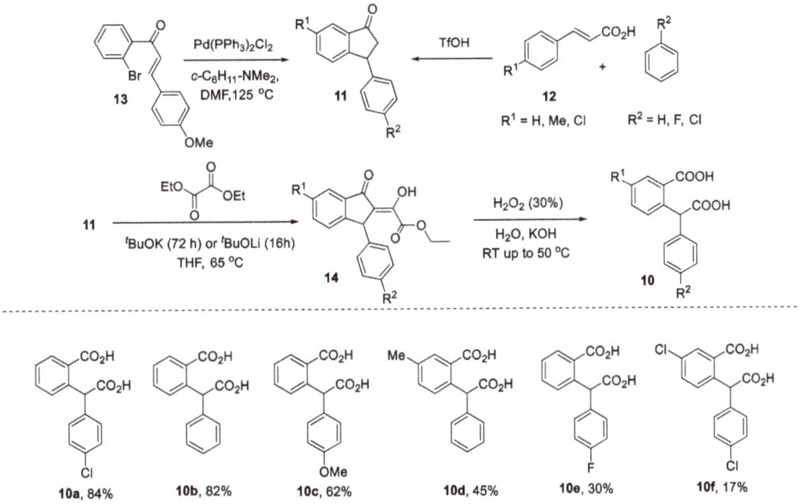

Scheme 1. Synthesis of substituted homophthalic acids **10**.

For the prospective employment of homophthalic acids in the CCR, anhydrides **8** were prepared immediately before the reaction using acetic anhydride as the cyclodehydrating agent and were used in the condensation with imines without further purification. For the preparation of anhydrides from homophthalic acids **10a–d,** the cyclodehydration was performed at room temperature in dichloromethane. For substrates **10e–f**, due to limited solubility in the latter conditions, the same reaction was performed in toluene at 80 °C.

Although the CCR of HPA can be conducted in a range of different solvents [19], after brief optimization, we found the reaction of anhydride derived from unsubstituted diacid **10b** to furnish an optimum 72% yield of THIQ cycloadduct **9a** as a single diastereomer after refluxing the reaction partners in acetonitrile over 18 h. The same reaction conducted in refluxing toluene gave lower (66%) yield. Interestingly, the reaction in acetonitrile also proceeded to completion at room temperature but with lower yield (55%) and lower diastereoselectivity (*dr* 5:1, *trans-*/*cis-*). Thus, the conditions involving refluxing acetonitrile were extended to anhydrides **8** of this and other homophthalic acids **10** in combination with various imines prepared from aromatic aldehydes (Scheme 2).

The yields of 4-aryl-substituted THIQ acids **9a–u** were generally good after simple evaporation of acetonitrile and trituration of the crude material with hexane and ether, with no need for chromatographic purification. The reactions were completely diastereoselective throughout except for those yielding products **9q–t**. The stereochemical identity of products **9a–u** was unequivocally confirmed as being *trans* with respect to the vicinal aryl groups by single-crystal X-ray analysis of compound **9a** (Figure 4, see ESI for details). The substituents in the homophthalic portion did not apparently influence the reaction outcome. The scope of the reaction was also quite broad with respect to the aromatic, aldehyde-derived group tolerating heterocyclic motifs as well as phenyl group with a nitro group. Likewise, the scope of amines, aromatic and aliphatic alike, was also fairly broad.

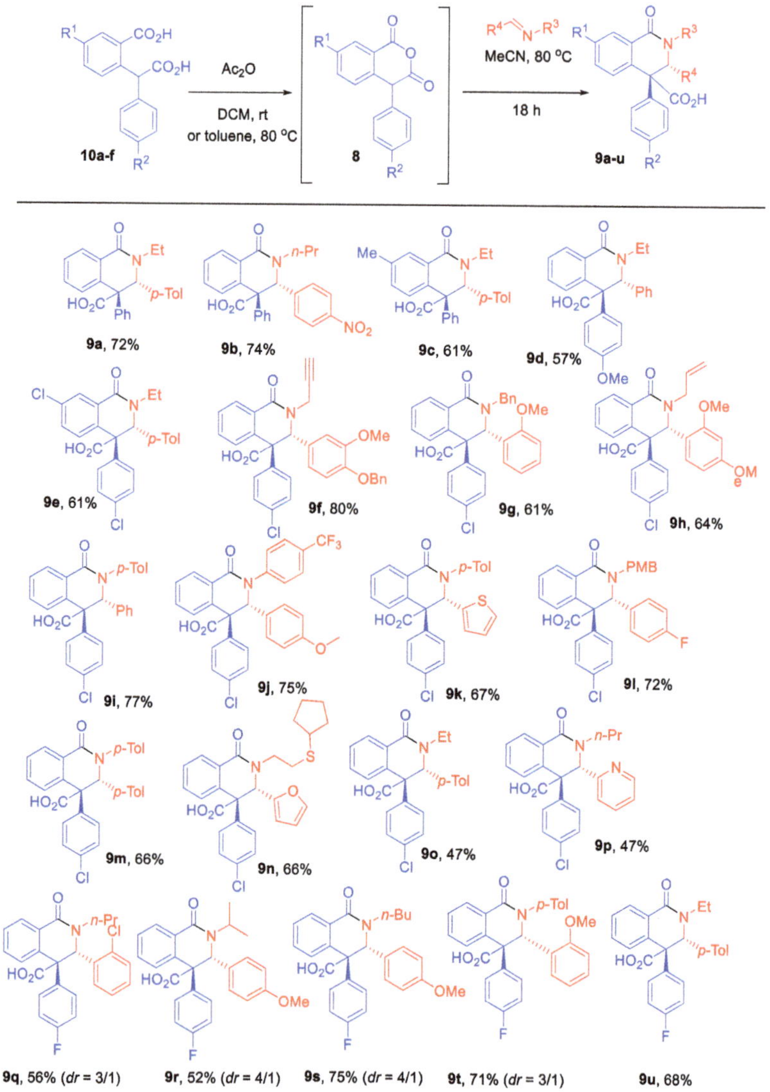

Scheme 2. The CCR of cyclic anhydrides **8** with imines.

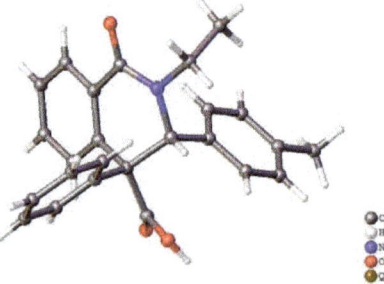

Figure 4. Crystallographic structure of compound **9a** (ORTEP plot, 50% probability level).

Despite our initial expectations of potentially lower reactivity of anhydrides **8** in the CCR due to increased steric bulk compared to HPA, the reactivity of these anhydrides was similar to that of HPA (considering the fact that the reaction also proceeded at room temperature, vide supra). This is in line with the observations by others for methyl- and benzyl-substituted versions of HPA [15].

In addition to dicarboxylic acids **10a–f**, we prepared 1,2,3-triazol-1-yl-substituted dicarboxylic acid **15** by copper-catalyzed [3 + 2] azide-alkyne cycloaddition of the known [20] azido-substituted homophthalic diethyl ester **16** and phenylacetylene followed by hydrolysis. Due to solubility issues, the cyclodehydration procedure to anhydride **17** was modified, and the reaction was performed in DMF using dicyclohexylcarbodiimide (DCC) as the cyclodehydrating agent. Anhydride **17** proved to be a competent substrate for the CCR; however, due to low solubility of **17** in acetonitrile, the reaction was conducted in DMF at room temperature. *Trans*-configured cycloadduct **18** was obtained as a single diastereomer in 50% yield, also with no need for chromatographic purification (Scheme 3).

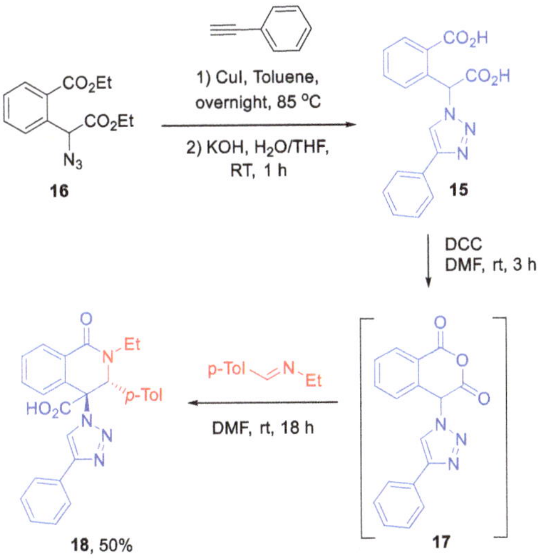

Scheme 3. Preparation and use of 1,2,3-triazol-1-yl-substituted cyclic anhydride **17** in the CCR.

3. Conclusions

We have described the synthesis of novel aryl-substituted homophthalic acids. Their cyclodehydration to the respective homophthalic anhydrides and the Castagnoli–Cushman reaction of the latter with a range of imines resulted in good yields and delivered hitherto undescribed 4-aryl-substituted tetrahydroisoquinolonic acids with remarkable diastereoselectivity, good yields and no need for chromatographic purification. These products are distinct in that they contain an all-carbon quaternary stereogenic centers in the α-position to the carboxylic acid. The cyclodehydration–Castagnoli–Cushman reaction protocol was found to be also transferrable to a novel 1,2,3-triazol-1-yl-substituted homophthalic acid. These findings significantly extend the range of cyclic anhydrides employable in the Castagnoli–Cushman reaction and signify access to a novel substitution pattern around the medicinally relevant tetrahydroisoquinolonic acid scaffold.

4. Materials and Methods

4.1. General Information

All reagents were obtained from commercial sources and used without further purification. Acetonitrile, toluene and N,N-dimethylformamide were distilled from suitable drying agents (CaH_2 or P_2O_5) and stored over MS 4Å. Mass spectra were recorded with a Bruker Maxis HRMS-ESI-qTOF spectrometer (Moscow, Russia) (electrospray ionization mode). NMR data were recorded with Bruker Avance 400/500 spectrometer (Moscow, Russia) (400.13 MHz for ^{1}H, 100.61 MHz and 125.73 MHz for ^{13}C and 376.50 MHz for ^{19}F) in DMSO-d_6 and were referenced to residual solvent proton peaks (δH = 2.51 ppm) and solvent carbon peaks (δC = 39.52 ppm). NMR and HRMS spectra are in the Supplementary Material.

4.2. Preparation of Arylhomophthalic Acids **10a–10f***: General Procedure* **1**

Step 1. Condensation of arylindanones with diethyl oxalate

Compounds **10a,b,d,e**: Corresponding indanone (9.6 mmol, 1 equiv.) and diethyl oxalate (4.2 g, 3.9 mL, 28.8 mmol, 3 equiv.) were dissolved in THF (10 mL, dry) in a round-bottom flask, and to the resulting solution a suspension of *t*-BuOK (3.23 g, 28.8 mmol, 3 equiv.) in THF (15 mL, dry) at room temperature was added dropwise. Next, the flask was stoppered, and the mixture was heated in a metal heating block at 65 °C for 72 h (conversion was estimated by TLC, using DCM as an eluent). After cooling to room temperature, the solvent was evaporated and the mixture was dissolved in CHCl$_3$ (30 mL), washed with 3% hydrochloric acid solution (1 × 15 mL), water (1 × 15 mL) and brine (1 × 15 mL), then organic layer was dried over anhydrous sodium sulfate. The solvent was evaporated, and the resulting mixture was used in the next step without purification. Compounds **10c,f** were obtained according to nearly the same procedure (but using *t*-BuOLi instead of *t*-BuOK), and the heating was performed for 16h.

Step 2. Oxidation

A solution of KOH (3.76 g, 67.2 mmol, 7 equiv.) in water (20 mL) was added to the product of the previous step in a round-bottom flask; the mixture was stirred for 20 min, then H$_2$O$_2$ (30%, 27.2 mL) was added dropwise. The solution was stirred overnight at room temperature, then heated in a metal heating block to 50 °C and stirred for two hours (until the mixture became transparent). Activated charcoal (12 g) (powder−100 particle size (mesh)) was added to the resulting chilled solution and intensively stirred for 15 min. The solution was filtered through zeolite, and a solution of concentrated hydrochloric acid was added to the filtrate at room temperature to reach pH 1. The precipitated acid was extracted into EtOAc (3 × 30 mL). The organic layer was combined, dried over anhydrous sodium sulfate and evaporated. The resulting acids **10a–e** did not require further purification. The acid **10f** was additionally crystallized from acetonitrile. Yields of compounds **10** were calculated for 2 steps.

4.2.1. 2-[Carboxy(4-chlorophenyl)methyl]benzoic Acid (**10a**)

Prepared according to the general procedure GP1 from 3-(4-chlorophenyl)-2,3-dihydro-1*H*-inden-1-one[21]. Yield 2.344 g, 84%. Colorless amorphous solid. ^{1}H NMR (400 MHz, DMSO-d_6) δ 12.89 (s, 2H), 8.02–7.80 (m, 1H), 7.55–7.48 (m, 1H), 7.46–7.34 (m, 3H), 7.31–7.23 (m, 2H), 7.15–7.10 (m, 1H), 5.99 (s, 1H). ^{13}C NMR (101 MHz, DMSO-d_6) δ 173.6, 169.0, 140.1, 138.5, 132.3, 132.1, 131.4, 131.0, 130.9, 130.2, 128.9, 127.5, 52.9. HRMS (ESI/Q-TOF) *m/z*: [M + Na$^+$]$^+$ Calcd for C$_{15}$H$_{11}$ClO$_4$Na$^+$ 313.0238; Found 313.0234.

4.2.2. 2-[Carboxy(phenyl)methyl]benzoic Acid (**10b**)

Prepared according to the general procedure GP1 from 3-phenyl-2,3-dihydro-1*H*-inden-1-one [16]. Yield 2.017 g, 82%. Colorless amorphous solid. ^{1}H NMR (400 MHz, DMSO-d_6) δ 12.81 (s, 2H), 7.95–7.82 (m, 1H), 7.53–7.44 (m, 1H), 7.41–7.32 (m, 3H), 7.32–7.22 (m, 3H), 7.13–7.04 (m, 1H), 5.97 (s, 1H). ^{13}C NMR (101 MHz, DMSO-d_6) δ 173.9, 169.1, 140.6, 139.5, 132.1, 131.1, 130.7, 130.3, 129.5, 129.0, 127.4, 127.3, 53.6. HRMS (ESI/Q-TOF) *m/z*: [M + Na$^+$]$^+$ Calcd for C$_{15}$H$_{12}$O$_4$Na$^+$ 279.0628; Found 279.0623.

4.2.3. 2-[Carboxy(4-methoxyphenyl)methyl]benzoic Acid (10c)

Prepared according to the general procedure GP1 from 3-(4-methoxyphenyl)-2,3-dihydro-1H-inden-1-one [21]. Yield 703 mg, 62%. Colorless amorphous solid. ^{1}H NMR (400 MHz, DMSO-d_6) δ 12.75 (s, 2H), 7.86 (dd, J = 7.7, 1.6 Hz, 1H), 7.46 (td, J = 7.5, 1.6 Hz, 1H), 7.33 (t, J = 7.5 Hz, 1H), 7.15 (d, J = 8.7 Hz, 2H), 7.08 (d, J = 7.7 Hz, 1H), 6.92 (d, J = 8.7 Hz, 2H), 5.88 (s, 1H), 3.74 (s, 3H). ^{13}C NMR (101 MHz, DMSO-d_6) δ 173.7, 168.7, 158.2, 140.6, 131.5, 130.9, 130.6, 130.2, 130.1, 129.7, 126.7, 113.9, 55.1, 52.3. HRMS (ESI/Q-TOF) m/z: [M-H]$^−$ Calcd for $C_{16}H_{13}O_5^−$ 285.0769; Found 285.0768.

4.2.4. 2-[Carboxy(phenyl)methyl]-5-methylbenzoic Acid (10d)

Prepared according to the general procedure GP1 from 6-methyl-3-phenyl-2,3-dihydro-1H-inden-1-one [16]. Yield 1.167 g, 45%. Colorless amorphous solid. ^{1}H NMR (400 MHz, DMSO-d_6) δ 12.71 (s, 2H), 7.71–7.65 (m, 1H), 7.38–7.32 (m, 2H), 7.30–7.26 (m, 2H), 7.25–7.19 (m, 2H), 7.05–6.89 (m, 1H), 5.91 (s, 1H), 2.31 (s, 3H). ^{13}C NMR (101 MHz, DMSO-d_6) δ 174.00, 169.17, 161.41, 139.66, 137.68, 136.58, 132.61, 131.11, 130.88, 130.24, 129.44, 128.92, 127.30, 53.20, 20.78. ^{13}C NMR (101 MHz, DMSO-d_6) δ 174.0, 169.2, 161.4, 139.7, 137.7, 136.6, 132.6, 131.1, 130.9, 130.2, 129.4, 128.9, 127.3, 53.2, 20.8. HRMS (ESI/Q-TOF) m/z: [M + Na]$^+$ Calcd for $C_{16}H_{14}O_4Na^+$ 293.0784; Found 293.0785.

4.2.5. 2-[Carboxy(4-fluorophenyl)methyl]benzoic Acid (10e)

Prepared according to the general procedure GP1 from 3-(4-fluorophenyl)-2,3-dihydro-1H-inden-1-one [16]. Yield 0.789 g, 30%. Colorless amorphous solid. ^{1}H NMR (400 MHz, DMSO-d_6) δ 12.88 (s, 2H), 7.94–7.83 (m, 1H), 7.53–7.45 (m, 1H), 7.40–7.33 (m, 1H), 7.31–7.25 (m, 2H), 7.23–7.09 (m, 3H), 6.06–5.94 (m, 1H). ^{13}C NMR (101 MHz, DMSO-d_6) δ 173.8, 169.1, 161.6 (d, J = 243.3 Hz), 140.4, 135.7 (d, J = 3.1 Hz), 132.2, 131.4 (d, J = 8.1 Hz), 131.1, 130.8, 130.1, 127.4, 115.7 (d, J = 21.3 Hz), 52.8. ^{19}F NMR (376 MHz, DMSO-d_6) δ −115.9. HRMS (ESI/Q-TOF) m/z: [M + Na$^+$]$^+$ Calcd for $C_{15}H_{11}FO_4Na^+$ 297.0534; Found 297.0528.

4.2.6. 2-[Carboxy(4-chlorophenyl)methyl]-5-chlorobenzoic Acid (10f)

Prepared according to the general procedure GP1 from 6-chloro-3-(4-chlorophenyl)-2,3-dihydro-1H-inden-1-one [22]. Yield 530 mg, 17%. Colorless amorphous solid. ^{1}H NMR (400 MHz, DMSO-d_6) δ 13.38 (s, 1H), 12.87 (s, 1H), 7.91–7.78 (m, 1H), 7.65–7.53 (m, 1H), 7.46–7.36 (m, 2H), 7.30–7.22 (m, 2H), 7.12–7.04 (m, 1H), 5.91 (s, 1H). ^{13}C NMR (101 MHz, DMSO-d_6) δ 173.3, 167.7, 139.1, 138.0, 133.0, 132.3, 132.3, 132.1, 132.0, 131.4, 130.3, 129.1, 52.6. HRMS (ESI/Q-TOF) m/z: [M + Na$^+$]$^+$ Calcd for $C_{15}H_{10}Cl_2O_4Na^+$ 346.9848; Found 346.9841.

4.3. General Procedure for Preparation of Tetrahydroisoquinonolones 9a–9u

Step 1. Anhydride synthesis.

Products 9a–c and f–u:

Diacid 10a–c,f (50 mg) was mixed with DCM (1 mL, dry.) in a screw-cap vial, after which acetic anhydride (6 equiv.) was added to the suspension and the reaction mixture was stirred overnight at room temperature. Then, the solvent was evaporated in vacuo. The resulting crude anhydride was used in the next step without purification or characterization.

For products 9d,e:

Diacid 10c,f (50 mg) was dissolved in toluene (3 mL, dry) in screw-cap vial, after which acetic anhydride (6 equiv.) was added to the suspension and the reaction mixture was stirred overnight at 80 °C in a metal heating box. Then, the solvent was evaporated in vacuo. The resulting crude anhydride was used in the next step without further purification.

Step 2. The Castagnoli–Cushman reaction

For products 9a–9u:

The resulting crude anhydride from the previous step was dissolved in MeCN (0.3 mL, dry) in a screw-cap vial, then imine (1.05 equiv.) dissolved in MeCN (0.2 mL, dry) was added with stirring. The reaction mixture was kept at 80 °C overnight in a metal heating block. Then, the solvent was evaporated. Next, the crude product was treated with diethyl ether (1 mL), after which pentane (3 mL) was added and the solid was thoroughly ground. After cooling to −20 °C for 20 min, the liquid was decanted. The resulting solid was dried in vacuo to give pure title compound.

Dr values were calculated from integrals of methine protons (^{1}H NMR spectra) from lactam ring.

4.3.1. (±)-(3R,4R)-2-Ethyl-1-oxo-4-phenyl-3-(p-tolyl)-1,2,3,4-tetrahydroisoquinoline-4-carboxylic Acid (**9a**)

Prepared according to the general procedure GP2 from **10b** and N-(4-methylbenzylidene)ethanamine. Yield 52 mg, 72%. Colorless amorphous solid. ^{1}H NMR (400 MHz, DMSO-d_6) δ 13.14 (s, 1H), 8.19–8.10 (m, 1H), 8.05–7.98 (m, 1H), 7.64 (s, 1H), 7.59–7.47 (m, 1H), 7.36–7.26 (m, 4H), 7.26–7.21 (m, 1H), 6.98 (s, 4H), 5.59 (s, 1H), 3.58 (dq, J = 14.0, 7.1 Hz, 1H), 3.20 (dq, J = 14.0, 7.1 Hz, 1H), 2.21 (s, 3H), 0.76 (t, J = 7.0 Hz, 3H). ^{13}C NMR (101 MHz, DMSO-d_6) δ 171.7, 162.3, 142.6, 137.8, 137.7, 136.0, 131.9, 130.4, 130.1, 129.0, 129.0, 128.5, 128.2, 128.0, 128.0, 127.6, 66.4, 59.0, 42.1, 21.0, 13.1. HRMS (ESI/Q-TOF) m/z: [M + H$^+$]$^+$ Calcd for $C_{25}H_{24}NO_3^+$ 386.1751; Found 386.1744.

Crystal Data for $C_{28.571429}H_{26.285714}N_{1.142857}O_{3.428571}$ (M = 440.51 g/mol): orthorhombic, space group Pbca (no. 61), a = 15.8652(2) Å, b = 14.6469(2) Å, c = 16.7176(2) Å, V = 3884.77(9) Å^3, Z = 7, T = 100.15 K, μ(CuKα) = 0.689 mm^{-1}, Dcalc = 1.318 g/cm^3, 41,022 reflections measured (9.774° ≤ 2Θ ≤ 152.44°), 4053 unique (R_{int} = 0.0439, R_{sigma} = 0.0168) which were used in all calculations. The final R_1 was 0.0408 (I > 2σ(I)) and wR_2 was 0.1105 (all data). Please see ESI (p.S2-5) for details.

4.3.2. (±)-(3R,4R)-3-(4-Nitrophenyl)-1-oxo-4-phenyl-2-propyl-1,2,3,4-tetrahydroisoquinoline-4-carboxylic Acid (**9b**)

Prepared according to the general procedure GP2 from **10b** and N-(4-nitrobenzylidene)propan-1-amine. Yield 62 mg, 74%. Colorless amorphous solid. ^{1}H NMR (400 MHz, DMSO-d_6) δ 13.46 (s, 1H), 8.12–8.05 (m, 3H), 8.03–7.98 (m, 1H), 7.72–7.66 (m, 1H), 7.60 (t, J = 7.5 Hz, 1H), 7.39 (d, J = 8.5 Hz, 2H), 7.29 (p, J = 6.6 Hz, 5H), 5.80 (s, 1H), 3.51 (ddd, J = 13.2, 8.9, 6.7 Hz, 1H), 2.97 (ddd, J = 13.6, 9.0, 5.0 Hz, 1H), 1.23 (dt, J = 8.1, 4.9 Hz, 1H), 1.15–0.99 (m, 1H), 0.47 (t, J = 7.3 Hz, 3H). ^{13}C NMR (101 MHz, DMSO-d_6) δ 171.7, 162.7, 147.6, 147.2, 142.3, 137.0, 132.4, 130.5, 130.3, 130.0, 128.7, 128.5, 128.3, 128.1, 127.8, 123.5, 66.3, 59.5, 48.3, 20.7, 11.4. HRMS (ESI/Q-TOF) m/z: [M + H$^+$]$^+$ Calcd for $C_{25}H_{23}N_2O_5^+$ 431.1601; Found 431.1606.

4.3.3. (±)-(3R,4R)-2-Ethyl-7-methyl-1-oxo-4-phenyl-3-(p-tolyl)-1,2,3,4-tetrahydroisoquinoline-4-carboxylic Acid (**9c**)

Prepared according to the general procedure GP2 from **10d** and N-(4-methylbenzylidene)ethanamine. Yield 45 mg, 61%. Colorless amorphous solid. ^{1}H NMR (400 MHz, DMSO-d_6) δ 13.06 (s, 1H), 8.05–7.99 (m, 1H), 7.83–7.80 (m, 1H), 7.47–7.41 (m, 1H), 7.35–7.26 (m, 4H), 7.25–7.19 (m, 1H), 7.02–6.92 (m, 4H), 5.55 (s, 1H), 3.57 (dq, J = 14.0, 7.1 Hz, 1H), 3.18 (dq, J = 14.0, 7.1 Hz, 1H), 2.42 (s, 3H), 2.21 (s, 3H), 0.74 (t, J = 7.1 Hz, 3H). ^{13}C NMR (126 MHz, DMSO-d_6) δ 171.8, 162.4, 142.8, 137.6, 137.5, 136.1, 134.9, 132.6, 130.2, 130.1, 129.0, 129.0, 128.5, 128.4, 128.0, 127.5, 66.5, 58.8, 42.1, 21.2, 21.0, 13.1. HRMS (ESI/Q-TOF) m/z: [M + Na$^+$]$^+$ Calcd for $C_{26}H_{25}NO_3Na^+$ 422.1727; Found 422.1718.

4.3.4. (±)-(3R,4R)-2-Ethyl-4-(4-methoxyphenyl)-1-oxo-3-(p-tolyl)-1,2,3,4-tetrahydroisoquinoline-4-carboxylic Acid (**9d**)

Prepared according to the general procedure GP2 from **10c** and N-(4-methylbenzylidene)ethanamine. Yield 41 mg, 57%. Colorless amorphous solid. ^{1}H NMR (400 MHz, DMSO-d_6) δ 13.04 (s, 1H), 8.17–8.09 (m, 1H), 8.05–7.94 (m, 1H), 7.64–7.59 (m, 1H), 7.54–7.46 (m, 1H), 7.25–7.19 (m, 2H), 7.02–6.95 (m, 4H), 6.88–6.82 (m, 2H), 5.54 (s, 1H), 3.70 (s, 3H), 3.57 (dq, J = 13.9, 7.2 Hz, 1H), 3.17 (dq, J = 13.9, 7.2 Hz, 1H), 2.20 (s, 3H), 0.77 (t, J = 7.1 Hz, 3H). ^{13}C NMR

(101 MHz, DMSO-d_6) δ 172.0, 162.4, 158.6, 138.1, 137.6, 136.0, 134.4, 131.8, 130.4, 130.1, 129.3, 129.0, 128.9, 128.1, 128.0, 113.8, 66.5, 58.4, 55.5, 42.1, 21.0, 13.1. HRMS (ESI/Q-TOF) *m/z*: [M + H$^+$]$^+$ Calcd for C$_{26}$H$_{26}$NO$_4^+$ 416.1856; Found 416.1848.

4.3.5. (±)-(3R,4R)-7-Chloro-4-(4-chlorophenyl)-2-ethyl-1-oxo-3-(p-tolyl)-1,2,3,4-tetrahydroisoquinoline-4-carboxylic acid (**9e**)

Prepared according to the general procedure GP2 from **10f** and *N*-(4-methylbenzylidene)ethanamine. Yield 43 mg, 61%. Colorless amorphous solid. ^{1}H NMR (400 MHz, DMSO-d_6) δ 13.48 (s, 1H), 8.26–8.13 (m, 1H), 8.01–7.93 (m, 1H), 7.79–7.55 (m, 1H), 7.52–7.27 (m, 4H), 7.10–6.85 (m, 4H), 5.65 (s, 1H), 3.58 (dq, *J* = 14.1, 7.1 Hz, 1H), 3.19 (dq, *J* = 14.1, 7.1 Hz, 1H), 2.21 (s, 3H), 0.78 (t, *J* = 7.1 Hz, 3H). ^{13}C NMR (101 MHz, DMSO-d_6) δ 171.0, 161.2, 141.0, 137.9, 136.4, 135.2, 133.4, 132.6, 132.6, 132.3, 131.9, 130.1, 129.2, 128.9, 128.5, 127.5, 66.3, 58.6, 42.3, 21.0, 13.0. HRMS (ESI/Q-TOF) *m/z*: [M + H$^+$]$^+$ Calcd for C$_{25}$H$_{22}$Cl$_2$NO$_3^+$ 454.0971; Found 454.0965.

4.3.6. (±)-(3R,4R)-3-(4-(Benzyloxy)-3-methoxyphenyl)-4-(4-chlorophenyl)-1-oxo-2-(prop-2-yn-1-yl)-1,2,3,4-tetrahydroisoquinoline-4-carboxylic Acid (**9f**)

Prepared according to the general procedure GP2 from **10b** and *N*-(4-(benzyloxy)-3-methoxybenzylidene)prop-2-yn-1-amine. Yield 76 mg, 80%. Colorless amorphous solid. ^{1}H NMR (400 MHz, DMSO-d_6) δ 13.37 (s, 1H), 8.29–8.18 (m, 1H), 8.08–7.96 (m, 1H), 7.74–7.65 (m, 1H), 7.58–7.26 (m, 10H), 6.92–6.82 (m, 1H), 6.67–6.61 (m, 1H), 6.59–6.50 (m, 1H), 5.75 (s, 1H), 4.98 (s, 2H), 4.65 (d, *J* = 17.4 Hz, 1H), 3.77 (d, *J* = 17.4 Hz, 1H), 3.51 (s, 3H). ^{13}C NMR (126 MHz, DMSO-d_6) δ 171.2, 162.3, 148.5, 148.3, 141.2, 137.9, 137.4, 136.7, 133.1, 132.7, 132.5, 130.2, 129.8, 129.3, 129.2, 128.9, 128.8, 128.6, 128.5, 128.4, 128.3, 128.3, 121.2, 113.1, 79.0, 76.1, 70.2, 58.4, 55.5. HRMS (ESI/Q-TOF) *m/z*: [M + H$^+$]$^+$ Calcd for C$_{33}$H$_{27}$ClNO$_5^+$ 552.1572; Found 552.1574.

4.3.7. (±)-(3R,4R)-2-Benzyl-4-(4-chlorophenyl)-3-(2-methoxyphenyl)-1-oxo-1,2,3,4-tetrahydroisoquinoline-4-carboxylic Acid (**9g**)

Prepared according to the general procedure GP2 from **10a** and *N*-(2-methoxybenzylidene)-1-phenylmethanamine. Yield 62 mg, 61%. Colorless amorphous solid. ^{1}H NMR (400 MHz, DMSO-d_6) δ 12.95 (s, 1H), 8.20–8.11 (m, 1H), 7.71–7.65 (m, 1H), 7.64–7.58 (m, 1H), 7.55 (d, *J* = 7.7 Hz, 1H), 7.27–7.19 (m, 2H), 7.13–7.07 (m, 2H), 7.05–7.00 (m, 3H), 6.83–6.60 (m, 6H), 5.79 (s, 1H), 5.19 (d, *J* = 14.6 Hz, 1H), 3.87 (s, 3H), 3.40 (d, *J* = 14.6 Hz, 1H). ^{13}C NMR (101 MHz, DMSO-d_6) δ 171.5, 162.9, 158.3, 142.0, 137.4, 136.7, 132.8, 132.2, 129.9, 129.7, 128.8, 128.5, 128.5, 128.3, 127.9, 127.4, 127.3, 126.8, 120.8, 111.5, 59.5, 59.2, 56.2, 48.4. HRMS (ESI/Q-TOF) *m/z*: [M + H$^+$]$^+$ Calcd for C$_{30}$H$_{25}$ClNO$_4^+$ 498.1467; Found 498.1471.

4.3.8. (±)-(3R,4R)-2-Allyl-4-(4-chlorophenyl)-3-(2,4-dimethoxyphenyl)-1-oxo-1,2,3,4-tetrahydroisoquinoline-4-carboxylic Acid (**9h**)

Prepared according to the general procedure GP2 from **10a** and *N*-(2,4-dimethoxybenzylidene)prop-2-en-1-amine. Yield 53 mg, 64%. Colorless amorphous solid. ^{1}H NMR (400 MHz, DMSO-d_6) δ 12.95 (s, 1H), 8.11–8.04 (m, 1H), 7.71–7.52 (m, 3H), 7.41–7.31 (m, 2H), 7.18–7.07 (m, 2H), 6.59–6.47 (m, 2H), 6.33–6.26 (m, 1H), 5.81 (s, 1H), 5.28–5.16 (m, 1H), 4.97 (d, *J* = 10.1 Hz, 1H), 4.88 (d, *J* = 17.1 Hz, 1H), 4.38 (dd, *J* = 15.8, 4.8 Hz, 1H), 3.82 (s, 3H), 3.69 (s, 3H), 3.21 (dd, *J* = 15.2, 7.6 Hz, 1H). ^{13}C NMR (126 MHz, DMSO-d_6) δ 171.6, 162.6, 160.7, 159.6, 142.1, 137.7, 133.0, 132.6, 132.3, 130.2, 130.0, 129.7, 128.6, 128.2, 128.1, 118.9, 118.7, 105.2, 98.7, 59.2, 59.0, 56.2, 55.6, 48.2. HRMS (ESI/Q-TOF) *m/z*: [M + H$^+$]$^+$ Calcd for C$_{27}$H$_{25}$ClNO$_5^+$ 478.1416; Found 478.1421.

4.3.9. (±)-(3R,4R)-4-(4-Chlorophenyl)-1-oxo-3-phenyl-2-(p-tolyl)-1,2,3,4-tetrahydroisoquinoline-4-carboxylic Acid (**9i**)

Prepared according to the general procedure GP2 from **10a** and *N*-benzylidene-4-methylaniline. Yield 62 mg, 77%. Colorless amorphous solid. ^{1}H NMR (400 MHz, DMSO-d_6) δ 13.48 (s, 1H), 8.17–8.03 (m, 2H), 7.77–7.69 (m, 1H), 7.66–7.59 (m, 1H), 7.50–7.41

(m, 2H), 7.37–7.30 (m, 2H), 7.27–7.17 (m, 3H), 7.11–7.03 (m, 4H), 6.67–6.49 (m, 2H), 5.75 (s, 1H), 2.25 (s, 3H). ^{13}C NMR (126 MHz, DMSO-d_6) δ 171.1, 162.7, 141.3, 139.5, 138.2, 137.1, 136.8, 132.8, 132.8, 130.6, 130.3, 129.9, 129.1, 128.9, 128.7, 128.7, 126.5, 70.4, 59.7, 21.0. HRMS (ESI/Q-TOF) m/z: [M + H$^+$]$^+$ Calcd for C$_{29}$H$_{23}$ClNO$_3{}^+$ 468.1361; Found 468.1368.

4.3.10. (±)-(3R,4R)-4-(4-Chlorophenyl)-3-(4-methoxyphenyl)-1-oxo-2-(4-(trifluoromethyl)phenyl)-1,2,3,4-tetrahydroisoquinoline-4-carboxylic Acid (**9j**)

Prepared according to the general procedure GP2 from **10a** and N-(4-methoxybenzylidene)-4-(trifluoromethyl)aniline. Yield 71 mg, 75%. Colorless amorphous solid. ^{1}H NMR (400 MHz, DMSO-d_6) δ 13.50 (s, 1H), 8.17–8.09 (m, 2H), 7.79–7.73 (m, 1H), 7.73–7.68 (m, 2H), 7.67–7.61 (m, 1H), 7.46–7.40 (m, 2H), 7.37–7.30 (m, 2H), 7.07–7.01 (m, 4H), 6.81–6.77 (m, 2H), 5.92 (s, 1H), 3.69 (s, 3H). ^{13}C NMR (101 MHz, DMSO-d_6) δ 170.9, 162.9, 159.5, 145.6, 141.1, 137.4, 133.1, 132.8, 130.4, 130.3, 130.1, 129.6, 128.9, 128.8, 128.7, 127.5 (q, J = 32.3 Hz), 127.2, 126.5 (q, J = 3.6 Hz), 124.4 (q, J = 271.9 Hz), 114.2, 69.0, 59.8, 55.5. ^{19}F NMR (376 MHz, DMSO-d_6) δ −60.9. HRMS (ESI/Q-TOF) m/z: [M + H$^+$]$^+$ Calcd for C$_{30}$H$_{22}$ClF$_3$NO$_4{}^+$ 552.1184; Found 552.1184.

4.3.11. (±)-(3S,4R)-4-(4-Chlorophenyl)-1-oxo-3-(thiophen-2-yl)-2-(p-tolyl)-1,2,3,4-tetrahydroisoquinoline-4-carboxylic Acid (**9k**)

Prepared according to the general procedure GP2 from **10a** and 4-methyl-N-(thiophen-2-ylmethylene)aniline. Yield 55 mg, 67%. Colorless amorphous solid. ^{1}H NMR (400 MHz, DMSO-d_6) δ 13.55 (s, 1H), 8.26–8.19 (m, 1H), 8.12–8.05 (m, 1H), 7.78–7.70 (m, 1H), 7.67–7.59 (m, 1H), 7.48–7.41 (m, 2H), 7.37–7.33 (m, 1H), 7.30–7.26 (m, 2H), 7.13–7.05 (m, 3H), 6.63–6.51 (m, 3H), 5.88 (s, 1H), 2.27 (s, 3H). ^{13}C NMR (101 MHz, DMSO-d_6) δ 171.2, 162.7, 140.8, 139.6, 139.4, 137.3, 136.7, 132.9, 132.8, 130.7, 130.1, 129.9, 129.8, 128.9, 128.8, 128.8, 128.0, 126.6, 126.3, 125.5, 66.2, 59.2, 21.0. HRMS (ESI/Q-TOF) m/z: [M + H$^+$]$^+$ Calcd for C$_{27}$H$_{21}$ClNO$_3$S$^+$ 474.0925; Found 474.0926.

4.3.12. (±)-(3R,4R)-4-(4-Chlorophenyl)-3-(4-fluorophenyl)-2-(4-methoxybenzyl)-1-oxo-1,2,3,4-tetrahydroisoquinoline-4-carboxylic Acid (**9l**)

Prepared according to the general procedure GP2 from **10a** and N-(4-fluorobenzylidene)-1-(4-methoxyphenyl)methanamine. Yield 64 mg, 72%. Colorless amorphous solid. ^{1}H NMR (400 MHz, DMSO-d_6) δ 13.38 (s, 1H), 8.16–8.09 (m, 1H), 8.02–7.95 (m, 1H), 7.72–7.64 (m, 1H), 7.62–7.57 (m, 1H), 7.16–6.99 (m, 6H), 6.96–6.87 (m, 4H), 6.77–6.65 (m, 2H), 5.38 (s, 1H), 5.04 (d, J = 14.3 Hz, 1H), 3.76 (s, 3H), 3.69 (d, J = 14.3 Hz, 1H). ^{13}C NMR (101 MHz, DMSO-d_6) δ 171.3, 162.5, 162.3 (d, J = 244.5 Hz), 159.1, 140.9, 136.7, 134.1 (d, J = 3.1 Hz), 132.6, 132.3, 131.2 (d, J = 8.2 Hz), 130.3, 130.0, 129.6, 128.8, 128.6, 128.5, 128.2, 115.4 (d, J = 21.3 Hz), 114.0, 65.5, 58.8, 55.5, 48.2. ^{19}F NMR (376 MHz, DMSO-d_6) δ −114.0. HRMS (ESI/Q-TOF) m/z: [M + H$^+$]$^+$ Calcd for C$_{30}$H$_{24}$ClFNO$_4{}^+$ 516.1372; Found 516.1373.

4.3.13. (±)-(3R,4R)-4-(4-Chlorophenyl)-1-oxo-2,3-di-p-tolyl-1,2,3,4-tetrahydroisoquinoline-4-carboxylic Acid (**9m**)

Prepared according to the general procedure GP2 from **10a** and 4-methyl-N-(4-methylbenzylidene)aniline. Yield 55 mg, 66%. Colorless amorphous solid. ^{1}H NMR (400 MHz, DMSO-d_6) δ 13.44 (s, 1H), 8.12–8.06 (m, 2H), 7.76–7.68 (m, 1H), 7.66–7.57 (m, 1H), 7.49–7.38 (m, 2H), 7.35–7.29 (m, 2H), 7.12–7.07 (m, 2H), 7.04–6.89 (m, 2H), 6.65–6.58 (m, 2H), 5.72 (s, 1H), 2.25 (s, 3H), 2.22 (s, 3H). ^{13}C NMR (101 MHz, DMSO-d_6) δ 171.1, 162.7, 141.3, 139.6, 137.9, 137.2, 136.7, 135.2, 132.7, 132.7, 130.6, 130.3, 129.8, 129.3, 129.0, 128.8, 128.7, 126.4, 70.1, 59.7, 21.0, 21.0. HRMS (ESI/Q-TOF) m/z: [M + H$^+$]$^+$ Calcd for C$_{30}$H$_{25}$ClNO$_3{}^+$ 482.1517; Found 482.1518.

4.3.14. (±)-(3S,4R)-4-(4-Chlorophenyl)-2-(2-(cyclopentylthio)ethyl)-3-(furan-2-yl)-1-oxo-1,2,3,4-tetrahydroisoquinoline-4-carboxylic Acid (**9n**)

Prepared according to the general procedure GP2 from **10a** and 2-(cyclopentylthio)-N-(furan-2-ylmethylene)ethanamine. Yield 56 mg, 66%. Colorless amorphous solid. ^{1}H NMR (400 MHz, DMSO-d_6) δ 13.52 (s, 1H), 8.20–8.14 (m, 1H), 8.00–7.94 (m, 1H), 7.73–7.64 (m,

1H), 7.59–7.50 (m, 1H), 7.48–7.43 (m, 1H), 7.42–7.35 (m, 2H), 7.32–7.23 (m, 2H), 6.27 (dd, J = 3.3, 1.9 Hz, 1H), 5.87 (d, J = 3.3 Hz, 1H), 5.84 (s, 1H), 3.68 (td, J = 13.2, 11.2, 5.0 Hz, 1H), 3.37 (td, 1H), 3.10 (p, J = 7.1 Hz, 1H), 2.33 (td, J = 12.8, 11.1, 5.0 Hz, 1H), 2.03 (ddd, J = 12.8, 11.1, 5.0 Hz, 1H), 1.92 (dt, J = 13.9, 6.9 Hz, 2H), 1.70–1.57 (m, 2H), 1.60–1.48 (m, 2H), 1.32 (dt, J = 13.9, 6.9 Hz, 2H). ^{13}C NMR (101 MHz, DMSO-d_6) δ 171.4, 163.1, 152.4, 143.0, 140.3, 132.8, 132.3, 130.0, 130.0, 129.1, 128.6, 128.5, 128.5, 110.8, 109.0, 61.2, 57.8, 47.7, 43.4, 33.9, 28.2, 24.8, 24.7. HRMS (ESI/Q-TOF) m/z: [M + H$^+$]$^+$ Calcd for $C_{27}H_{27}ClNO_4S^+$ 496.1344; Found 496.1344.

4.3.15. (±)-(3R,4R)-4-(4-Chlorophenyl)-2-ethyl-1-oxo-3-(p-tolyl)-1,2,3,4-tetrahydroisoquinoline-4-carboxylic acid (**9o**)

Prepared according to the general procedure GP2 from **10a** and *N*-(4-methylbenzylidene)ethanamine. Yield 34 mg, 47%. Colorless amorphous solid. ^{1}H NMR (400 MHz, DMSO-d_6) δ 13.06 (s, 1H), 8.09–7.97 (m, 1H), 7.86–7.79 (m, 1H), 7.49–7.42 (m, 1H), 7.34–7.26 (m, 4H), 7.25–7.19 (m, 1H), 7.04–6.92 (m, 4H), 5.55 (s, 1H), 3.57 (dq, J = 14.0, 7.1 Hz, 1H), 3.18 (dq, J = 14.0, 7.1 Hz, 1H), 2.42 (s, 3H), 2.21 (s, 3H), 0.74 (t, J = 7.1 Hz, 3H). ^{13}C NMR (101 MHz, DMSO-d_6) δ 171.3, 162.3, 141.5, 137.8, 137.3, 135.7, 132.4, 132.1, 130.4, 130.2, 130.1, 129.1, 128.9, 128.5, 128.4, 128.1, 66.4, 58.9, 42.0, 21.0, 13.1. HRMS (ESI/Q-TOF) m/z: [M + Na$^+$]$^+$ Calcd for $C_{25}H_{22}ClNO_3Na^+$ 442.1180; Found 442.1175.

4.3.16. (±)-(3S,4R)-4-(4-Chlorophenyl)-1-oxo-2-propyl-3-(pyridin-2-yl)-1,2,3,4-tetrahydroisoquinoline-4-carboxylic Acid (**9p**)

Prepared according to the general procedure GP2 from **10a** and *N*-(pyridin-3-ylmethylene)propan-1-amine. Yield 34 mg, 47%. Colorless amorphous solid. ^{1}H NMR (400 MHz, DMSO-d_6) δ 13.50 (s, 1H), 8.50–8.37 (m, 2H), 8.13–8.02 (m, 1H), 8.02–7.93 (m, 1H), 7.72–7.65 (m, 1H), 7.64–7.58 (m, 1H), 7.43–7.37 (m, 2H), 7.35–7.26 (m, 3H), 7.24–7.18 (m, 1H), 5.68 (s, 1H), 3.55–3.45 (m, 1H), 3.04–2.93 (m, 1H), 1.32–1.21 (m, 1H), 1.15–1.03 (m, 1H), 0.49 (t, J = 7.3 Hz, 3H). ^{13}C NMR (101 MHz, DMSO-d_6) δ 171.4, 162.6, 150.5, 149.6, 141.1, 136.5, 135.9, 134.7, 132.7, 132.5, 130.4, 130.1, 130.1, 128.9, 128.5, 128.4, 123.6, 64.7, 59.1, 48.2, 20.8, 11.3. HRMS (ESI/Q-TOF) m/z: [M + H$^+$]$^+$ Calcd for $C_{24}H_{22}ClN_2O_3^+$ 421.1313; Found 421.1315.

4.3.17. (±)-(3S,4R)-3-(2-Chlorophenyl)-4-(4-fluorophenyl)-1-oxo-2-propyl-1,2,3,4-tetrahydroisoquinoline-4-carboxylic Acid (**9q**)

Prepared according to the general procedure GP2 from **10e** and *N*-(2-chlorobenzylidene)propan-2-amine. Yield 45 mg, 56% (dr = 3/1). Colorless amorphous solid. Major isomer: ^{1}H NMR (400 MHz, DMSO-d_6) δ 13.34 (s, 1H), 8.17–8.10 (m, 1H), 7.75–7.69 (m, 1H), 7.68–7.60 (m, 2H), 7.51–7.45 (m, 1H), 7.30–7.24 (m, 1H), 7.22–7.07 (m, 5H), 6.80–6.73 (m, 1H), 5.90 (s, 1H), 3.71–3.58 (m, 1H), 2.62–2.54 (m, 1H), 1.28–1.13 (m, 1H), 1.09–0.96 (m, 1H), 0.38 (t, J = 7.5 Hz, 3H). ^{13}C NMR (126 MHz, DMSO-d_6) δ 171.3, 161.9, 161.2 (d, J = 244.8 Hz), 158.7, 138.0 (d, J = 2.9 Hz), 136.2, 131.6, 131.1, 131.1, 130.4 (d, J = 8.2 Hz), 130.1, 130.0, 128.1, 127.7, 114.6 (d, J = 21.5 Hz), 113.3, 64.1, 59.2, 55.0, 19.9, 19.6. ^{19}F NMR (376 MHz, DMSO-d_6) δ -115.5. HRMS (ESI/Q-TOF) m/z: [M + H$^+$]$^+$ Calcd for $C_{25}H_{22}ClFNO_3^+$ 438.1267; Found 438.1271. Minor isomer, partial data: ^{1}H NMR (400 MHz, DMSO) δ 7.44 (d, J = 7.9 Hz, 1H), 7.40–7.33 (m, 2H), 7.04 (t, J = 7.3 Hz, 2H), 6.63 (dd, J = 7.9, 1.6 Hz, 1H), 6.32 (t, J = 7.8 Hz, 1H), 5.98 (d, J = 1.9 Hz, 1H). ^{19}F NMR (376 MHz, DMSO) δ -108.77.

4.3.18. (±)-(3R,4R)-4-(4-Fluorophenyl)-2-isopropyl-3-(4-methoxyphenyl)-1-oxo-1,2,3,4-tetrahydroisoquinoline-4-carboxylic Acid (**9r**)

Prepared according to the general procedure GP2 from **10e** and *N*-(4-methoxybenzylidene)propan-2-amine. Yield 41 mg, 52% (dr = 4/1). Colorless amorphous solid. Major isomer: ^{1}H NMR (400 MHz, DMSO-d_6) δ 13.21 (s, 1H), 8.12–8.04 (m, 1H), 7.98–7.88 (m, 1H), 7.66–7.59 (m, 1H), 7.61–7.51 (m, 1H), 7.38–7.30 (m, 2H), 7.18–7.07 (m, 2H), 7.06–7.00 (m, 2H), 6.77–6.66 (m, 2H), 5.51 (s, 1H), 4.27 (hept, J = 6.7 Hz, 1H), 3.67 (s, 3H), 0.83 (dd, J = 20.2, 6.7 Hz, 6H). ^{13}C NMR (126 MHz, DMSO-d_6) δ 171.7, 162.3, 161.6 (d, J = 244.8 Hz), 159.1, 138.4 (d, J = 2.9 Hz), 136.6, 132.0, 131.5, 131.5, 130.8 (d, J = 8.2 Hz), 130.5, 130.4, 128.5, 128.1, 115.0 (d, J = 21.5 Hz), 113.7,

64.5, 59.6, 55.4, 48.8, 20.3, 20.0. ^{19}F NMR (376 MHz, DMSO-d_6) δ −115.1. HRMS (ESI/Q-TOF) m/z: [M + H$^+$]$^+$ Calcd for C$_{26}$H$_{25}$FNO$_4^+$ 434.1762; Found 434.1768. Minor isomer, partial data: ^{1}H NMR (400 MHz, DMSO-d_6) δ 7.23 (dd, J = 12.4, 8.1 Hz, 1H), 6.95–6.81 (m, 2H), 5.35 (s, 1H). ^{19}F NMR (376 MHz, DMSO-d_6) δ −108.81.

4.3.19. (±)-(3R,4R)-2-Butyl-4-(4-fluorophenyl)-3-(4-methoxyphenyl)-1-oxo-1,2,3,4-tetrahydroisoquinoline-4-carboxylic Acid (**9s**)

Prepared according to the general procedure GP2 from **10e** and *N*-(4-methoxybenzylidene)butan-1-amine. Yield 61 mg, 75% (*dr* = 4/1). Colorless amorphous solid. Major isomer: ^{1}H NMR (400 MHz, DMSO-d_6) δ 13.20 (s, 1H), 8.13–7.98 (m, 2H), 7.69–7.62 (m, 1H), 7.59–7.53 (m, 1H), 7.39–7.28 (m, 2H), 7.17–7.08 (m, 2H), 7.07–6.97 (m, 2H), 6.78–6.71 (m, 2H), 5.49 (s, 1H), 3.68 (s, 3H), 3.65–3.52 (m, 1H), 3.04–2.89 (m, 1H), 1.32–1.18 (m, 1H), 1.16–1.05 (m, 1H), 0.96–0.85 (m, 1H), 0.85–0.73 (m, 1H), 0.67 (t, J = 7.2 Hz, 3H). ^{13}C NMR (126 MHz, DMSO-d_6) δ 171.6, 162.6, 160.6, 159.3, 138.6 (d, J = 3.1 Hz), 137.3, 132.1, 130.4 (d, J = 5.9 Hz), 130.4, 130.3, 128.5, 128.1, 115.1 (d, J = 21.2 Hz), 113.8, 66.7, 58.9, 55.4, 46.0, 29.5, 19.8, 14.1. ^{19}F NMR (376 MHz, DMSO-d_6) δ −115.7. HRMS (ESI/Q-TOF) m/z: [M + H$^+$]$^+$ Calcd for C$_{27}$H$_{27}$FNO$_4^+$ 448.1919; Found 448.1924. Minor isomer, partial data: ^{1}H NMR (400 MHz, DMSO-d_6) δ 7.72 (dd, J = 8.1, 2.9 Hz, 2H), 7.25–7.18 (m, 1H), 7.06 (d, J = 8.2 Hz, 1H), 6.88 (d, J = 8.4 Hz, 2H), 5.31 (s, 1H), 3.24–3.12 (m, 1H). ^{19}F NMR (376 MHz, DMSO-d_6) δ −108.66.

4.3.20. (±)-(3R,4R)-4-(4-Fluorophenyl)-3-(2-methoxyphenyl)-1-oxo-2-(p-tolyl)-1,2,3,4-tetrahydroisoquinoline-4-carboxylic Acid (**9t**)

Prepared according to the general procedure GP2 from **10e** and 2-methoxy-*N*-(4-methylbenzylidene)aniline. Yield 52 mg, 71% (dr = 3/1). Colorless amorphous solid. ^{1}H NMR (400 MHz, DMSO-d_6) δ 12.98 (s, 1H), 8.22–8.11 (m, 1H), 7.77–7.67 (m, 1H), 7.66–7.58 (m, 1H), 7.48–7.41 (m, 1H), 7.25–7.18 (m, 3H), 7.16–7.09 (m, 2H), 7.05–6.98 (m, 2H), 6.96–6.88 (m, 2H), 6.84–6.65 (m, 1H), 6.49–6.37 (m, 2H), 6.09 (s, 1H), 3.68 (s, 3H), 2.21 (s, 3H). ^{13}C NMR (101 MHz, DMSO-d_6) δ 171.6, 162.9, 161.6 (d, J = 244.7 Hz), 158.0, 140.0, 139.6 (d, J = 2.6 Hz), 137.8, 136.8, 133.0, 130.7 (d, J = 8.3 Hz), 130.3, 130.0, 129.8, 129.6, 128.8, 128.4, 127.7, 127.0, 126.3, 120.7, 115.1 (d, J = 21.2 Hz), 111.3, 65.2, 60.5, 55.9, 21.0. ^{19}F NMR (376 MHz, DMSO-d_6) δ −115.5. HRMS (ESI/Q-TOF) m/z: [M + H$^+$]$^+$ Calcd for C$_{30}$H$_{25}$FNO$_4^+$ 482.1762; Found 482.1763. Minor isomer, partial data: ^{1}H NMR (400 MHz, DMSO-d_6) δ 7.37–7.26 (m, 2H), 6.36 (d, J = 8.0 Hz, 2H), 6.27 (d, J = 2.8 Hz, 1H), 3.64 (s, 3H). ^{19}F NMR (376 MHz, DMSO-d_6) δ −108.30.

4.3.21. (±)-(3R,4R)-2-Ethyl-4-(4-fluorophenyl)-1-oxo-3-(p-tolyl)-1,2,3,4-tetrahydroisoquinoline-4-carboxylic acid (**9u**)

Prepared according to the general procedure GP2 from **10e** and *N*-(4-methylbenzylidene)ethanamine. Yield 50 mg, 68%. Colorless amorphous solid. ^{1}H NMR (400 MHz, DMSO-d_6) δ 13.25 (s, 1H), 8.17–8.06 (m, 1H), 8.07–8.00 (m, 1H), 7.69–7.59 (m, 1H), 7.58–7.48 (m, 1H), 7.40–7.30 (m, 2H), 7.17–7.06 (m, 2H), 6.98 (s, 4H), 5.58 (s, 1H), 3.59 (dq, J = 13.9, 7.0 Hz, 1H), 3.15 (dq, J = 14.0, 6.8 Hz, 1H), 2.21 (s, 3H), 0.76 (t, J = 7.1 Hz, 3H). ^{13}C NMR (101 MHz, DMSO-d_6) δ 171.6, 162.3, 161.5 (d, J = 244.5 Hz), 138.8 (d, J = 3.4 Hz), 137.7, 135.8, 132.0, 130.4, 130.3, 130.2, 130.2, 129.0, 128.9, 128.3, 128.0, 115.2 (d, J = 21.2 Hz), 66.7, 58.7, 42.0, 21.0, 13.1. ^{19}F NMR (376 MHz, DMSO-d_6) δ −115.7. HRMS (ESI/Q-TOF) m/z: [M + Na$^+$]$^+$ Calcd for C$_{25}$H$_{22}$FNO$_3$Na$^+$ 426.1476; Found 426.1469.

4.4. 2-(Carboxy(4-phenyl-1H-1,2,3-triazol-1-yl)methyl)benzoic Acid (15)

Ethyl 2-(1-azido-2-methoxy-2-oxoethyl)benzoate[20] **16** (526 mg, 2 mmol, 1 equiv.) and phenylacetylene (206 mg, 1 equiv.) were added to a suspension of CuI (27 mg, 7 mol. %) in dry toluene. The reaction mixture was stirred at 85 °C overnight. The solvent was evaporated, and the title compound was extracted with ethyl acetate (30 mL). The organic layer was washed with water (20 mL × 2) and brine (20 mL × 1) and then dried over Na$_2$SO$_4$. The solvent was evaporated and the resulting compound (ethyl 2-(2-ethoxy-2-oxo-1-(4-phenyl-1H-1,2,3-triazol-1-yl)ethyl)benzoate) was used in the next step without further purification. The obtained ester and KOH (560 mg, 5 equiv.) were dissolved in 30 mL of

30% aq.THF and stirred for 1 h at room temperature. Activated charcoal (**12g**) (powder -100 particle size (mesh)) was added to the resulting mixture and intensively stirred at room temperature for 0.5 h. Next, the solution was filtered through a layer of zeolite, and 3 N HCl was added to it until pH = 1. The target compound (**15**) was extracted into diethyl ether, and the organic layer was combined and dried over Na_2SO_4 and was evaporated.

Yield 601 mg, 93% (2 steps). Colorless amorphous solid. ^{1}H NMR (400 MHz, DMSO-d_6) δ 13.55 (s, 2H), 8.72 (s, 1H), 8.07–8.00 (m, 1H), 7.94–7.82 (m, 2H), 7.71–7.61 (m, 1H), 7.60–7.51 (m, 2H), 7.50–7.42 (m, 2H), 7.39–7.30 (m, 1H), 7.29–7.20 (m, 1H). ^{13}C NMR (101 MHz, DMSO-d_6) δ 169.0, 168.5, 146.6, 135.7, 133.1, 131.4, 130.9, 130.6, 129.5, 129.4, 129.2, 128.5, 125.7, 122.9, 63.7. HRMS (ESI/Q-TOF) m/z: [M + H$^+$]$^+$ Calcd for $C_{17}H_{14}N_3O_4^+$ 324.0979; Found 324.0974.

*4.5. (±)-(3R,4S)-2-Ethyl-1-oxo-4-(4-phenyl-1H-1,2,3-triazol-1-yl)-3-(p-tolyl)-1,2,3,4-tetrahydroisoquinoline-4-carboxylic Acid (**18**)*

The diacid **15** (50 mg, 0.15 mmol) was dissolved in DMF (0.5 mL, dry) in a screw-cap, and DCC (1.1 equiv.) was added with stirring. After 3 h, N-(4-methylbenzylidene)ethanamine (1.1 equiv.) was added, and the reaction mixture was kept for a day at room temperature. The solution was then filtered through celite, EtOAc (15 mL) and 10 mL of brine were added to the filtrate, the precipitate formed was filtered off, and then organic layer of the filtrate was washed with brine (10 mL × 3), dried over sodium sulfate and evaporated. The residue was treated with diethyl ether (1 mL), after which pentane (3 mL) was added and the solid was thoroughly ground. After cooling to -20 °C for 20 min, the liquid was decanted. The resulting solid was dried in vacuo to give pure title compound. The substance undergoes decarboxylation easily and is therefore unstable in solutions even at room temperature.

Yield 35 mg, 50%. Colorless amorphous solid. ^{1}H NMR (400 MHz, DMSO-d_6) δ 14.25 (s, 1H), 8.83–8.72 (m, 1H), 8.17–8.07 (m, 1H), 7.88–7.79 (m, 3H), 7.75–7.62 (m, 2H), 7.50–7.40 (m, 2H), 7.39–7.31 (m, 1H), 7.16–6.90 (m, 4H), 6.05 (s, 1H), 3.69 (dq, J = 14.0, 7.0 Hz, 1H), 3.18 (dq, J = 14.0, 7.0 Hz, 1H), 2.24 (s, 3H), 0.90 (t, J = 7.0 Hz, 3H). ^{13}C NMR (101 MHz, DMSO-d_6) δ 167.7, 161.6, 146.2, 138.4, 133.3, 132.4, 131.5, 130.9, 130.4, 130.0, 129.4, 129.4, 128.8, 128.5, 127.9, 125.6, 122.1, 72.2, 65.6, 42.1, 21.1, 13.3. HRMS (ESI/Q-TOF) m/z: [M + Na$^+$]$^+$ Calcd for $C_{27}H_{24}N_4O_3Na^+$ 475.1741; Found 475.1730.

Supplementary Materials: The following supporting information can be downloaded at: https://www.mdpi.com/article/10.3390/molecules27238462/s1. Copies of NMR and HRMS spectra. X-ray data [23–25].

Author Contributions: Conceptualization, O.B. and D.D.; methodology, O.B. and D.D.; investigation, N.M. and A.K.; data curation, O.B.; writing—original draft preparation, M.K.; writing—review and editing, O.B. and D.D.; supervision, M.K.; funding acquisition, O.B. All authors have read and agreed to the published version of the manuscript.

Funding: We gratefully acknowledge the financial support from the Russian Foundation for Basic Research (grant# 20-03-00922) and Megagrant of the Government of Russian Federation (# 075-15-2021-637).

Institutional Review Board Statement: Not applicable.

Informed Consent Statement: Not applicable.

Data Availability Statement: Data available from the corresponding authors upon reasonable request.

Acknowledgments: We thank the Research Centre for Magnetic Resonance and the Center for Chemical Analysis and Materials Research and the Centre for X-ray Diffraction Methods of Saint Petersburg State University Research Park for obtaining the analytical data.

Conflicts of Interest: The authors declare no conflict of interest. The funders had no role in the design of the study; in the collection, analyses, or interpretation of data; in the writing of the manuscript; or in the decision to publish the results.

Sample Availability: Samples of the compounds **9** are available from the authors upon reasonable request.

References

1. Howard, S.Y.; Di Maso, M.J.; Shimabukuro, K.; Burlow, N.P.; Tan, D.Q.; Fettinger, J.C.; Malig, T.C.; Hein, J.E.; Shaw, J.T. Mechanistic Investigation of Castagnoli–Cushman Multicomponent Reactions Leading to a Three-Component Synthesis of Dihydroisoquinolones. *J. Org. Chem.* **2021**, *86*, 11599–11607. [CrossRef] [PubMed]
2. Krasavin, M.; Dar'in, D. Current diversity of cyclic anhydrides for the Castagnoli–Cushman-type formal cycloaddition reactions: Prospects and challenges. *Tetrahedron Lett.* **2016**, *57*, 1635–1640. [CrossRef]
3. Lepikhina, A.; Dar'in, D.; Bakulina, O.; Chupakhin, E.; Krasavin, M. Skeletal Diversity in Combinatorial Fashion: A New Format for the Castagnoli–Cushman Reaction. *ACS Comb. Sci.* **2017**, *19*, 702–707. [CrossRef] [PubMed]
4. Chupakhin, E.; Dar'in, D.; Krasavin, M. The Castagnoli–Cushman reaction in a three-component format. *Tetrahedron Lett.* **2018**, *59*, 2595–2599. [CrossRef]
5. Bakulina, O.; Chizhova, M.; Dar'in, D.; Krasavin, M. A General Way to Construct Arene-Fused Seven-Membered Nitrogen Heterocycles. *Eur. J. Org. Chem.* **2018**, *2018*, 362–371. [CrossRef]
6. Adamovskyi, M.I.; Avramenko, M.M.; Volochnyuk, D.M.; Ryabukhin, S.V. Fluoral Hydrate: A Perspective Substrate for the Castagnoli–Cushman Reaction. *ACS Omega* **2020**, *5*, 20932–20942. [CrossRef]
7. Bayles, T.; Guillou, C. Trifluoroethanol Promoted Castagnoli–Cushman Cycloadditions of Imines with Homophthalic Anhydride. *Molecules* **2022**, *27*, 844. [CrossRef]
8. Jarvis, C.L.; Hirschi, J.S.; Vetticatt, M.J.; Seidel, D. Catalytic Enantioselective Synthesis of Lactams through Formal [4 + 2] Cycloaddition of Imines with Homophthalic Anhydride. *Angew. Chem.* **2017**, *56*, 2670–2674. [CrossRef]
9. Ivashchenko, A.V.; Tkachenko, S.Y.; Okun, I.M.R.S.A.; Vladimirovich, K.D.; Khvat, A.V. Pharmaceutical Composition, Method for the Production and the Use Thereof. Patent WO 2007133108, 22 November 2007.
10. Okun, I.; Balakin, K.V.; Tkachenko, S.E.; Ivachtchenko, A.V. Caspase activity modulators as anticancer agents. *Anticancer Agents Med. Chem.* **2008**, *8*, 322–341. [CrossRef]
11. Pereira, G.A.N.; da Silva, E.B.; Braga, S.F.P.; Leite, P.G.; Martins, L.C.; Vieira, R.P.; Soh, W.T.; Villela, F.S.; Costa, F.M.R.; Ray, D.; et al. Discovery and characterization of trypanocidal cysteine protease inhibitors from the 'malaria box'. *Eur. J. Med. Chem.* **2019**, *179*, 765–778. [CrossRef]
12. Chen, Y.; Zhu, F.; Hammill, J.; Holbrook, G.; Yang, L.; Freeman, B.; White, K.L.; Shackleford, D.M.; O'Loughlin, K.G.; Charman, S.A.; et al. Selecting an anti-malarial clinical candidate from two potent dihydroisoquinolones. *Malar. J.* **2021**, *20*, 107. [CrossRef] [PubMed]
13. Guranova, N.; Golubev, P.; Bakulina, O.; Dar'in, D.; Kantin, G.; Krasavin, M. Unexpected formal [4 + 2]-cycloaddition of chalcone imines and homophthalic anhydrides: Preparation of dihydropyridin-2(1H)-ones. *Org. Biomol. Chem.* **2021**, *19*, 3829–3833. [CrossRef] [PubMed]
14. Laws, S.W.; Moore, L.C.; Di Maso, M.J.; Nguyen, Q.N.N.; Tantillo, D.J.; Shaw, J.T. Diastereoselective Base-Catalyzed Formal [4 + 2] Cycloadditions of N-Sulfonyl Imines and Cyclic Anhydrides. *Org. Lett.* **2017**, *19*, 2466–2469. [CrossRef] [PubMed]
15. Ng, P.Y.; Tang, Y.; Knosp, W.M.; Stadler, H.S.; Shaw, J.T. Synthesis of diverse lactam carboxamides leading to the discovery of a new transcription-factor inhibitor. *Angew. Chem.* **2007**, *46*, 5352–5355. [CrossRef]
16. Rendy, R.; Zhang, Y.; McElrea, A.; Gomez, A.; Klumpp, D.A. Superacid-catalyzed reactions of cinnamic acids and the role of superelectrophiles. *J. Org. Chem.* **2004**, *69*, 2340–2347. [CrossRef]
17. Püschl, A.; Rudbeck, H.C.; Faldt, A.; Confante, A.; Kehler, J. Versatile Synthesis of 3-Arylindan-1-ones by Palladium-Catalyzed Intramolecular Reductive Cyclization of Bromochalcones. *Synthesis* **2005**, *2005*, 291–295. [CrossRef]
18. Safrygin, A.; Zhmurov, P.; Dar'in, D.; Silonov, S.; Kasatkina, M.; Zonis, Y.; Gureev, M.; Krasavin, M. Three-component Castagnoli–Cushman reaction with ammonium acetate delivers 2-unsubstituted isoquinol-1-ones as potent inhibitors of poly(ADP-ribose) polymerase (PARP). *J. Enzym. Inhib. Med. Chem.* **2021**, *36*, 1916–1921. [CrossRef]
19. Liu, J.; Wang, Z.; Levin, A.; Emge, T.J.; Rablen, P.R.; Floyd, D.M.; Knapp, S. N-methylimidazole promotes the reaction of homophthalic anhydride with imines. *J. Org. Chem.* **2014**, *79*, 7593–7599. [CrossRef]
20. Scorzelli, F.; Di Mola, A.; De Piano, F.; Tedesco, C.; Palombi, L.; Filosa, R.; Waser, M.; Massa, A. A systematic study on the use of different organocatalytic activation modes for asymmetric conjugated addition reactions of isoindolinones. *Tetrahedron* **2017**, *73*, 819–828. [CrossRef]
21. Yue, G.; Lei, K.; Hirao, H.; Zhou, J.S. Palladium-catalyzed asymmetric reductive heck reaction of aryl halides. *Angew. Chem.* **2015**, *54*, 6531–6535. [CrossRef]
22. Bogeso, K.P. Neuroleptic activity and dopamine-uptake inhibition in 1-piperazino-3-phenylindans. *J. Med. Chem.* **1983**, *26*, 935–947. [CrossRef] [PubMed]
23. Dolomanov, O.V.; Bourhis, L.J.; Gildea, R.J.; Howard, J.A.K.; Puschmann, H. OLEX2: A complete structure solution, refinement and analysis program. *J. Appl. Crystallogr.* **2009**, *42*, 339–341. [CrossRef]
24. Sheldrick, G.M. SHELXT—Integrated space-group and crystal-structure determination. *Acta Crystallogr. A Found Adv.* **2015**, *71*, 3–8. [CrossRef] [PubMed]
25. Sheldrick, G.M. Crystal structure refinement with SHELXL. *Acta Crystallogr. C Struct. Chem.* **2015**, *71*, 3–8. [CrossRef]

Review

Recent Advances in Gold(I)-Catalyzed Approaches to Three-Type Small-Molecule Scaffolds via Arylalkyne Activation

Lu Yang [1,2,3], Hongwei Su [1,3], Yue Sun [1,2,3], Sen Zhang [1,2,3], Maosheng Cheng [1,3,*] and Yongxiang Liu [1,2,3,*]

[1] Key Laboratory of Structure-Based Drug Design and Discovery of Ministry of Education, Shenyang Pharmaceutical University, Shenyang 110016, China
[2] Wuya College of Innovation, Shenyang Pharmaceutical University, Shenyang 110016, China
[3] Institute of Drug Research in Medicine Capital of China, Shenyang Pharmaceutical University, Benxi 117000, China
* Correspondence: mscheng@syphu.edu.cn (M.C.); yongxiang.liu@syphu.edu.cn (Y.L.)

Abstract: Gold catalysts possess the advantages of water and oxygen resistance, with the possibility of catalyzing many novel chemical transformations, especially in the syntheses of small-molecule skeletons, in addition to achieving the rapid construction of multiple chemical bonds and ring systems in one step. In this feature paper, we summarize recent advances in the construction of small-molecule scaffolds, such as benzene, cyclopentene, furan, and pyran, based on gold-catalyzed cyclization of arylalkyne derivatives within the last decade. We hope that this review will serve as a useful reference for chemists to apply gold-catalyzed strategies to the syntheses of related natural products and active molecules, hopefully providing useful guidance for the exploration of additional novel gold-catalyzed approaches.

Keywords: gold(I)-catalyzed; arylalkyne; benzene derivatives; cyclopentene derivatives; furan and pyran derivatives

1. Introduction

Gold was long considered an inert precious metal that cannot be used in catalyzing chemical reactions until Bond and Ito discovered that gold exhibits excellent catalytic activity in nanoparticle form or as soluble complexes [1,2], opening the door for the subsequent development and application of gold-catalyzed chemical reactions [3]. The oxidation states of gold include Au(0), Au(I), and Au(III). Au(I) alone is unstable in solution and is generally used in linear complexes with phosphine ligands, carbene ligands, etc. (Figure 1a) [4]. The counterions of gold catalysts include trifluoromethanesulfonate (OTf$^-$), tetrafluoroborate (BF$_4^-$), hexafluoroantimonate (SbF$_6^-$), tetraphenylboron (BAr$_4^-$), etc. (Figure 1b). A reactive Au(I) complex is formed through counterion exchange with various silver salts (AgX) or with sodium tetra-aryl borate (NaBAr$_4$) and potassium tetra-aryl borate (KBAr$_4$) (Figure 1c).

In homogeneous gold-catalyzed reactions, gold, as a soft acid, is highly nucleophilic to the π-electron system in alkynes, alkenes, and allenes, promoting a series of chemical transformations. In 1998, the Teles group first reported the hydrofunctionalization of alkynes by a Au(I)-phosphine complex, after which the great potential of homogeneous gold catalysis in organic synthesis was gradually explored [5]. Over the past two decades, many subtle gold-catalyzed methodologies have been developed, including cycloaddition reactions, cycloisomerization reactions, and cascade cyclization reactions.

Gold catalysts are characterized by high catalytic reactivity, good chemical selectivity, mild reaction conditions, and high tolerance to water and air. The most common application of gold catalysts in organic synthesis is the cyclization reaction, which can be used

to synthesize a benzene ring, indole ring, quinoline ring, imidazole ring, oxazole ring, etc. [6–21]. Arylalkyne-containing building blocks are easily prepared and can undergo a variety of cyclization reactions, offering unique advantages with respect to the construction of small-molecule skeletons, such as benzenes, cyclopentenes, furans, and pyrans under the influence of gold catalysis (Figure 2). Therefore, we attempted to systematically summarize the building-block-directed construction of specific small-molecule scaffolds with arylalkyne substrates under gold(I)-catalyzed conditions within the last decade, and any works missed were unintentional.

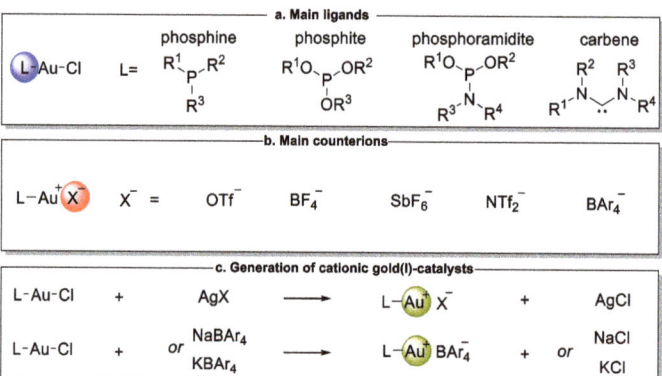

Figure 1. The main ligands, counterions, and generation of cationic gold(I) catalysts.

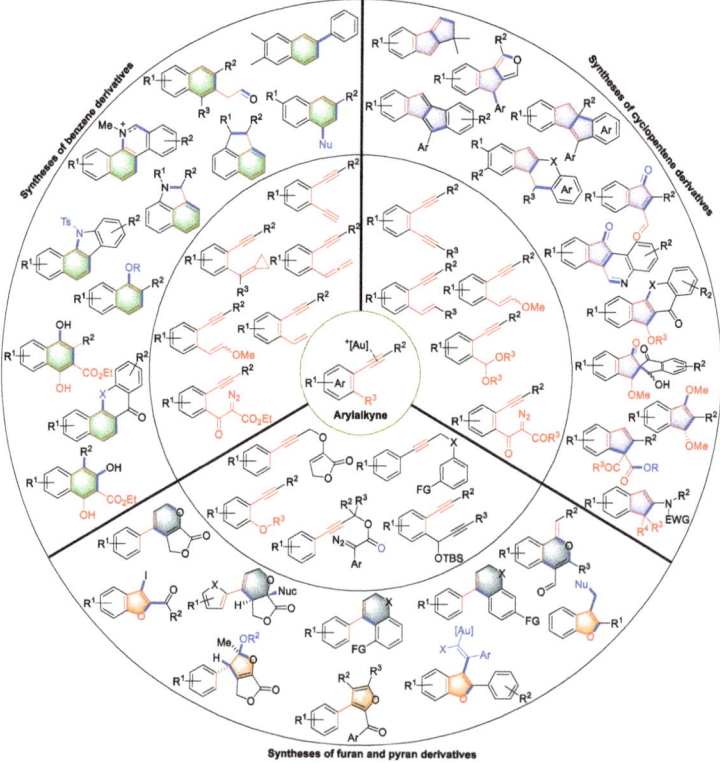

Figure 2. Arylalkyne blocks and the three corresponding types of products.

In this feature paper, studies are classified according to the structural characteristics of small-molecule skeletons, highlighting the development of strategies and the scope of research on gold-catalyzed cyclization of arylalkyne derivatives, including arene–diynes, arene–enynes, aryne–enolether, aryne–acetals, etc.

2. Syntheses of Benzene Derivatives

Many important natural products and drugs contain aromatic units, such as benzenes, naphthalenes, and biaryls; thus, the construction of benzene rings is significant in organic synthesis. The synthesis methods for benzene rings using [2+2+2] or [4+2] cycloaddition reactions usually require harsh conditions. However, Au(I)-catalyzed cyclization of arylalkyne substrates represents a mild and versatile approach to the construction of benzene rings. In this chapter, we summarize previous works on the synthesis of benzene derivatives based on the type of arylalkyne.

2.1. Arene-Diyne Substrates

In 2012, Hashmi and colleagues reported a double gold(I)-activated cyclization of arene-diynes to construct benzene rings for the synthesis of β-substituted naphthalene derivatives, which was achieved through an unexpected reaction pathway (Scheme 1) [22]. First, one terminal alkyne of arene-diyne (**1**) was activated by a gold catalyst to form a Au–C-σ bond through catalyst transfer, and the other terminal alkyne was activated to produce a double-activated intermediate (**2**). Subsequently, the activated triple bond was attacked by the β carbon of gold acetylide due to π coordination, which induced the formation of a five-membered ring to generate gold–vinylidene (**3**). Next, intermediate **4** was formed by a solvent attack (benzene) and a *1,3-H* shift, which was subsequently transformed into intermediate **5** via a ring expansion. Finally, after the elimination of the gold(I) catalyst and protonation, a β-substituted naphthalene product (**6**) was obtained. The reaction pathway was clearly verified through X-ray crystal structure analysis of the key intermediates and controlled experiments. The strategy of double gold activation had a significant influence on the later development of gold chemistry.

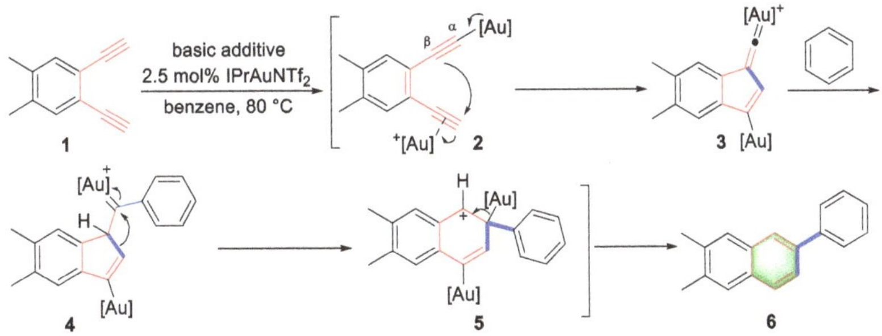

Scheme 1. Double gold(I)-catalyzed syntheses of β-substituted naphthalene derivatives.

In the same year, the Ohno group described a gold(I)-catalyzed tandem approach to 1,3-disubstituted naphthalenes using arene-diynes with 14 examples, achieving a quantitative yield (Scheme 2) [23]. This strategy mainly involves intermolecular nucleophilic addition and intramolecular nucleophilic addition reactions. Gold(I)-activated terminal alkyne was first attacked by nucleophilic reagents, such as ROH, RR'NH, and Ar-H, to generate intermediates (**8**) that were immediately converted to enolether/enamine-type intermediates (**9**) by protodeauration. A subsequent 6-*endo*-dig cyclization yielded intermediates (**10**) that underwent subsequent aromatization and protonation to provide naphthalene derivatives (**11**). The above reaction path was verified in detail by the syntheses of silyl enolether intermediates (**9**) and related deuteration experiments.

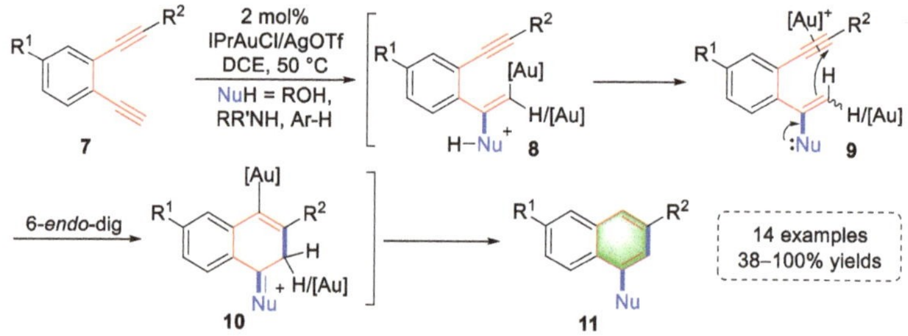

Scheme 2. Gold(I)-catalyzed syntheses of substituted naphthalene derivatives.

2.2. Arene-Enyne Substrates

Gold-catalyzed cyclization of arene-enynes is an important strategy for building small-molecule carbocyclic skeletons that has inspired many excellent methods to be reported. In 2017, Shi et al., developed a gold(I)-catalyzed tandem cyclization–oxidation strategy to access aryl acetaldehyde derivatives using alkylidene–cyclopropane and pyridine N-oxide (Scheme 3) [24]. First, coordination of the triple bond by [Au]$^+$ triggered the 6-endo-dig cyclization to form intermediates (13), and benzylic carbocation was stabilized by electron-rich cyclopropane and the benzene ring. Subsequently, the 3,5-dibromo-pyridine N-oxide acted as a nucleophile to attack cyclopropane and produce intermediates (14). Finally, aryl acetaldehyde derivatives (15) were generated by Kornblum-type oxidation with the simultaneous release of 3,5-dibromo-pyridine. The scope of application of the above strategy was examined using 27 examples with 36–93% yields. It is worth noting that when R^2 was a substituted phenyl group, the aryl acetaldehyde derivative could be further modified to polycyclic aromatic hydrocarbons (PAHs) under the catalysis of In(OTf)$_3$.

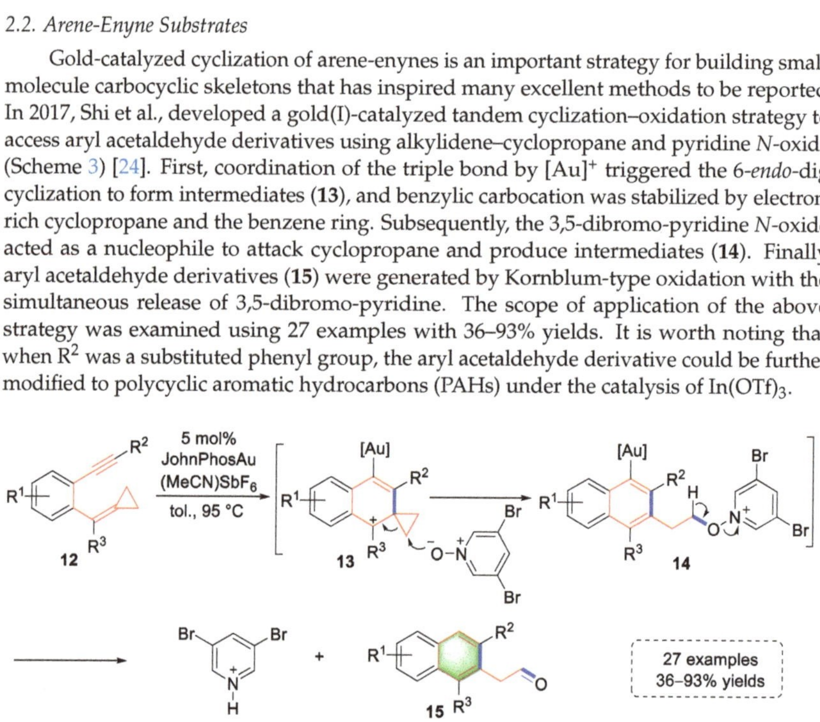

Scheme 3. Gold(I)-catalyzed syntheses of aryl acetaldehyde derivatives.

In 2019, the Ohno lab described a gold(I)-catalyzed cascade cyclization strategy for the syntheses of cyclopropanes derivatives, with 11 examples and yields of up to 96% (Scheme 4) [25]. The activation of the allenyl moiety of 1-allenyl-2-ethynyl-3-alkylbenzene substrates (16) by the gold complex induced a nucleophilic attack of alkyne to yield vinyl cationic intermediates (17). Then, a 1,5-H shift occurred to generate benzylic carbocation intermediates (18). Subsequent carbocation cyclization provided acenaphthene derivatives (19) after aromatization and protodeauration. In addition, a series of 1-(naphth-1-yl)cyclopropa-[b]benzofuran derivatives was successfully prepared when phenylene-tethered allenynes and benzofurans were subjected to the same gold-catalyzed conditions.

In 2021, the Ohno lab reported the syntheses of benzo[cd]indole skeletons by gold-catalyzed tandem cyclization based on their previous work (Scheme 5) [26]. In this approach, a series of amino-allenyne substrates (20) were designed and prepared. First, the activated allene was attacked by the electron-rich alkyne to form vinyl cationic intermediates (21). The vinyl cation was captured by the neighboring amine group to yield tricyclic fused indoles (**22**), which underwent an isomerization to furnish pyrrolonaphthalenes (**23**). The resulting tricyclic derivatives could be transformed into nitrogen-containing polycyclic aromatic compounds (*N*-PACs) with special photophysical properties through *N*-arylation or Friedel–Crafts acylation.

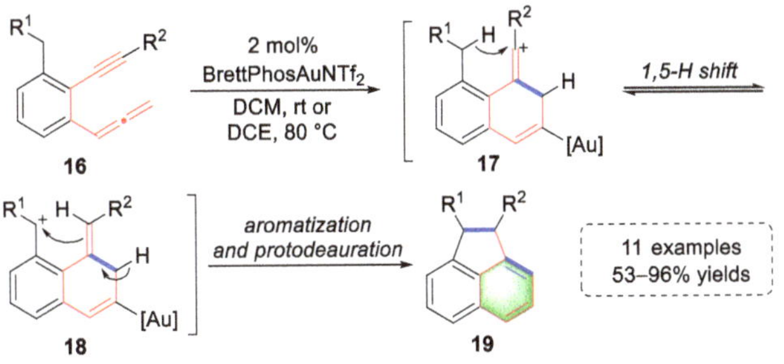

Scheme 4. Gold(I)-catalyzed syntheses of acenaphthene derivatives.

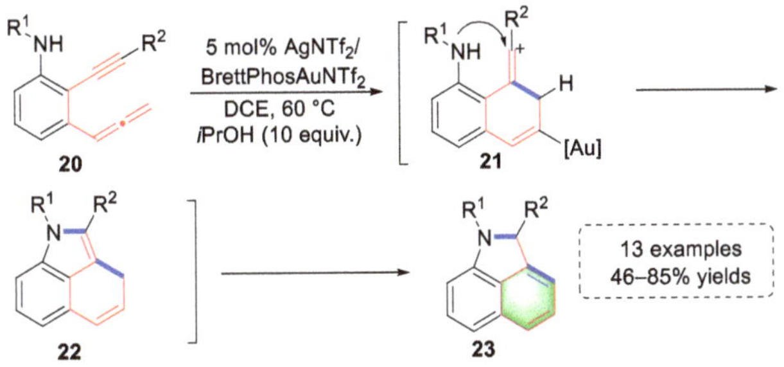

Scheme 5. Gold(I)-catalyzed syntheses of pyrrolonaphthalene derivatives.

Recently, Liu and colleagues achieved the gold(I)-catalyzed construction of benzene derivatives using arene–enyne substrates, which was applied to the total syntheses of eight natural products (Scheme 6) [27]. Coordination of alkyne in substrates (**24**) promoted a 6-*endo*-dig cyclization to yield intermediates (**25**) that were converted into iodonaphthalenes (**26**) in situ in the presence of *N*-iodosuccinimide (NIS). The intermediates (**26**) were used as key moieties to synthesize benzo[c]phenanthridine alkaloids in a pot-economic approach. Moreover, the cytotoxicities of these alkaloids were investigated, indicating the future potential of these molecules for anticancer research.

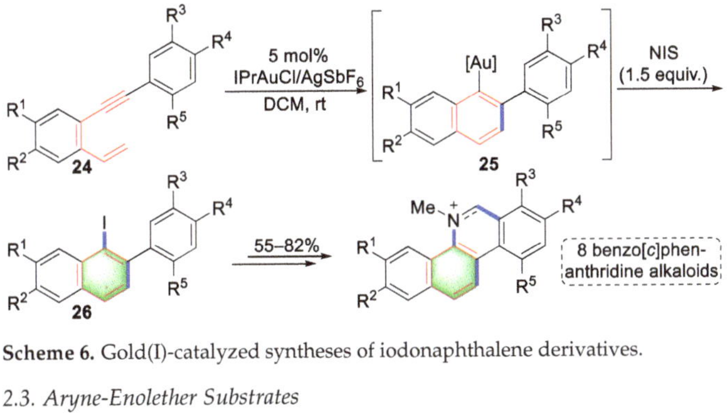

Scheme 6. Gold(I)-catalyzed syntheses of iodonaphthalene derivatives.

2.3. Aryne-Enolether Substrates

Enolether showed versatile properties in gold-catalyzed reactions, making it suitable for use not only as a nucleophile but also as an electrophile to be coordinated by [Au]$^+$. The combination of enolether with alkyne derivatives to form building blocks containing 1,5-enyne showed unique advantages in gold-catalyzed tandem cyclization for the syntheses of benzene derivatives. Accordingly, the Liu lab has reported a number of such studies in recent years.

In 2014, a gold(I)-catalyzed cycloisomerization of arylalkyne-enolether for the construction of multisubstituted naphthalenes was developed by the Liu group (Scheme 7) [28]. First, the triple bonds of the substrates (27) were activated by the gold species, which induced an intermolecular nucleophilic addition by alcohol to yield dienol ether intermediates (28). Coordination of [Au]$^+$ to the electron-rich enolether promoted cycloisomerization to provide multisubstituted naphthalenes (29) via the release of methanol and protodeauration. The scope of this strategy was examined by synthesizing 20 alkyne–enolether substrates with 38–88% yields.

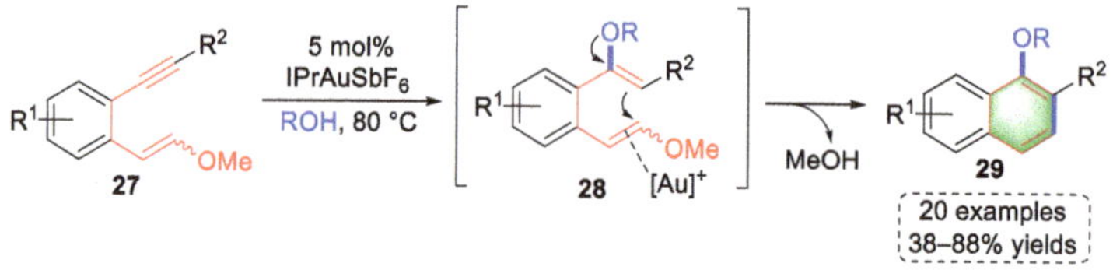

Scheme 7. Gold(I)-catalyzed syntheses of multisubstituted naphthalenes.

In 2017, Liu and colleagues achieved gold(I)-catalyzed tandem cyclization for the syntheses of benzo[a]carbazole derivatives using arylalkyne-enolether substrates (Scheme 8) [29]. The authors modulated the electronic properties of the triple bond through the substituent of the right benzene ring, which further tuned the cyclization order. When there were sulfonamide substituents on the appropriate benzenes, the α-position of the alkyne activated by [Au]$^+$ induced a 5-*endo*-dig cyclization to produce indole intermediates (31). The enolether was then activated by a gold(I) complex and attacked by the electron-rich indole to promote the second cyclization, yielding benzo[a]carbazoles (32) by elimination of methanol and protodeauration. The above reaction mechanism was verified by capturing the intermediates and further supported by DFT calculations. Notably, when the appropriate benzene rings of the substrates were substituted with amine groups, the order of cyclization was changed to yield indeno-[1,2-c]quinoline derivatives, which are described in detail in later sections.

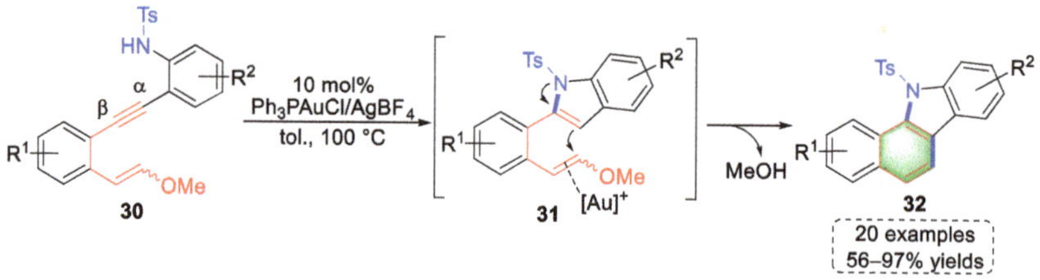

Scheme 8. Gold(I)-catalyzed syntheses of benzo[a]carbazole derivatives.

One year later, another cascade cyclization strategy was reported by the Liu group as an ongoing study on the gold-catalyzed cyclization of enolether-involved substrates in the construction of small-molecule scaffolds. The authors achieved the syntheses of xanthone and acridone derivatives by designing a series of alkyne–enolether substrates with 25 examples and up to 98% yield (Scheme 9) [30]. Initially, the triple bonds of the substrates (33) were chelated by the gold(I) species, which promoted an intramolecular Michael addition to obtain intermediates (34) after protodeauration. Then, gold(I)-activated enolether was attacked by newly generated enolethers or enamines to undergo a 6-*endo*-trig cyclization. Finally, xanthone or acridone derivatives (35) were achieved via a similar pathway as reported previously.

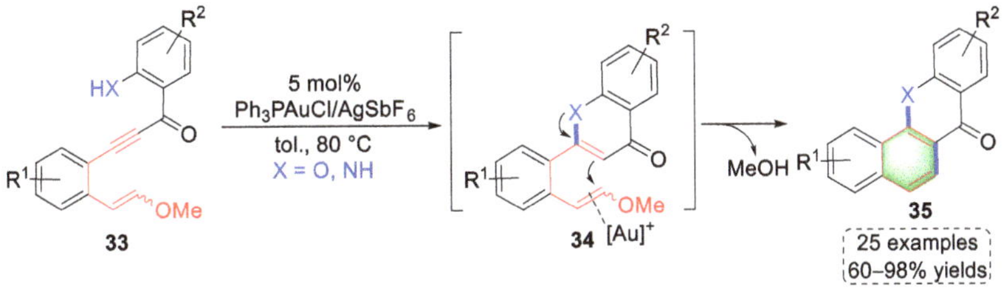

Scheme 9. Gold(I)-catalyzed syntheses of xanthone and acridone derivatives.

In 2022, Liu and colleagues reported a gold(I)-catalyzed 6-*endo*-dig cyclization of arylalkyne–enolethers (**36**) to construct 2-(naphthalen-2-yl)aniline derivatives (Scheme 10) [31]. The authors found that the amine group on the right-hand benzene ring benefited 6-*endo*-dig cyclization via an electron-donating effect to generate naphthalenes (**37**) after isomerization and protodeauration. In addition, several important heterocycles (**38–41**) were synthesized in a divergent manner from that of naphthalene derivatives (**37**).

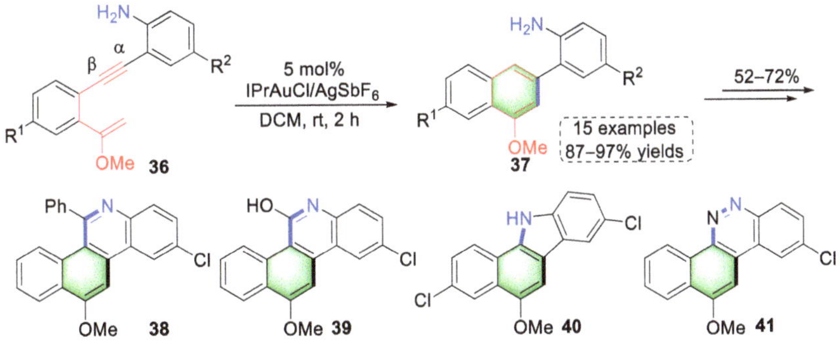

Scheme 10. Gold(I)-catalyzed syntheses of 2-(naphthalen-2-yl)aniline derivatives.

2.4. Other Arylalkyne Substrates

In 2013, Ye et al., described a gold(I)/acid-catalyzed methodology for the syntheses of anthracenes using *o*-alkynyl diarylmethanes with 21 examples and 58–80% yields (Scheme 11) [32]. Coordination of alkynes by gold catalysts triggered the attack of electron-rich benzene rings to furnish vinyl–gold intermediates (**43**) via 6-*exo*-dig cyclization. After protodeauration and [Au]$^+$/H$^+$ promoted isomerization, anthracenes (**45**) were obtained. An alternative pathway was also proposed; the alkyne of the substrates (**42**) was hydrolyzed under gold-catalyzed conditions to yield intermediates (**44**) that were converted to products (**45**) by an acid-mediated cyclodehydration. In addition, the products (**45**) were further modified into a variety of potentially valuable anthracene derivatives.

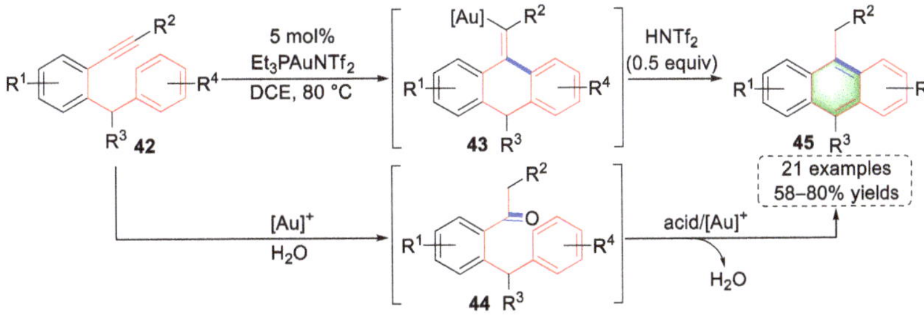

Scheme 11. Gold(I)-catalyzed syntheses of anthracene derivatives.

In 2017, a gold(I)-catalyzed tandem cycloisomerization, Diels–Alder, and retro-Diels–Alder reactions were reported by the Liu lab (Scheme 12) [33]. Activation of alkyne in substrates (**46**) initiated the first cycloisomerization to yield furopyran intermediates (**47**). A subsequent Diels–Alder reaction of dienes (**47**) and dienophiles occurred to form highly strained intermediates (**48**), which underwent a retro-Diels–Alder reaction to provide biaryl products (**49**) by releasing acetaldehyde (HCHO). The above pathways were reasonably explained by density functional theory (DFT).

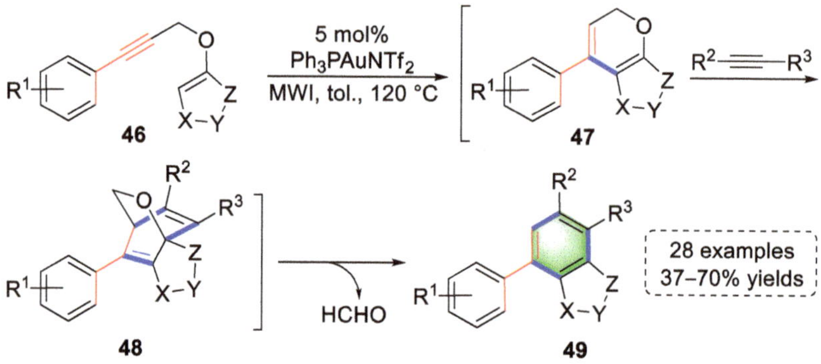

Scheme 12. Gold(I)-catalyzed syntheses of biaryl derivatives.

In 2021, the Hashmi group reported the syntheses of *meta*- and *para*dihydroxynaphthalenes based on diazoalkynes through a regiodivergent gold-catalyzed cyclization (Scheme 13) [34]. The activated triple bonds of substrates (**50**) were attacked by diazocarbon to generate intermediates (**51**), followed by the formation of quinoid gold carbene intermediates (**52**) via the release of nitrogen. At this stage, two different reaction paths occurred via the addition of water or Et$_3$N(HF)$_3$. Under the condition of water as an additive (path a), *meta*dihydroxynaphthalenes (**54**) were produced via carbene insertion of water after protodeauration. When H$_2$O and Et$_3$N(HF)$_3$ were used as additives, *para*dihydroxynaphthalenes (**56**) were obtained via Michael-type addition of quinoid carbene species, 1,2-phenyl migration, and protodeauration. Moreover, when only Et$_3$N(HF)$_3$ was used as an additive, "F$^-$" was inserted instead of water for gold carbene to generate the α-fluoronaphthalenes.

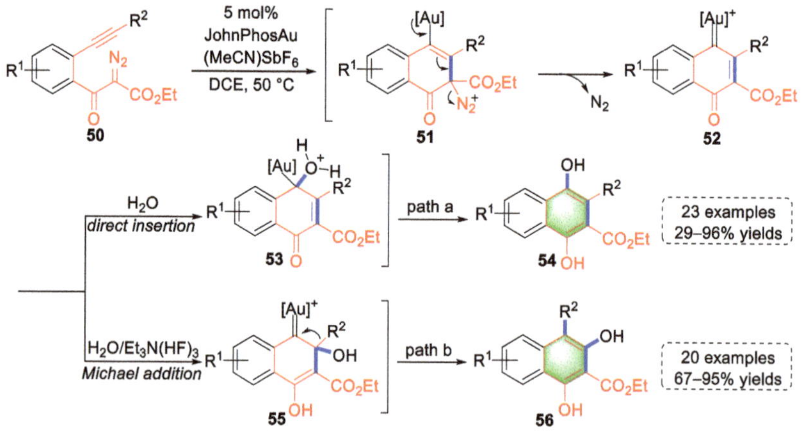

Scheme 13. Gold(I)-catalyzed syntheses of *meta*- and *para*dihydroxynaphthalene derivatives.

Gold(I)-catalyzed arylalkyne annulations provide abundant strategies for the syntheses of benzene derivatives, including the strategies shown in this chapter and several other intramolecular or intermolecular strategies [35–47].

3. Construction of Cyclopentene Derivatives

Small-molecule skeletons containing cyclopentene are important components of many natural products and pharmaceutical intermediates. The syntheses of useful cyclopentene derivatives have attracted a great deal of interest among chemists. Gold(I)-catalyzed annulations of a variety of arylacetylene substrates provide a range of versatile synthetic methods for the syntheses of benzocyclopentene derivatives.

3.1. Arene-Diyne Substrates

In 2012, the Hashmi group achieved the preparation of benzofulvene derivatives based on their previous strategy of double activation of diynes containing terminal alkynes (Scheme 14) [48]. Under the catalysis of a gold catalyst, dual σ/π-activated intermediates (**58**) were formed, which were rapidly transformed into gold vinylidenes (**59**) as a result of double activation. A 1,5-H shift to electrophilic vinylidene carbon occurred, leading to intermediates (**60**). After the trapping of the carbocation by the vinyl–gold species, benzofulvene products (**61**) were synthesized in association with the elimination of the gold catalyst. The applicability of the strategy was examined by 10 examples and up to 92% yield. The above strategy was characterized by easy preparation of the substrate and a novel reaction mechanism.

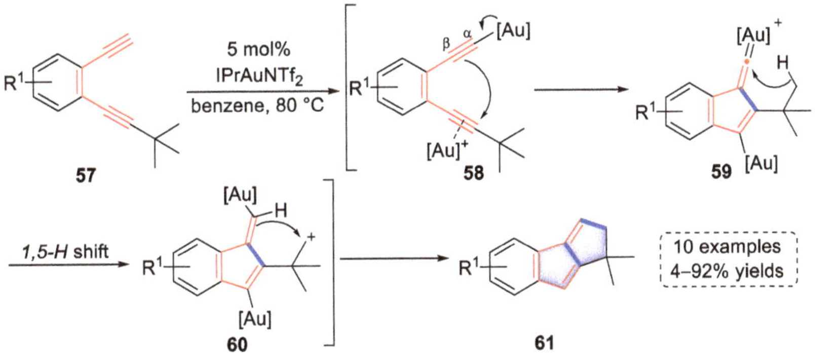

Scheme 14. Gold(I)-catalyzed syntheses of benzofulvene derivatives.

In 2017, the Hashmi group the construction of aryl-substituted dibenzopentalene derivatives using terminally aromatic substituted 1,5-diyne substrates under gold-catalyzed conditions (Scheme 15) [49]. One of the triple bonds was coordinated by [Au]$^+$, resulting in the attack of another electron-rich triple bond to form vinyl cation intermediates (**63**). The vinyl cation was trapped by the neighboring electron-rich benzene to produce intermediates (**64**), followed by rearomatization and protodeauration to yield intermediates (**65**). Ultimately, dibenzopentalene products (**66**) were obtained by ligand exchange of gold species. It is worth noting that benzene as a solvent was not involved in the above process to trap the vinyl cation.

In 2021, the Hashmi group developed a gold-catalyzed cycloisomerization of substituted 1,5-diynes to synthesize indeno[1,2-c]furan derivatives. The functional group tolerance was systematically examined using 29 examples with 16–81% yields (Scheme 16) [50]. Vinyl cationic intermediates (**68**) were formed through similar paths a those described previously in Schemes 14 and 15. Subsequently, a second annulation occurred immediately to yield oxonium intermediates (**69**). Intermediates (**71**) were produced via the release of benzyl carbocation, followed by [5,5]-sigmatropic rearrangement. Finally, indeno[1,2-c]furan derivatives (**73**) were obtained by rearomatization, the elimination of gold species, and proton transfer mediated by *p*-toluenesulfonic acid (PTSA). The authors fully explained the above reaction mechanism using DFT calculations, and the high regioselectivity of [5,5]-sigmatropic rearrangement was also reasonably illustrated.

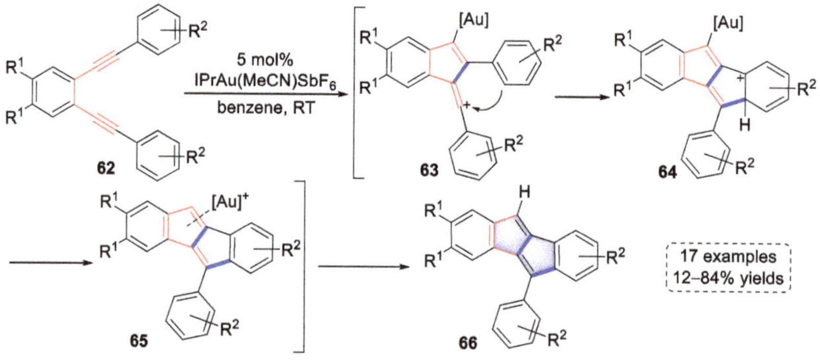

Scheme 15. Gold(I)-catalyzed syntheses of aryl-substituted dibenzopentalene derivatives.

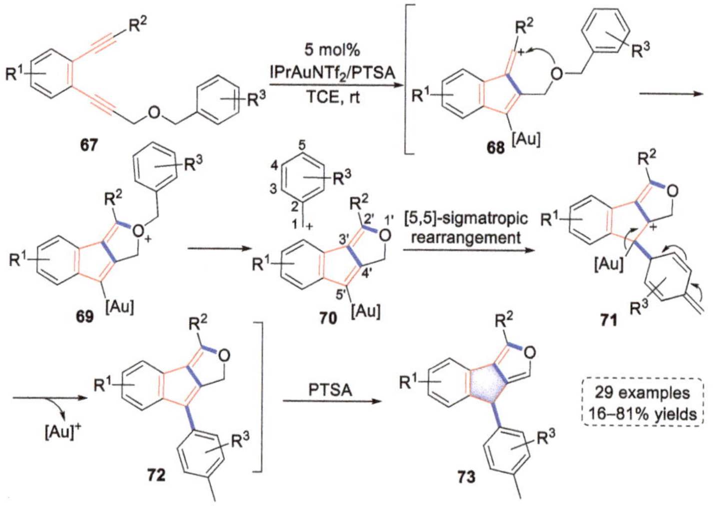

Scheme 16. Gold(I)-catalyzed syntheses of indeno[1,2-*c*]furan derivatives.

3.2. Arene-Enyne Substrates

In 2016, Sanz et al., reported a gold(I)-catalyzed tandem reaction using β,β-diaryl-*o*-(alkynyl)styrenes to synthesize dihydroindeno[2,1-*a*]indene derivatives (Scheme 17) [51]. A 5-*endo*-cyclization was induced by the activation of [Au]$^+$ to the alkyne, which produced carbocationic intermediates (**75**). After proton elimination and protodeauration, benzofulvene intermediates (**77**) were generated. The diene units in intermediates (**77**) were then activated by the gold species to generate allylic carbocationic intermediates (**78**), which were trapped by the neighboring electron-rich aryl group to access products (**79**). In addition, under the condition of 0 °C in DCM, benzofulvene intermediates (**77**) were isolated as products.

In 2022, the Sanz lab disclosed a gold-catalyzed domino method for the syntheses of indeno[2,1-*b*]thiochromene derivatives with 21 examples and 70–88% yields (Scheme 18) [52]. Activation of S/Se-substituted alkynes by [Au]$^+$ triggered the cyclization of alkene to afford cationic intermediates (**81**), the carbocations of which were trapped by the vinyl–gold to produce cyclopropyl gold carbenes (**82**). The cyclopropanes of **82** were attacked by electron-rich aromatic groups to form ring-opening intermediates (83) after rearomatization. Indeno[2,1-b]thiochromene derivatives (84) containing sulfur or selenium were ultimately obtained by protodeauration. Importantly, when (S)-DM-SEGPHOS(AuCl)$_2$, a chiral ligand, was used in the gold-catalyzed reaction, an enantioselective transformation was achieved in up to 80% ee.

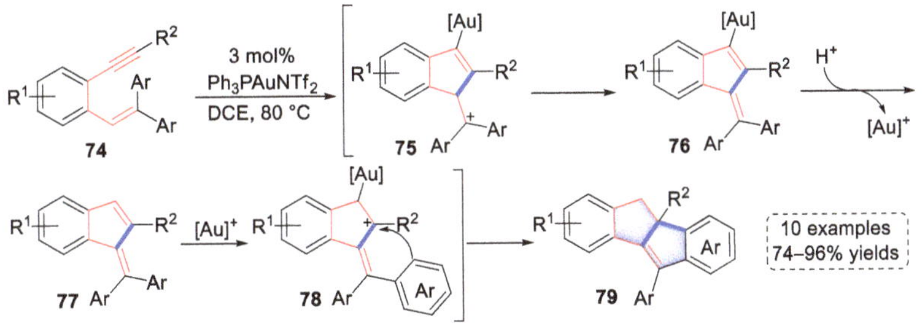

Scheme 17. Gold(I)-catalyzed syntheses of dihydroindeno[2,1-a]indene derivatives.

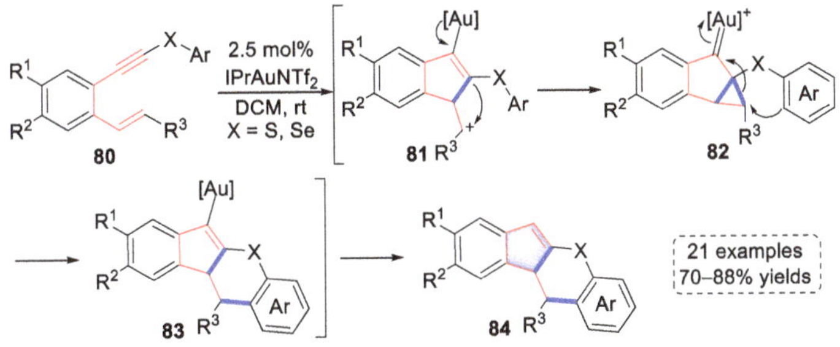

Scheme 18. Gold(I)-catalyzed syntheses of indeno[2,1-b]thiochromene derivatives.

3.3. Aryne-Enolether Substrates

In 2017, a strategy of synthesizing indeno[1,2-c]quinoline derivatives was developed by the Liu group through gold(I)-catalyzed cascade cyclization with 18 examples and up to 99% yield (Scheme 19) [29]. The coordination of gold species to the β position of the triple bond initiated an attack of the enolether to generate indene intermediates (86). Intermediates (87) were produced by the activation of double bonds in the conjugated enolether with [Au]$^+$, which were converted to aromatic intermediates (88) via intramolecular condensation with the release of MeOH after protodemetalation. In oxygen, the intermediates (88) were further oxidized to a more stable indeno[1,2-c]quinoline product (89). Notably, the electron-donating effect of the amine on the right benzene ring played a crucial role in the initiation of the above transformation.

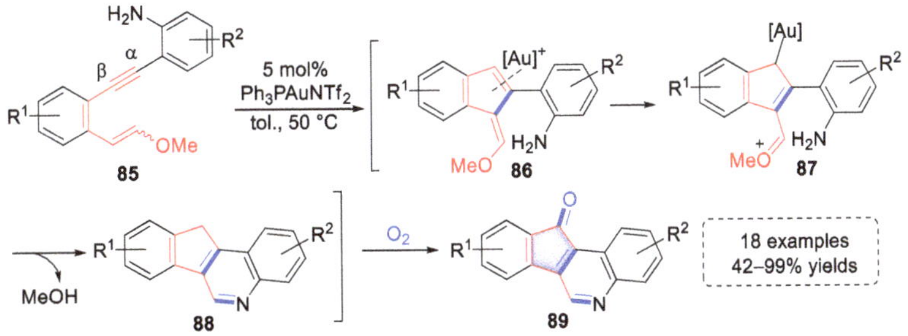

Scheme 19. Gold(I)-catalyzed syntheses of indeno[1,2-c]quinoline derivatives.

In 2018, Liu et al., used 1,5-enyne substrates to synthesize a series of 2-aryl indenone derivatives in the catalysis of a gold catalyst (Scheme 20) [53]. Intermediates (**92**) were formed via a gold-catalyzed cycloisomerization. An O_2-mediated radical addition to intermediates **92** afforded intermediates (**93**) that underwent aromatization to yield peroxy intermediates (**94**), which were subsequently transformed into oxonium intermediates (**95**) through the cleavage of the peroxide bond with the coordination of $[Au]^+$. Finally, indenone products (**96**) were achieved by the hydrolysis of oxonium with the release of MeOH. The above free radical reaction process was verified via control experiments and heavy atom labeling.

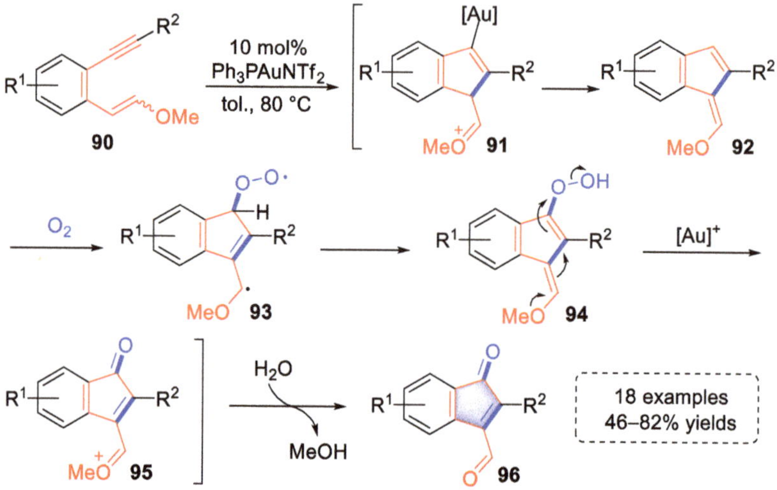

Scheme 20. Gold(I)-catalyzed syntheses of 2-aryl indenone derivatives.

3.4. Aryne-Acetal Substrates

Arylalkynes containing acetal moieties as useful building blocks exhibited excellent reactivity in the gold-catalyzed syntheses of cyclopentene derivatives. In 2013, in pioneering work, the Toste group developed a gold-catalyzed strategy for the enantioselective syntheses of β-alkoxy indanone derivatives using this kind of substrate (Scheme 21) [54]. It was proposed that the activation of triple bonds by gold complexes triggered the migration of an alkoxy group to the alkyne, generating oxonium intermediates (**99**) via intermediates (**98**). An enantioselective annulation then occurred to form oxonium intermediates (**100**), which were transformed into products (**101**) after isomerization. The use of $[Au]^+$ with chiral ligands ensured enantioselective cyclization with up to 98% *ee*. In addition, the β-alkoxy indanone derivatives could be further hydrolyzed to corresponding 3-methoxycyclopentenone derivatives under PTSA conditions with wet DCM.

In 2016, Liu et al., described a gold(I)-catalyzed hydrogen-bond-regulated tandem cyclization for the syntheses of indeno-chromen-4-one and indeno-quinolin-4-one derivatives by introducing a Michael acceptor in the substrates (Scheme 22) [55]. The double activation of a hydrogen bond and gold catalyst promoted methoxy migration to generate vinyl–gold intermediates (**103**), followed by an intramolecular annulation to produce intermediates (**104**) after isomerization. With conformational changing, intramolecular Michael addition occurred to yield indeno-chromen-4-one or indeno-quinolin-4-one derivatives (**105**) after the elimination of alkoxy groups.

In 2020, Sajiki and colleagues developed a gold(I)-catalyzed approach for the preparation of indenone derivatives using arylalkyne substrates containing cyclic acetals (Scheme 23) [56]. The triple bonds were first activated by the gold complex to produce vinyl–gold intermediates (**107**), which initiated the migration of benzylic hydride to generate oxonium cationic intermediates (**108**). Cyclized gold(I)–carbene intermediates (**109**) were then formed by intramolecular nucleophilic addition. At this stage, when the system contained water, a carbene insert process occurred to yield intermediates (**110**), followed by a [Au]⁺-activated dehydration reaction to produce indenone derivatives (**112**). Alternatively, products (**112**) were generated directly from the cyclized gold(I)–carbene intermediates (**109**) through a *1,2-H* shift and elimination of gold species. The key 1,5-hydride shift was verified by deuterium-labeled experiments and 2D NMR analysis.

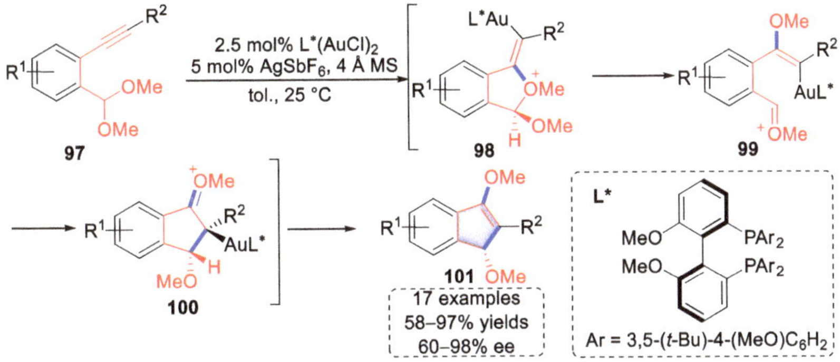

Scheme 21. Gold(I)-catalyzed syntheses of β-alkoxy indanone derivatives.

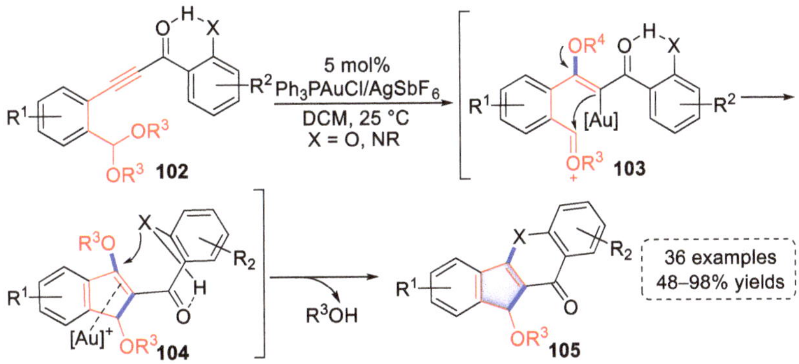

Scheme 22. Gold(I)-catalyzed syntheses of indeno-chromen-4-one and indeno-quinolin-4-one derivatives.

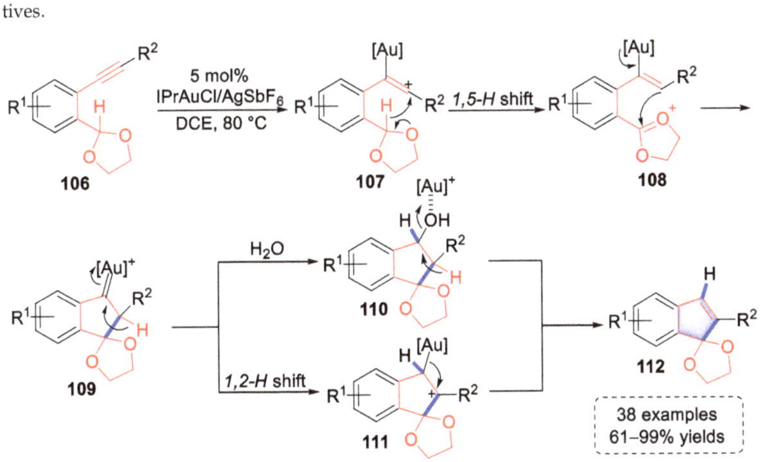

Scheme 23. Gold(I)-catalyzed syntheses of indenone derivatives.

In 2020, the Liu group reported a gold(I)-catalyzed domino reaction to construct benzo[*b*]indeno[1,2-*e*][1,4]diazepine derivatives using *o*-phenylenediamines and ynones (Scheme 24) [57]. The coordination of the gold species with a triple bond induced a series of transformations into intermediates (**115**), which was similar to the generation of intermediates (**104**) shown in Scheme 22. The intermediates (**115**) underwent Michael addition with exogenous *o*-phenylenediamine to produce intermediates (**117**) after the elimination of MeOH. Ultimately, benzo[*b*]indeno[1,2-*e*][1,4]diazepine derivatives (**118**) were synthesized by intramolecular condensation and aromatization accompanied by the elimination of MeOH and H$_2$O. Controlled experiments were further conducted to determine the rationality of the above reaction.

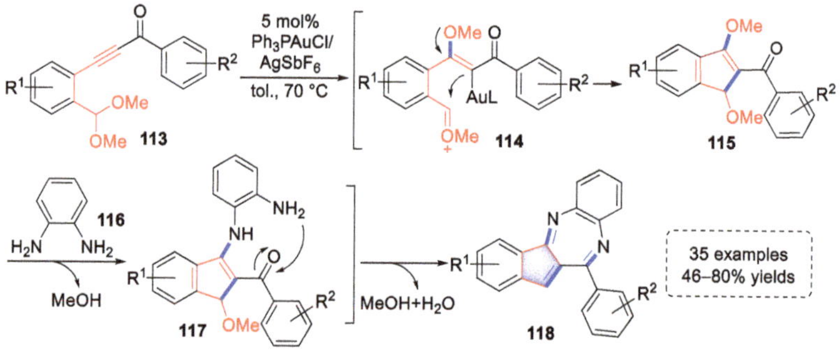

Scheme 24. Gold(I)-catalyzed syntheses of benzo[*b*]indeno[1,2-*e*][1,4]diazepine derivatives.

Recently, the Liu group developed a synthetic strategy for 2,2′-spirobi[indene] derivatives using arylalkyne–acetal substrates based on their previous research (Schemes 22 and 24), mainly involving methoxylation and aldol condensation (Scheme 25) [58]. Intermediates (**120**) were easily produced by the activation of [Au]$^+$/H$^+$ and converted into intermediates (**121**) through an intramolecular aldol reaction. After releasing MeOH, 2,2′-spirobi[indene] derivatives were obtained. It should be noted that the reversible equilibrium of aldol/retro-aldol reactions led to the isomerization of the hydroxyl group.

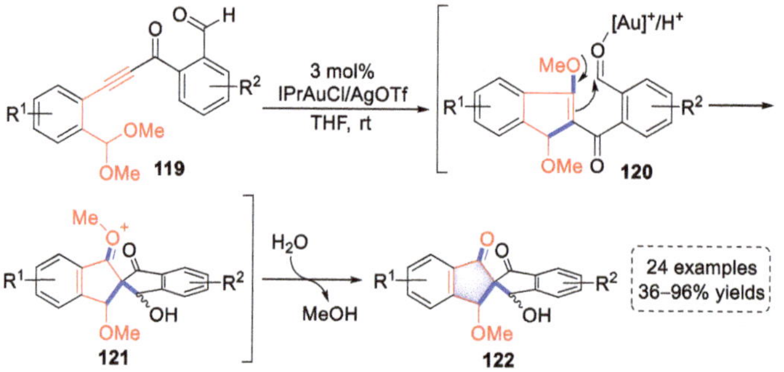

Scheme 25. Gold(I)-catalyzed syntheses of 2,2′-spirobi[indene] derivatives.

3.5. Other Arylalkyne Substrates

In 2021, Xu and colleagues achieved a cascade strategy for the syntheses of indene derivatives involving gold(I)-catalyzed Wolff rearrangement and ketene C=C dual functionalization (Scheme 26) [59]. Diazoketone substrates (**123**) were activated by a gold complex to form gold carbine, which as converted to ketene intermediates (**124**) by Wolff rearrangement. The ketene units of **124** were then attacked by nucleophiles (ROH) to form enol intermediates (**125**). Activation of a triple bond by gold(I) species initiated a C-5-*endo*-dig cyclization to obtain indene products (**126**). In addition, when nucleophiles such as indoles or pyrroles were used, O-7-*endo*-dig cyclization occurred to generate benzo[*d*]oxepine derivatives. The scope of the above strategy was studied in detail with 46 examples and up to 88% yield, and the related reaction pathways were explained by DFT calculations.

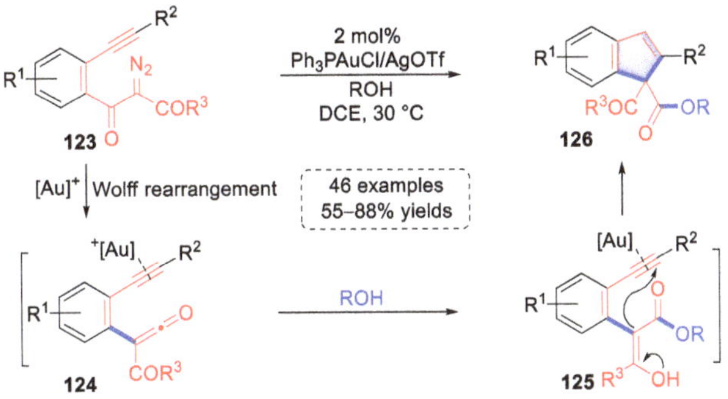

Scheme 26. Gold(I)-catalyzed syntheses of indene derivatives.

In 2021, a strategy for the syntheses of indene derivatives based on the cyclization of ynamides was developed by the Evano group (Scheme 27) [60]. Gold–keteniminium ions (**128**) were formed upon the coordination of [Au]⁺ to the triple bond in ynamide, followed by a *1,5-H* shift, resulting in carbocation intermediates (**129**). Subsequently, the carbocations of **129** were trapped by vinyl–gold to trigger a cyclization, producing intermediates (**130**). After a *1,2-H* shift and elimination of [Au]⁺, indene products (**131**) were achieved. Alternatively, indene products (**131**) could be formed by the elimination of a proton and protodeauration. This method is associated with a wide range of substrates and was systematically studied using 20 examples with 40–96% yields.

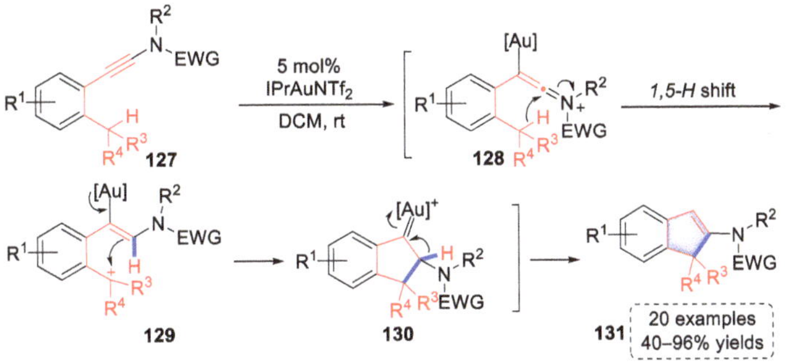

Scheme 27. Gold(I)-catalyzed syntheses of polysubstituted indene derivatives.

Recently, the Ohno lab reported a gold(I)-catalyzed cascade acetylenic Schmidt reaction/ *1,5-H* shift/*N*- or *C*-cyclization method producing indole[*a*]- and [*b*]-fused polycycle derivatives (Scheme 28) [61]. The isotopic labeling experiment showed that the reaction started with an acetylenic Schmidt reaction activated by gold species, which resulted in the formation of α-imino gold carbenes (**133**), followed by a *1,5-H* shift to yield carbocationic intermediates (**134**), which were in reversible equilibrium with aromatized intermediates (**135** and **137**). Finally, *C*-cyclization products (**136**) were generated via aromatized intermediates (**135**), and the *N*-cyclization products (**13b**) were yielded via aromatized intermediates (**137**) with a bond rotation. Notably, the selectivity of the *N* and *C*-cyclization products could be tuned through the electron density of the left benzene ring, the stability of the carbocation, and the effect of the counterion. Moreover, the above strategy is excellent example of benzylation of benzylic C(sp³)-H functionalizations, providing a concise method for the syntheses of indole[*a*]- and [*b*]-fused polycycle derivatives.

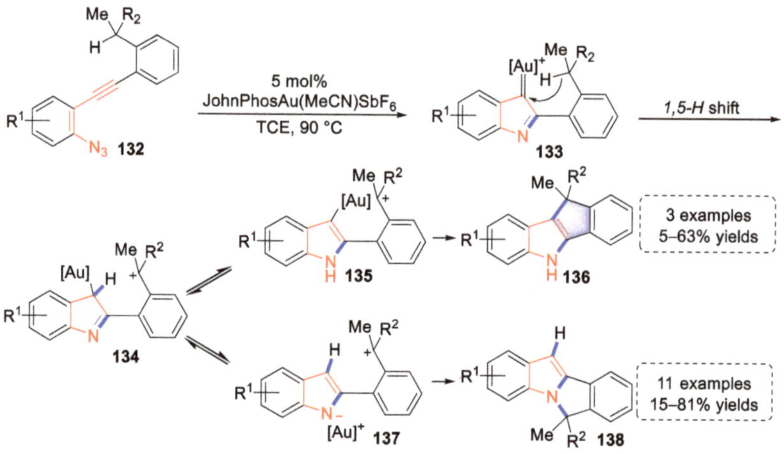

Scheme 28. Gold(I)-catalyzed syntheses of indole[*a*]- and [*b*]-fused polycycle derivatives.

Based on the cases summarized in this chapter, it seems that the gold(I)-catalyzed tandem approach using a variety of arylalkyne substrates could be used to synthesize corresponding cyclopentene derivatives, such as benzofulvenes, dibenzopentalenes, 2,2′-spirobi[indene], indenes, etc. These structurally diverse cyclopentene derivatives can provide further possibilities for the discovery of bioactive lead compounds and provide strategic support for the syntheses of related bioactive molecules.

4. Construction of Furan and Pyran Derivatives

Furan and pyran derivatives are valuable heterocyclic skeletons and important intermediates for the syntheses of drugs and lead compounds. For example, benzofuran derivatives exhibited excellent inhibition of both drug-sensitive and drug-resistant pathogens through a unique antitubercular and antibacterial mechanism [62]. Gold(I)-catalyzed arylalkyne cyclization can be used to construct a variety of furan- or pyran-containing derivatives, such as polycyclic furans, polycyclic pyrans, benzofurans, and benzopyrans. In this chapter, we discuss in detail the synthetic strategies and the scope of furan and pyran derivatives depending on the arylene substrates.

4.1. 1,5-Enyne Substrates

1,5-enyne is an important building block in the gold(I)-catalyzed construction of small-molecule heterocycles. In 2016, Liu and colleagues reported a gold(I)-catalyzed tandem strategy for the syntheses of furopyran derivatives involving Claisen rearrangement and 6-*endo*-trig cyclization, the regioselectivity of which was mainly controlled by the angle strain of propargyl γ-butyrolactone-2-enol ethers (**139**) (Scheme 29) [63]. A 6-*endo*-dig cyclization was initiated by the coordination of the gold catalyst to the triple bond to form intermediates (**140**) that were rearranged into β-allenic ketones (**141**). Intermediates (**143**) were produced by keto–enol tautomerism, and angle strain controlled 6-*endo*-dig cyclization. After demetallation, furopyran derivatives (**144**) were successfully obtained. The reason for the change in regioselectivity from 5-*exo*-trig to 6-*endo*-trig was explained by DFT calculation.

In the same year, Liu and colleagues achieved the syntheses of multisubstituted furofuran derivatives based on the studies represented in Scheme 29 by trapping key intermediates (**141**) (Scheme 30a) [64]. Alkynes of substrates (**145**) were activated by gold species to induce a rearrangement reaction and yield allene intermediates (**147**), consistent with the generation of intermediates (**141**). The terminal alkene of the allene was coordinated by a [Au]⁺ complex to enable the attack of nucleophiles, generating σ-allyl gold species (**148**). After SE′-type protodeauration of **148**, intermediates (**149**) were accessed, the enolether units of which were activated by gold species to trigger a 5-*exo*-trig cyclization. Finally, furofuran products (**150**) were delivered after protodeauration. In addition, multisubstituted furopyran derivatives were successfully produced when the propargyl terminal was substituted with thiophene or furan (Scheme 30b). Substrates (**151**) were converted to intermediates (**152**) under the activation of a gold catalyst, which was similar to the formation of intermediates (**140**) shown in Scheme 30. Intermediates (**152**) were not rearranged to β-allenic ketones due to the chelation of the heteroatom to the gold complex but were transformed to intermediates (**153**). Ultimately, furopyran products (**154**) were obtained via the nucleophilic addition of oxonium moiety after protodeauration. Thus, the authors achieved the syntheses of furofuran and furopyran derivatives by substituent modulation using propargyl vinyl ethers in the catalysis of gold(I) catalysts.

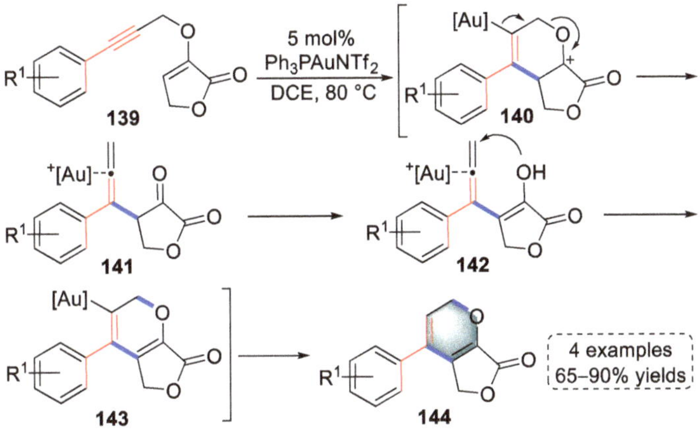

Scheme 29. Gold(I)-catalyzed syntheses of furopyran derivatives.

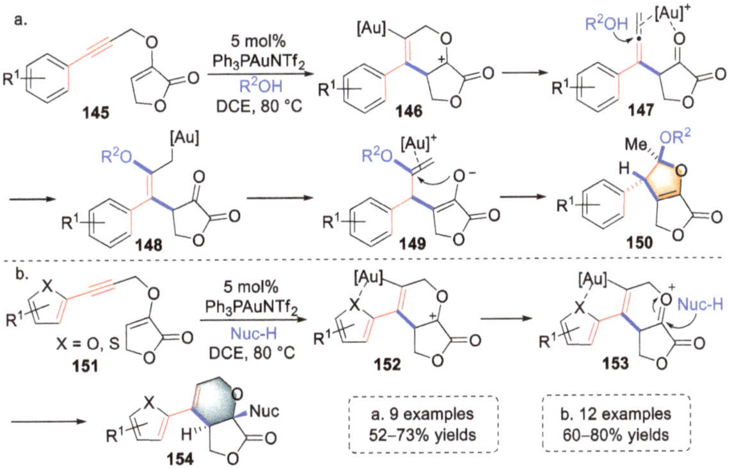

Scheme 30. Gold(I)-catalyzed syntheses of multisubstituted (**a**) furofuran and (**b**) furopyran derivatives.

In 2016, Jiang et al., developed a gold(I)-catalyzed, ligand-regulated cyclization for the syntheses furopyran or dihydroquinoline derivatives using 1,5-enyne substrates containing directing groups (Scheme 31) [65]. When using tris(2,4-di-*tert*-butylphenyl) phosphite (L1) in combination with trifluoromethanesulfonate (OTf⁻), gold(I)-π-alkyne intermediates (**156**) were formed by three coordinations, which were attributed to the increased electrophilicity of the gold center. The activated triple bond was attacked by the ortho position of the left aromatic ring, which overcame the steric hindrance. After protodeauration, furopyran and dihydroquinoline derivatives (**158**) were accessed (Scheme 31a). When a combination of Xphos ligand (L2) and NTf_2^- was used, intermediates (**160**) were produced, which were attributed to the decreased electrophilicity of the gold center. Next, the activated alkyne was attacked by the para position of the left aromatic ring, which yielded products (**162**) after protodeauration (Scheme 31b). The above regiodivergent cyclization depended mainly on the electronic and steric effects of the ligand in gold species. The authors systematically examined the scope of the above switchable strategy with moderate to excellent yields.

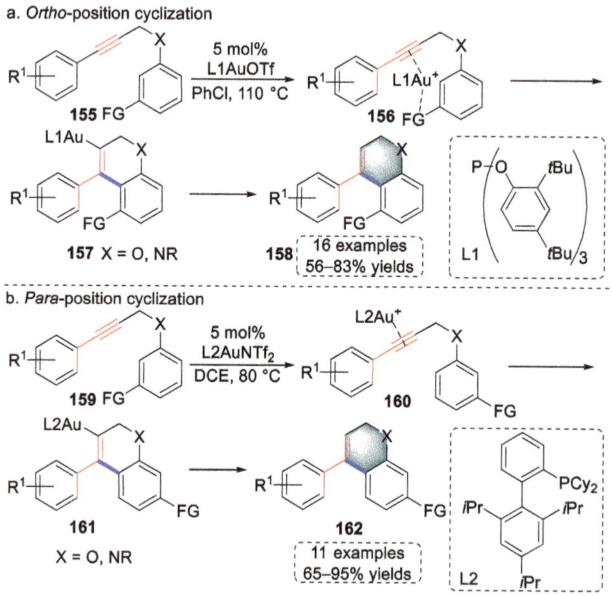

Scheme 31. Gold(I)-catalyzed syntheses of (**a**) furopyran and (**b**) dihydroquinoline derivatives.

4.2. Alkyne–Phenol Substrates

In 2016, a gold-catalyzed tandem cyclization to benzofuran derivatives was reported by Saito and colleagues (Scheme 32) [66]. The coordination of a gold complex to the triple bond initiated cyclization to generate intermediates (**164**), which were subsequently transformed into intermediates (**165**) with an α-alkoxy alkyl-shift. Under the influence of the activation of a gold catalyst, oxonium intermediates (**166**) were formed by releasing R^2OH, the α,β-enone moieties of which were attacked by the nucleophilic group to generate benzofuran products (**167**). Moreover, this strategy could be used for the construction of a larger number of small-molecule heterocyclic derivatives by regulating side chains in o-alkyl aryl ethers (**163**).

In 2019, the González lab achieved a gold(I)-catalyzed tandem cycloisomerization for the syntheses of benzofuran derivatives using 2-(iodoethynyl)-aryl esters with 15 examples and up to 85% yield (Scheme 33) [67]. The triple bonds of substrates (**167**) were activated by the gold complex to generate gold–vinylidene intermediates (**168**) via a 1,2-iodine shift. 3-iodo-2-acyl benzofuran products (**169**) were assembled by inserting gold carbine into the O-acyl bond. Importantly, the capture of intermediates (**168**) by silane through supplementary experiments implied a gold-catalyzed iodine rearrangement.

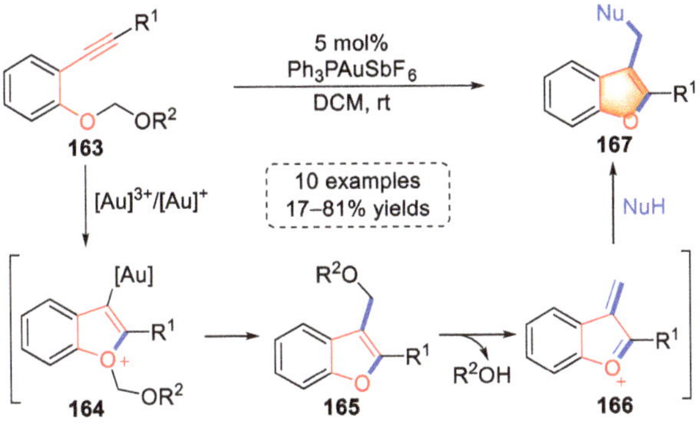

Scheme 32. Gold(I/III)-catalyzed syntheses of benzofuran derivatives.

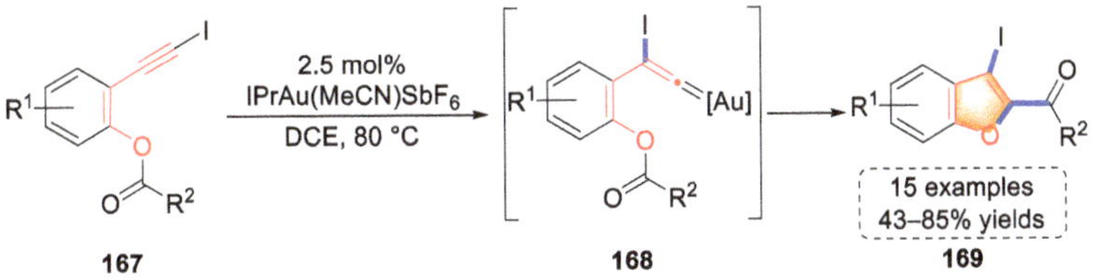

Scheme 33. Gold(I)-catalyzed syntheses of 3-iodo-2-acyl benzofuran derivatives.

A series of vinyl benzofuran derivatives was synthesized via a gold(I)-catalyzed cascade cyclization/hydroarylation method developed by the Xia group in 2022 (Scheme 34) [68]. With SIPrAuCl as catalyst and NaBARF as cocatalyst, benzofurans (**171**) were formed from *o*-alkyl phenol substrates (**170**). The triple bond of the haloalkyne was activated by the gold complex and thus attacked by the C3 position of the benzofuran through transition states (**172**). Then, cationic vinyl–gold intermediates were produced, which were then transformed into vinyl benzofurans (**173**) through a proton transfer. The authors demonstrated the reaction mechanism via experiments and computational calculations, and the functional group tolerance of the above strategy was examined with 20 examples and 19–98% yields.

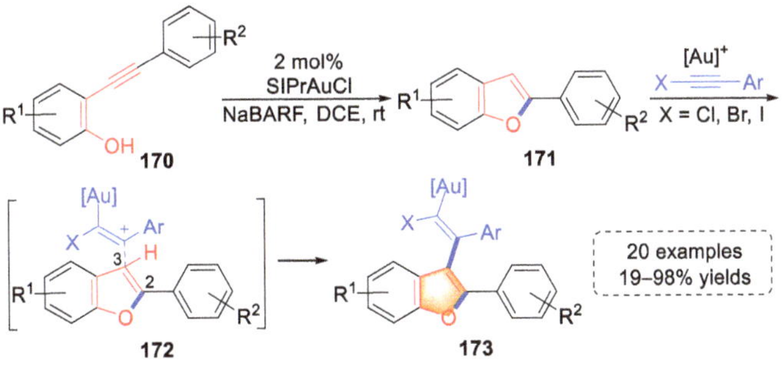

Scheme 34. Gold(I)-catalyzed syntheses of vinyl benzofuran derivatives.

4.3. Other Arylalkyne Substrates

In 2018, the Xu group synthesized furan derivatives using a series of propargyl diazoacetates through a gold(I)-catalyzed, water-involved tandem approach with 29 examples and up to 90% yield (Scheme 35) [69]. Initially, diazoacetate substrates (**174**) were transformed into gold carbene intermediates via the activation of the gold catalyst with the release of N_2, the gold carbene moieties of which were then attacked by H_2O to form oxonium ylides (**175**). After isomerization, enol intermediates (**176**) were produced by proton transfer, followed by a 6-*endo*-dig cyclization to yield cyclized intermediates (**177**). The carbonyl groups of **177** were nucleophilically attacked by the vinyl–gold to generate ring contraction intermediates (**178**). After the cleavage of cyclopropane, secondary carbene intermediates (**179**) were generated with the elimination of H_2O via an intramolecular H-bond-assisted pinacol rearrangement. When R^2 or R^3 was H, the final processes of β-H elimination and protodeauration yielded furan products (**180**). The authors demonstrated the formation of intermediates by interception experiments and verified the involvement of H_2O by isotope-labeled experiments.

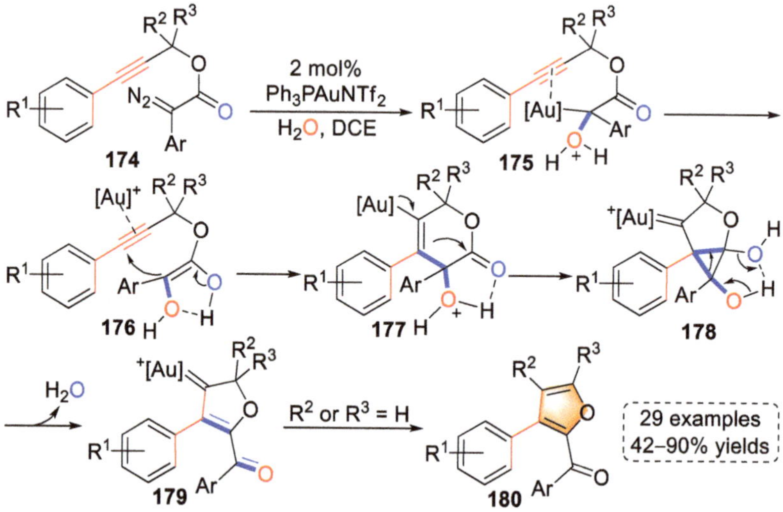

Scheme 35. Gold(I)-catalyzed syntheses of furan derivatives.

In the same year, Liu and colleagues reported a gold(I)-catalyzed tandem protocol involving oxidation, 1,2-enynyl migration, and 6-*exo*-dig cyclization to prepare 1*H*-isochromene derivatives (Scheme 36) [70]. The R^3-substituted alkyne of o-(alkynyl)-phenyl propargyl ether substrates (**181**) was coordinated by the gold catalyst to initiate an attack of N-oxide, followed by the elimination of the pyridine derivative to generate gold carbene intermediates (**182**). Next, a novel 1,2-enynyl migration resulted in the formation of oxonium ion intermediates (**183**), which were then converted into 1*H*-isochromene products (**184**) by 6-*exo*-dig cyclization after protodeauration. Notably, the reaction mechanism was supported by isotopic labeling experiments.

There are many excellent examples of the syntheses of furan and pyran derivatives reported, other than those listed in this chapter [71–75], including multicomponent, one-pot reactions [76,77]. Gold(I)-catalyzed tandem reactions are significant for the construction of small-molecule scaffolds containing furan or pyran. Furthermore, the development of gold(I)-catalyzed strategies also provides material support for the study of the bioactivity of furan and pyran derivatives.

In addition, the use of gold(I)-catalyzed alkyne cyclization to construct N-heterocyclic skeletons, e.g., pyrrole, indole, quinoline, pyridine, carbazole, is an important research direction. This class of reactions has been systematically summarized in recent reviews, so is not be described repeatedly in this feature paper [78–80].

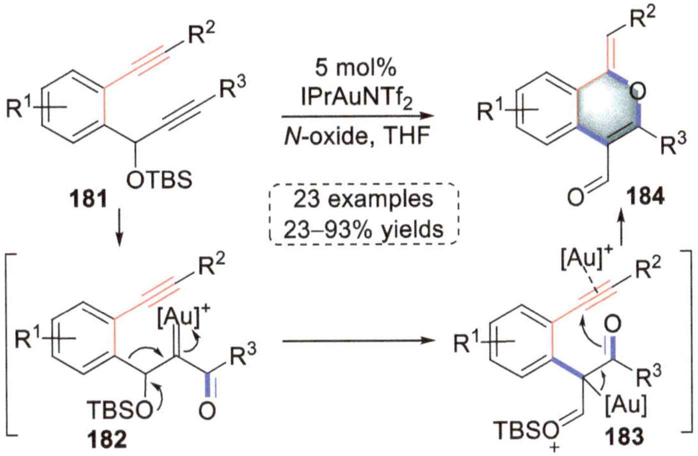

Scheme 36. Gold(I)-catalyzed syntheses of 1*H*-isochromene derivatives.

5. Conclusions

In the last decade, homogeneous gold(I)-catalyzed cyclization for the construction of small-molecule skeletons from arylalkyne substrates has been developed rapidly, owing to the ease of substrate preparation and the stability of gold catalysts.

In this feature paper, we systematically summarized the gold(I)-catalyzed syntheses of benzene, cyclopentene, furan, and pyran derivatives, which were carefully classified according to the type of arylalkyne substrate. Gold(I)-catalyzed tandem approaches for the construction of small-molecule scaffolds generally involve cyclization, isomerization, aromatization, migration, rearrangement, and other processes that are usually verified by controlled experiments and isotopic labeling experiments, as well as DFT calculations. In addition, the efficient strategies of gold catalysis were featured, with good functional group tolerance and reaction yield.

Although many excellent works have been reported with respect to gold catalysis for the syntheses of small-molecule skeletons, additional gold(I)-catalyzed asymmetric strategies are urgently required. Therefore, studies on the enantioselective construction of small-molecule scaffolds with the participation of chiral ligands will be further developed. In addition, dual gold/photoredox-catalyzed or dual gold/enzyme-catalyzed organic reactions can contribute to the development of this field [81].

Author Contributions: Conceptualization, L.Y. and Y.L.; software, H.S. and Y.S.; validation, L.Y. and S.Z.; formal analysis, H.S., S.Z. and Y.L.; investigation, L.Y., H.S., Y.S. and S.Z.; resources, L.Y. and Y.L.; data curation, L.Y.; writing—original draft preparation, L.Y. and Y.L.; writing—review and editing, L.Y., M.C. and Y.L.; visualization, L.Y. and M.C.; supervision, M.C. and Y.L.; project administration, M.C. and Y.L.; funding acquisition, L.Y. and Y.L. All authors have read and agreed to the published version of the manuscript.

Funding: This research was funded by the National Natural Science Foundation of China (No. 21977073), the Foundation of Liaoning Province Education Administration (No. LJKZ0908 and No. LJKQZ2022236), the Liaoning Provincial Foundation of Natural Science (No. 2022-MS-245), and the China Postdoctoral Science Foundation (No. 2022MD723807).

Institutional Review Board Statement: Not applicable.

Informed Consent Statement: Not applicable.

Data Availability Statement: Not applicable.

Acknowledgments: The authors acknowledge the Program for the Innovative Research Team of the Ministry of Education and the Program for the Liaoning Innovative Research Team in University.

Conflicts of Interest: The authors declare no conflict of interest.

References

1. Bond, G.C.; Sermon, P.A.; Webb, G.; Buchanan, D.A.; Wells, P.B. Hydrogenation over supported gold catalysts. *J. Chem. Soc. Chem. Commun.* **1973**, *13*, 444–445. [CrossRef]
2. Ito, Y.; Sawamura, M.; Hayashi, T. Catalytic asymmetric aldol reaction: Reaction of aldehydes with isocyanoacetate catalyzed by a chiral ferrocenylphosphine-gold(I) complex. *J. Am. Chem. Soc.* **1986**, *108*, 6405–6406. [CrossRef]
3. Gadon, V. Modern gold catalyzed synthesis. Edited by A. Stephen, K. Hashmi and F. Dean Toste. *Angew. Chem. Int. Ed.* **2012**, *51*, 11200. [CrossRef]
4. Li, Y.Y.; Li, W.B.; Zhang, J.L. Gold-catalyzed enantioselective annulations. *Eur. Chem. J.* **2017**, *23*, 467–512. [CrossRef]
5. Teles, J.H.; Brode, S.; Chabanas, M. Cationic gold(I) complexes: Highly efficient catalysts for the addition of alcohols to alkynes. *Angew. Chem., Int. Ed.* **1998**, *37*, 1415–1418. [CrossRef]
6. Hashmi, A.S.; Rudolph, M. Gold catalysis in total synthesis. *Chem. Soc. Rev.* **2008**, *37*, 1766–1775. [CrossRef]
7. Wittstock, A.; Bäumer, M. Catalysis by unsupported skeletal gold catalysts. *Acc. Chem. Res.* **2014**, *47*, 731–739. [CrossRef]
8. Friend, C.M.; Hashmi, A.S. Gold catalysis. *Acc. Chem. Res.* **2014**, *47*, 729–730. [CrossRef]
9. Hashmi, A.S.K. Dual Gold Catalysis. *Acc. Chem. Res.* **2014**, *47*, 864–876. [CrossRef]
10. Zhang, L.M. A non-diazo approach to α-oxo gold carbenes via gold-catalyzed alkyne oxidation. *Acc. Chem. Res.* **2014**, *47*, 877–888. [CrossRef]
11. Wang, Y.M.; Lackner, A.D.; Toste, F.D. Development of catalysts and ligands for enantioselective gold catalysis. *Acc. Chem. Res.* **2014**, *47*, 889–901. [CrossRef] [PubMed]
12. Obradors, C.; Echavarren, A.M. Gold-catalyzed rearrangements and beyond. *Acc. Chem. Res.* **2014**, *47*, 902–912. [CrossRef] [PubMed]
13. Zhang, D.H.; Tang, X.Y.; Shi, M. Gold-catalyzed tandem reactions of methylenecyclopropanes and vinylidenecyclopropanes. *Acc. Chem. Res.* **2014**, *47*, 913–924. [CrossRef] [PubMed]
14. Harris, R.J.; Widenhoefer, R.A. Gold carbenes, gold-stabilized carbocations, and cationic intermediates relevant to gold-catalysed enyne cycloaddition. *Chem. Soc. Rev.* **2016**, *45*, 4533–4551. [CrossRef]
15. Li, W.; Yu, B. Gold-catalyzed glycosylation in the synthesis of complex carbohydrate-containing natural products. *Chem. Soc. Rev.* **2018**, *47*, 7954–7984. [CrossRef]
16. Lu, Z.C.; Hammond, G.B.; Xu, B. Improving homogeneous cationic gold catalysis through a mechanism-based approach. *Acc. Chem. Res.* **2019**, *52*, 1275–1288. [CrossRef]
17. Greiner, L.C.; Matsuoka, J.; Inuki, S.; Ohno, H. Azido-alkynes in gold(I)-catalyzed indole syntheses. *Chem. Rev.* **2021**, *21*, 3897–3910. [CrossRef]
18. Li, D.Y.; Zang, W.Q.; Bird, M.J.; Hyland, C.J.T.; Shi, M. Gold-catalyzed conversion of highly strained compounds. *Chem. Rev.* **2021**, *121*, 8685–8755. [CrossRef]
19. Mato, M.; Franchino, A.; Garci, A.M.C.; Echavarren, A.M. Gold-catalyzed synthesis of small rings. *Chem. Rev.* **2021**, *121*, 8613–8684. [CrossRef]
20. Reyes, R.L.; Iwai, T.; Sawamura, M. Construction of medium-sized rings by gold catalysis. *Chem. Rev.* **2021**, *121*, 8926–8947. [CrossRef]
21. Ghosh, T.; Chatterjee, J.; Bhakta, S. Gold-catalyzed hydroarylation reactions: A comprehensive overview. *Org. Biomol. Chem.* **2022**, *20*, 7151–7187. [CrossRef]
22. Hashmi, A.S.K.; Braun, I.; Rudolph, M.; Rominger, F. The role of gold acetylides as a selectivity trigger and the importance of gem-diaurated species in the gold-catalyzed hydroarylating-aromatization of arene-diynes. *Organometallics* **2012**, *31*, 644–661. [CrossRef]
23. Naoe, S.; Suzuki, Y.; Hirano, K.; Inaba, Y.; Oishi, S.; Fujii, N.; Ohno, H. Gold(I)-catalyzed regioselective inter-/intramolecular addition cascade of di- and triynes for direct construction of substituted naphthalenes. *J. Org. Chem.* **2012**, *77*, 4907–4916. [CrossRef]
24. Yu, L.Z.; Wei, Y.; Shi, M. Synthesis of polysubstituted polycyclic aromatic hydrocarbons by gold-catalyzed cyclization–oxidation of alkylidenecyclopropane-containing 1,5-enynes. *ACS Catal.* **2017**, *7*, 4242–4247. [CrossRef]
25. Ikeuchi, T.; Inuki, S.; Oishi, S.; Ohno, H. Gold(I)-catalyzed cascade cyclization reactions of allenynes for the synthesis of fused cyclopropanes and acenaphthenes. *Angew. Chem. Int. Ed.* **2019**, *58*, 7792–7796. [CrossRef]
26. Komatsu, H.; Ikeuchi, T.; Tsuno, H.; Arichi, N.; Yasui, K.; Oishi, S.; Inuki, S.; Fukazawa, A.; Ohno, H. Construction of tricyclic nitrogen heterocycles by a gold(I)-catalyzed cascade cyclization of allenynes and application to polycyclic π-electron systems. *Angew. Chem. Int. Ed.* **2021**, *60*, 27019–27025. [CrossRef]
27. Fu, J.Y.; Li, B.B.; Wang, X.X.; Wang, H.; Deng, M.H.; Yang, H.L.; Lin, B.; Cheng, M.S.; Yang, L.; Liu, Y.X. Collective total syntheses of benzo[c]phenanthridine alkaloids via a sequential transition metal-catalyzed pot-economic approach. *Org. Lett.* **2022**, *24*, 8310–8315. [CrossRef]
28. Liu, Y.X.; Guo, J.; Liu, Y.; Wang, X.Y.; Wang, Y.S.; Jia, X.Y.; Wei, G.F.; Chen, L.Z.; Xiao, J.Y.; Cheng, M.S. Au(I)-catalyzed triple bond alkoxylation/dienolether aromaticity-driven cascade cyclization to naphthalenes. *Chem. Commun.* **2014**, *50*, 6243–6245. [CrossRef]

29. Peng, X.S.; Zhu, L.F.; Hou, Y.Q.; Pang, Y.D.; Li, Y.M.; Fu, J.Y.; Yang, L.; Lin, B.; Liu, Y.X.; Cheng, M.S. Access to benzo[a]carbazoles and indeno[1,2-c]quinolines by a gold(I)-catalyzed tunable domino cyclization of difunctional 1,2-diphenylethynes. *Org. Lett.* **2017**, *19*, 3402–3405. [CrossRef]
30. Xiong, Z.L.; Zhang, X.H.; Li, Y.M.; Peng, X.S.; Fu, J.Y.; Guo, J.; Xie, F.K.; Jiang, C.G.; Lin, B.; Liu, Y.X.; et al. Syntheses of 12H-benzo[a]xanthen-12-ones and benzo[a]acridin-12(7H)-ones through Au(I)-catalyzed Michael addition/6-endo-trig cyclization/aromatization cascade annulation. *Org. Biomol. Chem.* **2018**, *16*, 7361–7374. [CrossRef]
31. Fu, J.Y.; Li, B.B.; Wang, X.G.; Liang, Q.D.; Peng, X.S.; Yang, L.; Wan, T.; Wang, X.X.; Lin, B.; Cheng, M.S. Au(I)-catalyzed 6-endo-dig cyclizations of aromatic 1,5-enynes to 2-(naphthalen-2-yl)anilines leading to divergent syntheses of benzo[a]carbazole, benzo[c,h]cinnoline and dibenzo[i]phenanthridine derivatives. *Chin. J. Chem.* **2022**, *40*, 46–52. [CrossRef]
32. Shu, C.; Chen, C.B.; Chen, W.X.; Ye, L.W. Flexible and practical synthesis of anthracenes through gold-catalyzed cyclization of o-alkynyldiarylmethanes. *Org. Lett.* **2013**, *15*, 5542–5545. [CrossRef] [PubMed]
33. Jin, S.F.; Niu, Y.J.; Liu, C.J.; Zhu, L.F.; Li, Y.M.; Cui, S.S.; Xiong, Z.L.; Cheng, M.S.; Lin, B.; Liu, Y.X. Gold(I)-initiated cycloisomerization/Diels-Alder/retro-Diels-Alder cascade strategy to biaryls. *J. Org. Chem.* **2017**, *82*, 9066–9074. [CrossRef]
34. Zhang, C.; Sun, Q.; Rudolph, M.; Rominger, F.; Hashmi, A.S.K. Gold-catalyzed regiodivergent annulations of diazo-alkynes controlled by Et₃N(HF)₃. *ACS Catal.* **2021**, *11*, 15203–15211. [CrossRef]
35. Wang, C.Y.; Chen, Y.F.; Xie, X.; Liu, J.; Liu, Y.H. Gold-catalyzed furan/yne cyclizations for the regiodefined assembly of multisubstituted protected 1-naphthols. *J. Org. Chem.* **2012**, *77*, 1915–1921. [CrossRef]
36. Song, X.R.; Xia, X.F.; Song, Q.B.; Yang, F.; Li, Y.X.; Liu, X.Y.; Liang, Y.M. Gold-catalyzed cascade reaction of hydroxy enynes for the synthesis of oxanorbornenes and naphthalene derivatives. *Org. Lett.* **2012**, *14*, 3344–3347. [CrossRef]
37. Taguchi, M.; Tokimizu, Y.; Oishi, S.; Fujii, N.; Ohno, H. Synthesis of fused carbazoles by gold-catalyzed tricyclization of conjugated diynes via rearrangement of an N-propargyl group. *Org. Lett.* **2015**, *17*, 6250–6253. [CrossRef]
38. Kale, B.S.; Liu, R.S. Gold-catalyzed aromatizations of 3-ene-5-siloxy-1,6-diynes with nitrosoarenes to enable 1,4-N,O-functionalizations: One-pot construction of 4-hydroxy-3-aminobenzaldehyde cores. *Org. Lett.* **2019**, *21*, 8434–8438. [CrossRef]
39. Milian, A.; Garcia-Garcia, P.; Perez-Redondo, A.; Sanz, R.; Vaquero, J.J.; Fernandez-Rodriguez, M.A. Selective synthesis of phenanthrenes and dihydrophenanthrenes via gold-catalyzed cycloisomerization of biphenyl embedded trienynes. *Org. Lett.* **2020**, *22*, 8464–8469. [CrossRef]
40. Koshikawa, T.; Nagashima, Y.; Tanaka, K. Gold-catalyzed [3 + 2] annulation, carbenoid transfer, and C–H insertion cascade: Elucidation of annulation mechanisms via benzopyrylium intermediates. *ACS Catal.* **2021**, *11*, 1932–1937. [CrossRef]
41. Wang, H.F.; Guo, L.N.; Fan, Z.B.; Tang, T.H.; Zi, W.W. Gold-catalyzed formal hexadehydro-Diels-Alder/carboalkoxylation reaction cascades. *Org. Lett.* **2021**, *23*, 2676–2681. [CrossRef]
42. Bharath Kumar, P.; Raju, C.E.; Chandubhai, P.H.; Sridhar, B.; Karunakar, G.V. Gold(I)-catalyzed regioselective cyclization to access cyclopropane-fused tetrahydrobenzochromenes. *Org. Lett.* **2022**, *24*, 6761–6766. [CrossRef]
43. Fu, J.Y.; Li, B.B.; Zhou, Z.F.; Cheng, M.S.; Yang, L.; Liu, Y.X. Formal total synthesis of macarpine via a Au(I)-catalyzed 6-endo-dig cycloisomerization strategy. *Beilstein J. Org. Chem.* **2022**, *18*, 1589–1595. [CrossRef] [PubMed]
44. Samala, S.; Mandadapu, A.K.; Saifuddin, M.; Kundu, B. Gold-catalyzed sequential alkyne activation: One-pot synthesis of NH-carbazoles via cascade hydroarylation of alkyne/6-endo-dig carbocyclization reactions. *J. Org. Chem.* **2013**, *78*, 6769–6774. [CrossRef]
45. Li, N.; Lian, X.L.; Li, Y.H.; Wang, T.Y.; Han, Z.Y.; Zhang, L.; Gong, L.Z. Gold-catalyzed direct assembly of aryl-annulated carbazoles from 2-alkynyl arylazides and alkynes. *Org. Lett.* **2016**, *18*, 4178–4181. [CrossRef]
46. Li, X.S.; Xu, D.T.; Niu, Z.J.; Li, M.; Shi, W.Y.; Wang, C.T.; Wei, W.X.; Liang, Y.M. Gold-catalyzed tandem annulations of pyridylhomopropargylic alcohols with propargyl alcohols. *Org. Lett.* **2021**, *23*, 832–836. [CrossRef]
47. Pandit, Y.B.; Liu, R.S. Dynamic kinetic resolution in gold-catalyzed (4 + 2)-annulations between alkynyl benzaldehydes and allenamides to yield enantioenriched all-carbon diarylalkylmethane derivatives. *Org. Lett.* **2022**, *24*, 548–553. [CrossRef]
48. Hashmi, A.S.K.; Braun, I.; Nosel, P.; Schadlich, J.; Wieteck, M.; Rudolph, M.; Rominger, F. Simple gold-catalyzed synthesis of benzofulvenes-gem-diaurated species as "instant dual-activation" precatalysts. *Angew. Chem. Int. Ed.* **2012**, *51*, 4456–4460. [CrossRef]
49. Wurm, T.; Bucher, J.; Duckworth, S.B.; Rudolph, M.; Rominger, F.; Hashmi, A.S.K. Catalyzed generation of vinyl cations from 1,5-diynes. *Angew. Chem. Int. Ed.* **2017**, *56*, 3364–3368. [CrossRef]
50. Hu, C.; Farshadfar, K.; Dietl, M.C.; Cervantes-Reyes, A.; Wang, T.; Adak, T.; Rudolph, M.; Rominger, F.; Li, J.; Ariafard, A.; et al. Gold-catalyzed [5,5]-rearrangement. *ACS Catal.* **2021**, *11*, 6510–6518. [CrossRef]
51. Sanjuan, A.M.; Virumbrales, C.; Garcia-Garcia, P.; Fernandez-Rodriguez, M.A.; Sanz, R. Formal [4 + 1] cycloadditions of β,β-diaryl-substituted ortho-(alkynyl)styrenes through gold(I)-catalyzed cycloisomerization reactions. *Org. Lett.* **2016**, *18*, 1072–1075. [CrossRef] [PubMed]
52. Virumbrales, C.; El-Remaily, M.; Suarez-Pantiga, S.; Fernandez-Rodriguez, M.A.; Rodriguez, F.; Sanz, R. Gold(I) catalysis applied to the stereoselective synthesis of indeno[2,1-b]thiochromene derivatives and seleno analogues. *Org. Lett.* **2022**, *24*, 8077–8082. [CrossRef]
53. Guo, J.; Peng, X.S.; Wang, X.Y.; Xie, F.K.; Zhang, X.H.; Liang, G.D.; Sun, Z.H.; Liu, Y.X.; Cheng, M.S.; Liu, Y. A gold-catalyzed cycloisomerization/aaerobic oxidation cascade strategy for 2-aryl indenones from 1,5-enynes. *Org. Biomol. Chem.* **2018**, *16*, 9147–9151. [CrossRef]

54. Zi, W.W.; Toste, F.D. Gold(I)-catalyzed enantioselective carboalkoxylation of alkynes. *J. Am. Chem. Soc.* **2013**, *135*, 12600–12603. [CrossRef]
55. Jiang, C.G.; Xiong, Z.L.; Jin, S.F.; Gao, P.; Tang, Y.Z.; Wang, Y.S.; Du, C.; Wang, X.Y.; Liu, Y.; Lin, B.; et al. A Au(I)-catalyzed hydrogen bond-directed tandem strategy to synthesize indeno-chromen-4-one and indeno-quinolin-4-one derivatives. *Chem. Commun.* **2016**, *52*, 11516–11519. [CrossRef]
56. Yamada, T.; Park, K.; Tachikawa, T.; Fujii, A.; Rudolph, M.; Hashmi, A.S.K.; Sajiki, H. Gold-catalyzed cyclization of 2-alkynylaldehyde cyclic acetals via hydride shift for the synthesis of indenone derivatives. *Org. Lett.* **2020**, *22*, 1883–1888. [CrossRef]
57. Xie, F.K.; Zhang, B.; Chen, Y.Y.; Jia, H.W.; Sun, L.; Zhuang, K.T.; Yin, L.L.; Cheng, M.S.; Lin, B.; Liu, Y.X. A gold(I)-catalyzed tandem cyclization to benzo[b]indeno[1,2-e][1,4]diazepines from o-phenylenediamines and ynones. *Adv. Synth. Catal.* **2020**, *362*, 3886–3897. [CrossRef]
58. Zhang, J.F.; Zhang, S.; Ding, Z.X.; Hou, A.B.; Fu, J.Y.; Su, H.W.; Cheng, M.S.; Lin, B.; Yang, L.; Liu, Y.X. Gold(I)-catalyzed tandem intramolecular methoxylation/double aldol condensation strategy yielding 2,2′-spirobi[indene] derivatives. *Org. Lett.* **2022**, *24*, 6777–6782. [CrossRef]
59. Bao, M.; Chen, J.Z.; Pei, C.; Zhang, S.J.; Lei, J.P.; Hu, W.H.; Xu, X.F. Gold-catalyzed ketene dual functionalization and mechanistic insights: Divergent synthesis of indenes and benzo[d]oxepines. *Sci. Chin. Chem.* **2021**, *64*, 778–787. [CrossRef]
60. Thilmany, P.; Guarnieri-Ibáñez, A.; Jacob, C.; Lacour, J.; Evano, G. Straightforward synthesis of indenes by gold-catalyzed intramolecular hydroalkylation of ynamides. *ACS Org. Inorg. Au* **2021**, *2*, 53–58. [CrossRef]
61. Greiner, L.C.; Arichi, N.; Inuki, S.; Ohno, H. Gold(I)-catalyzed benzylic C(sp3)-H functionalizations: Divergent synthesis of indole[a]- and [b]-fused polycycles. *Angew. Chem. Int. Ed.* **2022**, e202213653. [CrossRef]
62. Xu, Z.; Zhao, S.J.; Lv, Z.S.; Feng, L.S.; Wang, Y.L.; Zhang, F.; Bai, L.Y.; Deng, J.L. Benzofuran derivatives and their anti-tubercular, anti-bacterial activities. *Eur. J. Med. Chem.* **2019**, *162*, 266–276. [CrossRef]
63. Jin, S.F.; Jiang, C.G.; Peng, X.S.; Shan, C.H.; Cui, S.S.; Niu, Y.Y.; Liu, Y.; Lan, Y.; Liu, Y.X.; Cheng, M.S. Gold(I)-catalyzed angle strain controlled strategy to furopyran derivatives from propargyl vinyl ethers: Insight into the regioselectivity of cycloisomerization. *Org. Lett.* **2016**, *18*, 680–683. [CrossRef]
64. Liu, Y.X.; Jin, S.F.; Wang, Y.S.; Cui, S.S.; Peng, X.S.; Niu, Y.Y.; Du, C.; Cheng, M.S. A gold(I)-catalyzed substituent-controlled cycloisomerization of propargyl vinyl ethers to multi-substituted furofuran and furopyran derivatives. *Chem. Commun.* **2016**, *52*, 6233–6236. [CrossRef]
65. Ding, D.; Mou, T.; Feng, M.G.; Jiang, X.F. Utility of ligand effect in homogenous gold catalysis: Enabling regiodivergent π-bond-activated cyclization. *J. Am. Chem. Soc.* **2016**, *138*, 5218–5221. [CrossRef]
66. Obata, T.; Suzuki, S.; Nakagawa, A.; Kajihara, R.; Noguchi, K.; Saito, A. Gold-catalyzed domino synthesis of functionalized benzofurans and tetracyclic isochromans via formal carboalkoxylation. *Org. Lett.* **2016**, *18*, 4136–4139. [CrossRef]
67. Fernandez-Canelas, P.; Rubio, E.; Gonzalez, J.M. Oxyacylation of iodoalkynes: Gold(I)-catalyzed expeditious access to benzofurans. *Org. Lett.* **2019**, *21*, 6566–6569. [CrossRef]
68. Wu, J.W.; Wei, C.B.; Zhao, F.; Du, W.Q.; Geng, Z.S.; Xia, Z.H. Gold(I)-catalyzed tandem cyclization/hydroarylation of o-alkynylphenols with haloalkynes. *J. Org. Chem.* **2022**, *87*, 14374–14383. [CrossRef]
69. Bao, M.; Qian, Y.; Su, H.; Wu, B.; Qiu, L.H.; Hu, W.H.; Xu, X.F. Gold(I)-catalyzed and H2O-mediated carbene cascade reaction of propargyl diazoacetates: Furan synthesis and mechanistic insights. *Org. Lett.* **2018**, *20*, 5332–5335. [CrossRef]
70. Zhao, J.D.; Xu, W.; Xie, X.; Sun, N.; Li, X.D.; Liu, Y.H. Gold-catalyzed oxidative cyclizations of {o-(alkynyl)phenyl propargyl} silyl ether derivatives involving 1,2-enynyl migration: Synthesis of functionalized *1H*-isochromenes and *2H*-pyrans. *Org. Lett.* **2018**, *20*, 5461–5465. [CrossRef]
71. Tang, Y.; Li, J.; Zhu, Y.; Li, Y.; Yu, B. Mechanistic insights into the gold(I)-catalyzed activation of glycosyl ortho-alkynylbenzoates for glycosidation. *J. Am. Chem. Soc.* **2013**, *135*, 18396–18405. [CrossRef]
72. Aparece, M.D.; Vadola, P.A. Gold-catalyzed dearomative spirocyclization of aryl alkynoate esters. *Org. Lett.* **2014**, *16*, 6008–6011. [CrossRef]
73. Mallampudi, N.A.; Reddy, G.S.; Maity, S.; Mohapatra, D.K. Gold(I)-catalyzed cyclization for the synthesis of 8-hydroxy-3-substituted isocoumarins: Total synthesis of exserolide F. *Org. Lett.* **2017**, *19*, 2074–2077. [CrossRef]
74. Hamada, N.; Yamaguchi, A.; Inuki, S.; Oishi, S.; Ohno, H. Gold(I)-catalyzed oxidative cascade cyclization of 1,4-diyn-3-ones for the construction of tropone-fused furan scaffolds. *Org. Lett.* **2018**, *20*, 4401–4405. [CrossRef]
75. Wagner, P.; Ghosh, N.; Gandon, V.; Blond, G. Solvent effect in gold(I)-catalyzed domino reaction: Access to furopyrans. *Org. Lett.* **2020**, *22*, 7333–7337. [CrossRef]
76. Li, J.; Liu, J.; Ding, D.; Sun, Y.X.; Dong, J.L. Gold(III)-catalyzed three-component coupling reaction (TCC) selective toward furans. *Org. Lett.* **2013**, *15*, 2884–2887. [CrossRef]
77. Hosseyni, S.; Su, Y.J.; Shi, X.D. Gold catalyzed synthesis of substituted furan by intermolecular cascade reaction of propargyl alcohol and alkyne. *Org. Lett.* **2015**, *17*, 6010–6013. [CrossRef]
78. Campeau, D.; León Rayo, D.F.; Mansour, A.; Muratov, K.; Gagosz, F. Gold-catalyzed reactions of specially activated alkynes, allenes, and alkenes. *Chem. Rev.* **2021**, *121*, 8756–8867. [CrossRef]
79. Shandilya, S.; Gogoi, M.P.; Dutta, S.; Sahoo, A.K. Gold-catalyzed transformation of ynamides. *Chem. Rec.* **2021**, *21*, 4123–4149. [CrossRef]

80. Bag, D.; Sawant, S.D. Heteroarene-tethered functionalized alkyne metamorphosis. *Chem. Eur. J.* **2021**, *27*, 1165–1218. [CrossRef]
81. Hopkinson, M.N.; Tlahuext-Aca, A.; Glorius, F. Merging visible light photoredox and gold catalysis. *Acc. Chem. Res.* **2016**, *49*, 2261–2272. [CrossRef]

Article

Copper-Catalyzed Asymmetric Sulfonylative Desymmetrization of Glycerol

Kosuke Yamamoto, Keisuke Miyamoto, Mizuki Ueno, Yuki Takemoto, Masami Kuriyama and Osamu Onomura *

Graduate School of Biomedical Sciences, Nagasaki University, 1-14 Bunkyo-machi, Nagasaki 852-8521, Japan
* Correspondence: onomura@nagasaki-u.ac.jp

Abstract: Glycerol is the main side product in the biodiesel manufacturing process, and the development of glycerol valorization methods would indirectly contribute the sustainable biodiesel production and decarbonization. Transformation of glycerol to optically active C3 units would be one of the attractive routes for glycerol valorization. We herein present the asymmetric sulfonylative desymmetrization of glycerol by using a CuCN/(R,R)-PhBOX catalyst system to provide an optically active monosulfonylated glycerol in high efficiency. A high degree of enantioselectivity was achieved with a commercially available chiral ligand and an inexpensive carbonate base. The optically active monosulfonylated glycerol was successfully transformed into a C3 unit attached with differentially protected three hydroxy moieties. In addition, the synthetic utility of the present reaction was also demonstrated by the transformation of the monosulfonylated glycerol into an optically active synthetic ceramide, sphingolipid E.

Keywords: copper catalysis; asymmetric reaction; desymmetrization; glycerol

1. Introduction

The use of renewable energies instead of conventional fossil fuels has become a global trend from the perspective of expected fossil fuel depletion and global climate change. Biodiesel fuel, which is produced by the transesterification of vegetable or animal fats with methanol, has emerged as a promising alternative to petroleum-derived diesel fuel [1,2]. The biodiesel manufacturing process inevitably provides 10 wt% of glycerol (1,2,3-propanetriol) as the main side product, and an oversupply of the crude glycerol would be projected with the growing biodiesel market [3]. Thus, great effort has been devoted to the development of an efficient process for the transformation of glycerol to value-added commodity chemicals, which would indirectly contribute to sustainable biodiesel production [4–7]. Among them, optically active glycerol derivatives would be an attractive target in the field of glycerol valorization. Chiral glycerol derivatives such as glyceraldehyde and glycidyl tosylate are utilized as valuable C3 building blocks in medicinal [8–12] and synthetic organic chemistry [13–18]. A chiral pool approach is a traditional strategy to access enantiopure glycerol derivatives, but the need for multi-step transformations may be a major drawback [19–23]. Asymmetric desymmetrization of glycerol would be one of the most straightforward methods for chiral glycerol derivatives production. Several types of enzymes, i.e., lipase, kinase, and dehydrogenase/oxidase, have been successfully applied to this strategy, affording optically active glycerols with various enantioselectivities [24–29]. On the other hand, despite the recent development of the enantioselective desymmetrization [30,31] of 1,2-diols [32–39] and 1,3-diols [40–48], including C2-substituted glycerols [49–53], the non-enzymatic direct desymmetrization of glycerol is still a challenging task presumably due to an extremely high hydrophilic nature of glycerol. In this context, the use of 2-O-protected glycerol derivatives would be the most common strategy for the chemical desymmetrization of glycerol (Scheme 1a) [54–57]. In

2013, Tan et al. developed the first non-enzymatic direct desymmetrization of glycerol through organocatalyzed enantioselective silylation (Scheme 1b) [58]. In their protocol, the high enantioselectivity was achieved through the secondary kinetic resolution on the initially formed monosilylated glycerol. Very recently, the copper-catalyzed sulfonylative desymmetrization of glycerol using a non-commercially available ligand with silver carbonate was described in the Chinese patent [59]. Although the desired product was obtained with high enantioselectivity under copper-catalyzed conditions, the use of a commercially available chiral ligand and a non-precious metal base would be desirable from a practical and economical point of view [60–62]. Herein, we report the asymmetric desymmetrization of glycerol through sulfonylation with a Cu/(R,R)-PhBOX complex and sodium carbonate affording the optically active monosulfonylated glycerol in an excellent yield and enantioselectivity (Scheme 1c) [63].

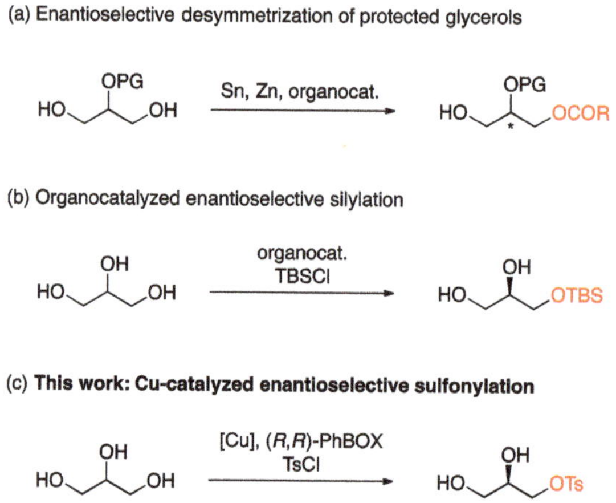

(a) Enantioselective desymmetrization of protected glycerols

(b) Organocatalyzed enantioselective silylation

(c) **This work: Cu-catalyzed enantioselective sulfonylation**

Scheme 1. Asymmetric desymmetrization of glycerols. (**a**) Enantioselective desymmetrization of protected glycerols. (**b**) Organocatalyzed enantioselective silylation. (**c**) This work: Cu-catalyzed enantioselective sulfonylation. The asterisk denotes the chiral center.

2. Results and Discussion

For the initial attempt to optimize the enantioselective desymmetrization of glycerol, compound **1** was treated with *p*-toluenesulfonyl chloride (TsCl) in the presence of copper trifluoromethanesulfonate (Cu(OTf)$_2$)/(R,R)-PhBOX and sodium carbonate in acetonitrile. Pleasingly, the desired monotosylated glycerol **2** was obtained in 91% yield with 83% ee (Table 1, entry 1). Using other carbonate salts, i.e., potassium carbonate and cesium carbonate, resulted in a decrease in both yield and enantioselectivity (entries 2 and 3). Organic bases were not suitable for the present transformation (entries 4 and 5). Next, other copper catalysts were examined to evaluate the catalytic activity in this reaction system. While CuCl provided **2** with a slightly lowered yield and enantioselectivity, CuBr and CuI exhibited a similar reactivity compared with Cu(OTf)$_2$ (entries 6–8). The use of CuCN led to the formation of **2** in 83% yield with higher enantioselectivity, and the reaction concentration was able to be doubled without significant changes regarding both yield and enantioselectivity (entries 9–10). Pleasingly, we found that acetone was a better solvent choice to afford the desired product in an excellent yield and enantioselectivity (96% yield, 94% ee), and the concentration of 0.25 M was found to be suitable for the present reaction (entries 11–12). The catalyst loading was able to be reduced to 5 mol% without a significant decrease in the yield and enantioselectivity (entry 13). We also examined the feasibility of the gram-scale preparation of **2**. The reaction with 6.0 mmol of glycerol successfully provided the desired product **2** in 88% yield (1.30 g) with 93% ee (entry 14).

Control experiments revealed that both the copper salt and the BOX ligand were essential to promote the tosylation of **1** (entries 15–16).

Table 1. Optimization of reaction conditions [1].

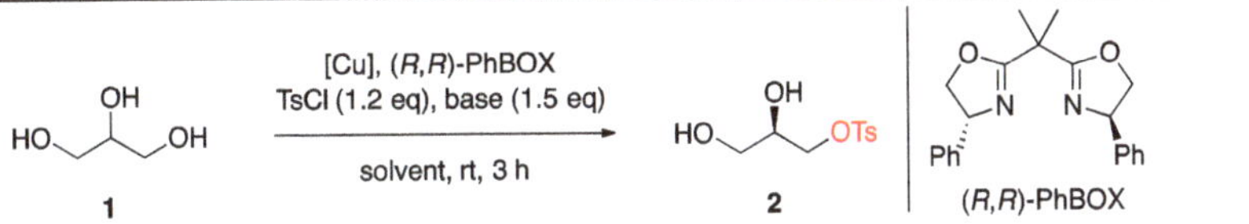

Entry	[Cu]	Base	Solvent	Yield (%) [2]	ee (%) [3]
1	Cu(OTf)$_2$	Na$_2$CO$_3$	CH$_3$CN	91	83
2	Cu(OTf)$_2$	K$_2$CO$_3$	CH$_3$CN	62	65
3 [4]	Cu(OTf)$_2$	Cs$_2$CO$_3$	CH$_3$CN	16	47
4 [4]	Cu(OTf)$_2$	pyridine	CH$_3$CN	41	rac
5 [4]	Cu(OTf)$_2$	DIPEA	CH$_3$CN	51	43
6	CuCl	Na$_2$CO$_3$	CH$_3$CN	73	79
7	CuBr	Na$_2$CO$_3$	CH$_3$CN	89	84
8	CuI	Na$_2$CO$_3$	CH$_3$CN	80	87
9	CuCN	Na$_2$CO$_3$	CH$_3$CN	83	90
10 [5]	CuCN	Na$_2$CO$_3$	CH$_3$CN	82	90
11 [5]	CuCN	Na$_2$CO$_3$	acetone	96	94
12 [6]	CuCN	Na$_2$CO$_3$	acetone	93	89
13 [5,7]	CuCN	Na$_2$CO$_3$	acetone	83	91
14 [5,8]	CuCN	Na$_2$CO$_3$	acetone	88	93
15 [5]	–	Na$_2$CO$_3$	acetone	trace	–
16 [5,9]	CuCN	Na$_2$CO$_3$	acetone	3	rac

[1] Reaction conditions: **1** (1.0 mmol), TsCl (1.2 mmol), [Cu] (0.1 mmol), (*R,R*)-PhBOX (0.1 mmol), base (1.5 mmol), solvent (0.125 M), rt, 3 h. [2] Isolated yield after column chromatography. [3] Determined by chiral HPLC analysis. [4] Reaction time (12 h). [5] Solvent (0.25 M). [6] Solvent (0.5 M). [7] CuCN (0.05 mmol), (*R,R*)-PhBOX (0.05 mmol), 10 h. [8] **1** (6.0 mmol), 9 h. [9] The reaction was carried out in the absence of (*R,R*)-PhBOX.

In order to gain insight into the chemoselectivity of the present reaction, we performed competition studies with alcohol additives (Table 2). The addition of *n*-propanol (1.0 eq) led to a slight decrease in both yield and enantioselectivity, but the formation of *n*-propyl tosylate was not detected (entry 1 vs. entry 2). Moreover, selective sulfonylation of glycerol (**1**) over 1,2- and 1,3-diols was observed under the present reaction conditions, and the desired monosulfonylated glycerol **2** was obtained without a significant loss of enantioselectivity (entry 1 vs. entries 3–4). In addition, the reaction of 2-*O*-benzylglycerol (**4**) provided the corresponding monotosylated product **5** in a low yield with poor enantioselectivity (Scheme 2). These results indicated that the present reaction system would be highly selective for the glycerol transformation even in the presence of other alcohols, and the presence of a free 2-hydroxy moiety would play a crucial role in accelerating the tosylation with high asymmetric induction.

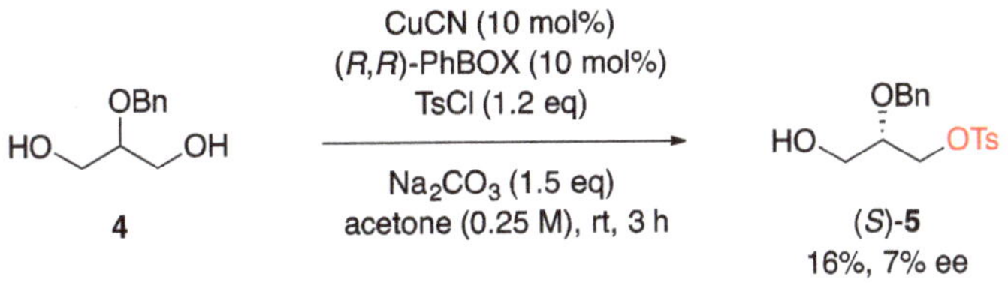

Scheme 2. Enantioselective desymmetrization of 2-*O*-benzylglycerol.

Table 2. Competition studies [1].

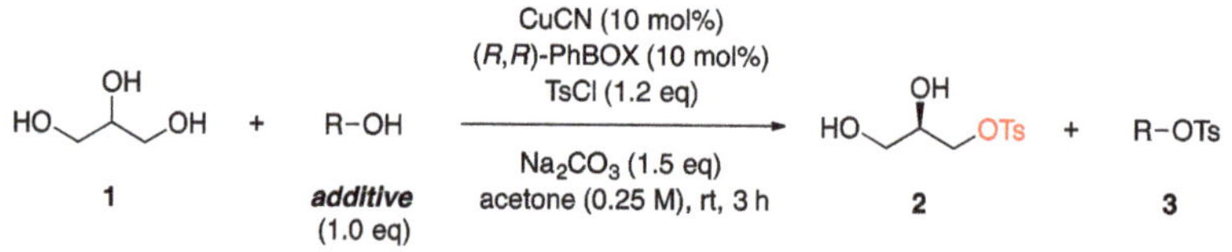

Entry	Additive	Yield of 2 (%) [2]	ee of 2 (%) [3]	Yield of 3 (%) [2]
1	none	96	94	n.d.
2	 ⌒OH	81	85	n.d.
3	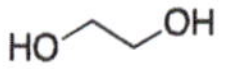 HO⌒OH	93	94	n.d.
4	 HO⌒⌒OH	91	95	n.d.

[1] Reaction conditions: **1** (1.0 mmol), additive (1.0 mmol), TsCl (1.2 mmol), CuCN (0.1 mmol), (R,R)-PhBOX (0.1 mmol), Na$_2$CO$_3$ (1.5 mmol), acetone (0.25 M), rt, 3 h. [2] Isolated yield after column chromatography. n.d. = not detected. [3] Determined by chiral HPLC analysis.

With successful asymmetric desymmetrization of glycerol achieved, we then investigated the synthetic applications of the obtained optically active glycerol. First, the site-selective protection of the remained hydroxy groups in (R)-**2** was examined (Scheme 3). The primary hydroxy moiety was selectively protected by using tert-butyldimethylsilyl chloride (TBSCl) with imidazole, affording the corresponding product (R)-**6**. Acetylation of the secondary hydroxy group with Ac$_2$O in the presence of a DMAP catalyst provided (R)-**7** in an excellent yield. Since each protective group would be removed by different deprotecting protocols, (R)-**7** would be potentially useful as a versatile chiral C3 building block.

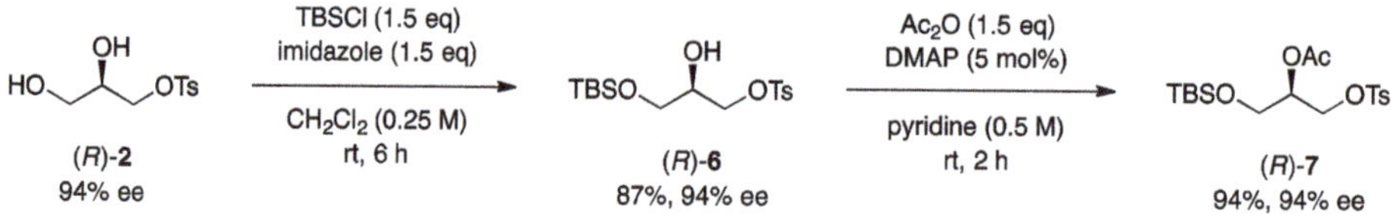

Scheme 3. Synthesis of optically active glycerol derivatives.

Next, we turned our attention to the application of the present transformation in the synthesis of an optically active synthetic ceramide. Ceramides are major components of the lamellar structure in stratum corneum lipids which protect the epidermis from excess transepidermal water loss and from the permeation of pathogens [64]. Interestingly, optically active natural ceramides showed different thermotropic behavior from racemic variants, and the lamellar liquid crystalline system containing optically active natural ceramides improved recovering effects of a water-holding ability and a barrier function of the skin [65]. Sphingolipid E (SLE) was a synthetic ceramide designed and synthesized by Kao Corporation as a structural analog of natural type 2 ceramide and was found to form a stable lamellar structure that exhibits a high water-holding ability [66]. Although SLE has been utilized as a racemic form, optically active SLE might affect the physicochemical property to form a lamellar structure and the interaction mode with water molecules. The synthesis of optically active SLE commenced with the introduction of nitrogen functionality to (R)-**2** (Scheme 4). The nucleophilic azide substitution of the tosyloxy group in (R)-**2** with sodium azide provided azide diol (S)-**8** in an excellent yield in the presence of 15-crown-5. The enantiomeric excess of (S)-**8** was determined after the tosylation of the primary hydroxy group, and no obvious racemization was observed in the azide substitution step. The boronic acid-catalyzed site-selective alkylation [67] of (S)-**8** with cetyl bromide followed by the Pd/C-catalyzed reduction of the azide group afforded the corresponding aminoalcohol (S)-**10**. N-Alkylated product (S)-**11** was successfully obtained by the reductive amination of (S)-**10** with TBS-protected glycolaldehyde using 2-picoline borane as a reductant. Finally,

(S)-11 was transformed into the optically active synthetic ceramide (S)-12 via the amidation with palmitoyl chloride followed by the removal of the TBS group.

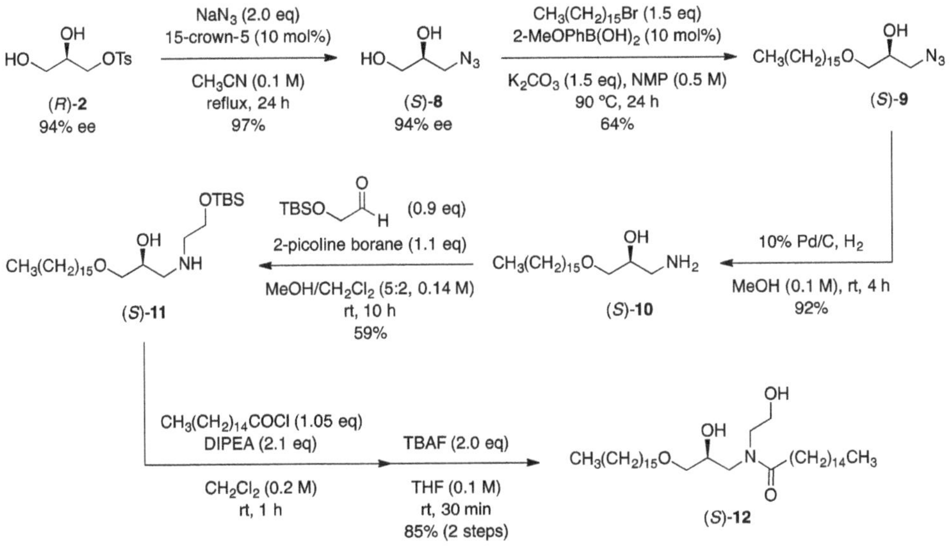

Scheme 4. Synthesis of an optically active synthetic ceramide.

In conclusion, we have developed the copper-catalyzed asymmetric sulfonylative desymmetrization of glycerol. The reaction smoothly proceeded under mild reaction conditions with a commercially available (R,R)-PhBOX ligand and an inexpensive inorganic base, providing the optically active monotosylated glycerol derivative in a high yield with high enantiomeric excess. The synthetic utility of the present transformation was demonstrated by the preparation of an enantio-enriched C3 building block with three different types of protective groups. Moreover, the synthesis of the optically active synthetic ceramide was also achieved from the monotosylated glycerol in six steps without a notable loss of enantiopurity.

3. Materials and Methods

3.1. General

Unless otherwise noted, all reactions were performed under an argon atmosphere, and all reagents and solvents were used as received without further purification. Column chromatography was performed on *Fuji silysia* Chromatorex 60B silica gel. Melting points (mp) were measured with a *Yanako* Micro Melting Point Apparatus MP-J3 and reported without correction. Infrared (IR) spectra were recorded on a *Shimadzu* IRAffinity-1 spectrometer and expressed as frequency of absorption (cm^{-1}). Optical rotations were measured with *JASCO* DIP-1000 or P-2200 spectrometers. ^1H and $^{13}C\{^1H\}$ NMR spectra (Supplementary Materials) were recorded on *JEOL* JNM-AL400, JNM-ECZ400R (400 MHz for ^1H NMR, 100 MHz for $^{13}C\{^1H\}$ NMR) or *Varian* NMR System 500PS SN (125 MHz for $^{13}C\{^1H\}$ NMR) spectrometers. Chemical shift values are expressed in parts per million (ppm) relative to internal TMS (δ 0.00 ppm for ^1H NMR) or deuterated solvent peaks (δ 77.0 ppm (CDCl$_3$) for $^{13}C\{^1H\}$ NMR). Abbreviations are as follows: s, singlet; d, doublet; t, triplet; q, quartet; m, multiplet; br, broad; app, apparent. Enantiomeric excess values were determined by chiral high-performance liquid chromatography (HPLC) analysis using *DAICEL* CHIRALPAK AD or AY-H columns. HPLC chromatograms (Supplementary Materials) were recorded on a CR8A CHROMATOPAC with an LC-20AT pump and SPD-20A UV detector (*Shimadzu*). High-resolution mass spectra (HRMS) were obtained on a *JEOL* JMS-700N (double-focusing magnetic sector mass analyzer) spectrometer with either the electron impact ionization (EI) or the fast atom bombardment (FAB) methods or a *JEOL* JMS-T100TD (TOF mass analyzer) spectrometer with either the direct analysis in real-time (DART) or the electrospray ionization (ESI) method.

3.2. Copper-Catalyzed Asymmetric Desymmetrization of Glycerol

A solution of (*R,R*)-PhBOX (33.4 mg, 0.10 mmol) and CuCN (8.96 mg, 0.10 mmol) in CH_2Cl_2 (4.0 mL) was stirred for 3 h at 40 °C. The reaction mixture was allowed to cool to room temperature and filtered into a round bottom flask using a cotton plug. After removal of the solvent under reduced pressure, the resulting solid was dried in vacuo for 30 min. To the round-bottom flask containing CuCN/(*R,R*)-PhBOX was added a solution of glycerol (92.1 mg, 1.0 mmol) in acetone (4.0 mL), and the resulting mixture was stirred for 10 min at room temperature. To the mixture was successively added Na_2CO_3 (159 mg, 1.5 mmol) and TsCl (229 mg, 1.2 mmol), and the mixture was stirred for 3 h at room temperature. The reaction was quenched with saturated aqueous NH_4Cl, and the resulting mixture was extracted with AcOEt. Combined organic layers were washed with brine, dried over Na_2SO_4, filtered, and concentrated under reduced pressure. The crude product was purified by silica gel column chromatography (hexane/AcOEt = 1/2) to afford (*R*)-**2** (236 mg, 0.96 mmol, 96% yield).

(*R*)-2,3-Dihydroxypropyl 4-methylbenzenesulfonate ((*R*)-**2**). White solid; mp = 56–58 °C; $[\alpha]_D^{23}$ −8.3 (*c* 1.00, MeOH) for 94% ee; ^{1}H NMR (400 MHz, $CDCl_3$): δ 7.81 (d, *J* = 8.5 Hz, 2H), 7.37 (d, *J* = 8.5 Hz, 2H), 4.13–4.06 (m, 2H), 3.99–3.93 (m, 1H), 3.74–3.69 (m, 1H), 3.66–3.60 (m, 1H), 2.51 (d, *J* = 5.5 Hz, 1H), 2.46 (s, 3H), 1.94 (t, *J* = 5.9 Hz, 1H); ^{13}C{^{1}H} NMR (100 MHz, $CDCl_3$): δ 145.2, 132.2, 130.0, 127.9, 70.7, 69.6, 62.7, 21.6; IR (ATR): 3368, 2926, 1350, 1171, 968, 812 cm^{-1}; HRMS (EI) *m/z*: $[M]^+$ calcd for $C_{10}H_{14}O_5S$ 246.0562, found 254.0562; HPLC analysis: Chiralpak AY-H, hexane/EtOH = 2/1, flow rate 1.0 mL/min, wavelength 254 nm, t_R 13.5 min (minor) and 15.6 min (major). The absolute configuration of **2** was established by comparing the sign of the specific rotation of **2** with the literature value ($[\alpha]_D^{22}$ −9.3 (*c* 4.99, MeOH) for (*R*)-**2**) [68].

3.3. Large-Scale Experiment

A solution of (*R,R*)-PhBOX (200 mg, 0.60 mmol) and CuCN (53.7 mg, 0.60 mmol) in CH_2Cl_2 (24 mL) was stirred for 3 h at 40 °C. The reaction mixture was allowed to cool to room temperature and filtered into a round bottom flask using a cotton plug. After removal of the solvent under reduced pressure, the resulting solid was dried in vacuo for 1 h. To the round-bottom flask containing CuCN/(*R,R*)-PhBOX was added a solution of glycerol (553 mg, 6.0 mmol) in acetone (24 mL), and the resulting mixture was stirred for 10 min at room temperature. To the mixture was successively added Na_2CO_3 (950 mg, 9.0 mmol) and TsCl (1.37 g, 7.2 mmol), and the mixture was stirred for 9 h at room temperature. The reaction was quenched with saturated aqueous NH_4Cl, and the resulting mixture was extracted with AcOEt. Combined organic layers were washed with brine, dried over Na_2SO_4, filtered, and concentrated under reduced pressure. The crude product was purified by silica gel column chromatography (hexane/AcOEt = 1/2) to afford (*R*)-**2** (1.30 g, 5.28 mmol, 88% yield, 93% ee).

3.4. Copper-Catalyzed Asymmetric Desymmetrization of 2-O-Benzylglycerol

The reaction was performed using 2-*O*-benzylglycerol (**4**, 182 mg, 1.0 mmol) [69] according to the procedure described in Section 3.2. Silica gel column chromatography (hexane/acetone = 3/1) to afford (*S*)-**5** (52.6 mg, 0.16 mmol, 16% yield).

(*S*)-2-(Benzyloxy)-3-hydroxypropyl 4-methylbenzenesulfonate ((*S*)-**5**). colorless oil; $[\alpha]_D^{21}$ −2.1 (*c* 1.00, $CHCl_3$) for 7% ee; ^{1}H NMR (400 MHz, $CDCl_3$): δ 7.78 (d, *J* = 8.2 Hz, 2H), 7.36–7.27 (m, 7H), 4.63 (d, *J* = 11.7 Hz, 1H), 4.55 (d, *J* = 11.7 Hz, 1H), 4.17–4.10 (m, 2H), 3.75–3.67 (m, 2H), 3.62–3.56 (m, 1H), 2.45 (s, 3H), 1.80 (t, *J* = 6.3 Hz, 1H); ^{13}C{^{1}H} NMR (125 MHz, $CDCl_3$): δ 145.0, 137.5, 132.6, 129.9, 128.5, 128.1, 128.0, 127.9, 76.7, 72.5,

68.7, 61.4, 21.7; IR (ATR): 3445, 3032, 2879, 1597, 1354, 1173, 974 cm^{-1}; HRMS (ESI) m/z: [M + Na]$^+$ calcd for $C_{17}H_{20}NaO_5S$ 359.0929, found 359.0922; HPLC analysis: Chiralpak AD, hexane/EtOH = 3/1, flow rate 1.0 mL/min, wavelength 254 nm, t_R 7.1 min (minor) and 8.8 min (major). The absolute configuration of **5** was established by comparing the sign of the specific rotation of **5** with the literature value ([α]$_D^{22}$ +29.5 (*c* 1.01, CHCl$_3$) for (*R*)-**5**) [70].

3.5. Synthesis of Optically Active Glycerol Derivatives

(R)-3-((tert-Butyldimethylsilyl)oxy)-2-hydroxypropyl 4-methylbenzenesulfonate ((R)-6). To a solution of (*R*)-**2** (134 mg, 0.54 mmol) in CH$_2$Cl$_2$ (2.2 mL) was successively added imidazole (55.8 mg, 0.82 mmol) and TBSCl (123 mg, 0.82 mmol) at room temperature. After stirring for 6 h at the same temperature, the reaction was quenched with H$_2$O. The resulting mixture was extracted with AcOEt. The combined organic layers were dried over Na$_2$SO$_4$, filtered, and concentrated under reduced pressure. The residue was purified by silica gel column chromatography (hexane/AcOEt = 2/1) to afford (*R*)-**6** (103 mg, 0.287 mmol, 87% yield) as a colorless oil; [α]$_D^{27}$ −12.4 (*c* 1.01, AcOEt) for 94% ee; ^{1}H NMR (400 MHz, CDCl$_3$): δ 7.82–7.79 (m, 2H), 7.37–7.34 (m, 2H), 4.07 (dd, *J* = 9.9, 5.6 Hz, 1H), 4.01 (dd, *J* = 9.9, 5.6 Hz, 1H), 3.87–3.83 (m, 1H), 3.66–3.59 (m, 2H), 2.45 (s, 3H), 2.40 (d, *J* = 6.2 Hz, 1H), 0.85 (s, 9H), 0.039 (s, 3H), 0.035 (s, 3H); ^{13}C{^{1}H} NMR (100 MHz, CDCl$_3$): δ 145.0, 132.6, 129.9, 128.0, 70.0, 69.2, 62.8, 25.7, 21.6, 18.1, −5.6; IR (ATR): 3537, 2930, 2857, 1360, 1252, 1175, 980, 833 cm^{-1}; HRMS (EI) m/z: [M]$^+$ calcd for $C_{16}H_{28}O_5SSi$ 360.1427, found 360.1429. HPLC analysis: Chiralpak AY-H, hexane/*i*-PrOH = 10/1, flow rate 1.0 mL/min, wavelength 254 nm, t_R 22.4 min (minor) and 25.3 min (major).

(R)-1-((tert-Butyldimethylsilyl)oxy)-3-(tosyloxy)propan-2-yl acetate ((R)-7). To a solution of (*R*)-**6** (144 mg, 0.40 mmol) in pyridine (0.80 mL) was successively added DMAP (2.44 mg, 0.020 mmol) and Ac$_2$O (49.0 mg, 0.48 mmol) at room temperature. After stirring for 2 h at the same temperature, the reaction mixture was diluted with toluene (5.0 mL) and concentrated under reduced pressure. The residue was purified by silica gel column chromatography (hexane/AcOEt = 5/1) to afford (*R*)-**7** (151 mg, 0.378 mmol, 94% yield) as a colorless oil. [α]$_D^{27}$ −2.2 (*c* 1.24, AcOEt) for 94% ee; ^{1}H NMR (400 MHz, CDCl$_3$): δ 7.79 (d, *J* = 8.2 Hz, 2H), 7.35 (d, *J* = 8.0 Hz, 2H), 4.96–4.91 (m, 1H), 4.22 (dd, *J* = 10.7, 3.7 Hz, 1H), 4.17 (dd, *J* = 10.9, 5.4 Hz, 1H), 3.70–3.63 (m, 2H), 2.45 (s, 3H), 2.00 (s, 3H), 0.83 (s, 9H), 0.01 (s, 6H); ^{13}C{^{1}H} NMR (100 MHz, CDCl$_3$): δ 170.0, 144.9, 132.7, 129.8, 128.0, 71.3, 67.8, 60.5, 25.6, 21.6, 20.8, 18.1, −5.58, −5.62; IR (ATR): 1744, 1362, 1233, 1175, 988, 833 cm^{-1}; HRMS (EI) m/z: [M]$^+$ calcd for $C_{18}H_{30}O_6SSi$ 402.1532, found 402.1533; HPLC analysis: Chiralpak AD, hexane/EtOH = 20/1, flow rate 1.0 mL/min, wavelength 254 nm, t_R 4.8 min (major) and 6.5 min (minor).

3.6. Synthesis of an Optically Active Synthetic Ceramide

(S)-3-Azidopropane-1,2-diol ((S)-8). To a solution of (*R*)-**2** (246 mg, 1.0 mmol) in CH$_3$CN (10 mL) was successively added 15-crown-5 (22.1 mg, 0.10 mmol) and NaN$_3$ (130 mg, 2.0 mmol) at room temperature. After refluxing for 24 h, the reaction mixture was filtered using celite, and the filtrate was concentrated under reduced pressure. The residue was purified by silica gel column chromatography (CH$_2$Cl$_2$/MeOH = 12/1) to afford (*S*)-**8** (114 mg, 0.974 mmol, 97% yield) as a colorless oil; [α]$_D^{23}$ −16.2 (*c* 1.10, MeOH); ^{1}H NMR (400 MHz, CDCl$_3$): δ 3.92–3.86 (m, 1H), 3.74–3.72 (m, 1H), 3.65–3.61 (m, 1H), 3.47–3.38 (m, 2H), 2.49 (d, *J* = 4.1 Hz, 1H), 1.95 (br s, 1H); ^{13}C{^{1}H} NMR (100 MHz, CDCl$_3$): δ 70.9, 63.9, 53.4; IR (ATR): 3333, 2924, 2855, 2093, 1443, 1272, 1103, 1038, 928 cm^{-1}; HRMS (EI) m/z: [M]$^+$ calcd for $C_3H_7N_3O_2$ 117.0538, found 117.0546. The absolute configuration of **8** was established by comparing the sign of the specific rotation of **8** with the literature value ([α]$_D^{20}$ −17.4 (*c* 1.00, MeOH) for (*S*)-**8**) [71].

(S)-1-Azido-3-(hexadecyloxy)propan-2-ol ((S)-9). To a solution of (*S*)-**8** (46.8 mg, 0.40 mmol) in *N*-methylpyrrolidone (0.80 mL) was successively added 2-methoxyphenylboronic acid (6.0 mg, 0.040 mmol) and K$_2$CO$_3$ (82.9 mg, 0.60 mmol). The resulting mixture was stirred

for 30 min at room temperature, and then cetyl bromide (183 mg, 0.60 mmol) was added. After stirring for 24 h at 95 °C, the reaction mixture was diluted with H_2O and extracted with AcOEt. The combined organic layers were washed with brine, dried over Na_2SO_4, filtered, and concentrated under reduced pressure. The residue was purified by silica gel column chromatography (hexane/Et_2O = 8/2) to afford (S)-**9** (87.3 mg, 0.256 mmol, 64% yield) as a white solid; mp = 37–39 °C; $[\alpha]_D^{24}$ −11.8 (c 1.00, MeOH); ^{1}H NMR (400 MHz, CDCl$_3$): δ 3.97–3.91 (m, 1H), 3.51–3.33 (m, 6H), 2.42 (d, J = 5.0 Hz, 1H), 1.61–1.54 (m, 2H), 1.32–1.26 (m, 26H), 0.88 (t, J = 6.9 Hz, 3H); ^{13}C{^{1}H} NMR (100 MHz, CDCl$_3$): δ 71.8, 71.7, 69.6, 53.5, 31.9, 29.68, 29.66, 29.64, 29.59, 29.57, 29.5, 29.4, 29.3, 26.1, 22.7, 14.1; IR (ATR): 3429, 2914, 2876, 2846, 2088, 1466, 1337, 1290, 1113, 989 cm^{-1}; HRMS (DART) m/z: [M + H]$^+$ calcd for $C_{19}H_{40}N_3O_2$ 342.3121, found 342.3171.

(S)-1-Amino-3-(hexadecyloxy)propan-2-ol ((S)-**10**). To a reaction vessel charged with 10% Pd/C (24.9 mg, 10% w/w) was added a solution of (S)-**9** (249 mg, 0.73 mmol) in MeOH (7.3 mL) at room temperature under argon atmosphere. The reaction vessel was charged with H_2 gas, and then the mixture was stirred for 4 h at room temperature. The reaction mixture was filtered using celite, and then the filtrate was concentrated under reduced pressure. The residue was dissolved in 10% aqueous HCl and washed with AcOEt. The aqueous layer was basified with saturated aqueous NaHCO$_3$ and then extracted with CHCl$_3$. The combined organic layers were dried over Na$_2$SO$_4$, filtered, and concentrated under reduced pressure to afford (S)-**10** (211 mg, 0.67 mmol, 92% yield) as a white solid; mp = 60–61 °C; $[\alpha]_D^{24}$ −3.2 (c 0.50, CHCl$_3$); ^{1}H NMR (400 MHz, CDCl$_3$): δ 3.76–3.70 (m, 1H), 3.50–3.43 (m, 3H), 3.38 (dd, J = 9.5, 6.5 Hz, 1H), 2.83 (dd, J = 12.7, 3.1 Hz, 1H), 2.72 (dd, J = 12.5, 6.7 Hz, 1H), 1.61–1.54 (m, 2H), 1.32–1.26 (m, 26H), 0.88 (t, J = 6.7 Hz, 3H); ^{13}C{^{1}H} NMR (100 MHz, CDCl$_3$): δ 73.0, 71.7, 71.1, 44.4, 31.9, 29.7, 29.64, 29.59, 29.58, 29.5, 29.3, 26.1, 22.7, 14.1; IR (ATR): 2912, 2827, 1470, 1130, 1032, 924 cm^{-1}; HRMS (FAB) m/z: [M + H]$^+$ calcd for $C_{19}H_{42}NO_2$ 316.3216, found 316.3200.

(S)-2,2,3,3-Tetramethyl-4,11-dioxa-7-aza-3-silaheptacosan-9-ol ((S)-**11**). To a solution of 2-(tert-butyldimethylsilyloxy)acetaldehyde (34.9 mg, 0.20 mmol) [72] in MeOH/CH$_2$Cl$_2$ (5:2, 1.4 mL) was added (S)-**10** (69.4 mg, 0.22 mmol) at room temperature. After stirring for 10 min at the same temperature, 2-picoline borane (256 mg, 0.24 mmol) was added, and then the reaction mixture was stirred for an additional 10 h. The reaction mixture was diluted with H$_2$O and extracted with CH$_2$Cl$_2$. The combined organic layers were dried over Na$_2$SO$_4$, filtered, and concentrated under reduced pressure. The residue was purified by silica gel column chromatography (CH$_2$Cl$_2$/MeOH = 12/1) to afford (S)-**11** (55.8 mg, 0.117 mmol, 59% yield) as a colorless amorphous; $[\alpha]_D^{25}$ −3.6 (c 1.00, CHCl$_3$); ^{1}H NMR (400 MHz, CDCl$_3$): δ 3.86–3.81 (m, 1H), 3.75–3.67 (m, 2H), 3.49–3.39 (m, 4H), 2.78–2.64 (m, 4H), 1.60–1.54 (m, 2H), 1.33–1.25 (m, 26H), 0.90–0.86 (m, 12H), 0.06 (s, 6H); ^{13}C{^{1}H} NMR (100 MHz, CDCl$_3$): δ 73.3, 71.7, 68.7, 62.2, 51.7, 51.5, 31.9, 29.7, 29.62, 29.58, 29.5, 29.3, 26.1, 25.9, 22.7, 18.3, 14.1, −5.4; IR (ATR): 2914, 2849, 1472, 1464, 1256, 1119, 1080, 968, 937, 831 cm^{-1}; HRMS (FAB) m/z: [M + H]$^+$ calcd for $C_{27}H_{60}NO_3Si$ 473.4342, found 473.4300.

(S)-N-(3-(Hexadecyloxy)-2-hydroxypropyl)-N-(2-hydroxyethyl)palmitamide ((S)-**12**). To a solution of (S)-**11** (135 mg, 0.28 mmol) and i-Pr$_2$NEt (77.0 mg, 0.60 mmol) in CH$_2$Cl$_2$ (1.4 mL) was added palmitoyl chloride (81.9 mg, 0.30 mmol) at room temperature. After stirring for 1 h at the same temperature, all volatile was removed under reduced pressure. The residue was dissolved in THF (2.8 mL), and then a 1.0 M solution of TBAF in THF (0.57 mL) was added at room temperature. After stirring for 30 min at the same temperature, the reaction mixture was diluted with H$_2$O and extracted with CHCl$_3$. The combined organic layers were dried over Na$_2$SO$_4$, filtered, and concentrated under reduced pressure. The residue was purified by silica gel column chromatography (hexane/AcOEt = 1/1) to afford (S)-**12** (144 mg, 0.241 mmol, 85% yield) as a white solid; mp = 67–68 °C; $[\alpha]_D^{25}$ −4.4 (c 1.00, CHCl$_3$); ^{1}H NMR (400 MHz, CDCl$_3$, mixture of rotamers): δ 4.16–3.93 (m, 1H), 3.85–3.74 (m, 2H), 3.67–3.25 (m, 8H), 2.46–2.30 (m, 2H), 1.67–1.52 (m, 4H), 1.28 (m, 50H), 0.88 (app t, J = 6.9 Hz, 6H). ^{13}C{^{1}H} NMR (100 MHz, CDCl$_3$): δ 175.8, 72.4, 72.1, 71.8, 71.6, 69.8, 69.4, 61.7, 60.6, 53.3, 52.5, 51.4, 51.1, 33.6, 33.5, 31.9, 29.7, 29.62, 29.60, 29.59, 29.56, 29.53, 29.47,

29.44, 29.42, 29.3, 26.1, 26.0, 25.29, 25.26, 22.7, 14.1; IR (ATR): 3320, 2916, 2849, 1611, 1464, 1437, 1375, 1306, 1290, 1261, 1206, 1165, 1109, 1094, 1059, 1040, 955, 845, 814 cm^{-1}; HRMS (FAB) m/z: [M + H]$^+$ calcd for $C_{37}H_{76}NO_4$ 598.5773, found 598.5800.

3.7. Tosylation of Azide Diol (S)-8

(*S*)-3-*Azido-2-hydroxypropyl 4-methylbenzenesulfonate* ((*S*)-**8'**). To a solution of (*S*)-**8** (23.4 mg, 0.20 mmol) in CH$_3$CN (0.80 mL) was added pyridine (23.7 mg, 0.30 mmol) and TsCl (57.2 mg, 0.30 mmol) at room temperature. After stirring for 10 h at the same temperature, the reaction was quenched with H$_2$O, and the resulting mixture was extracted with AcOEt. The combined organic layers were washed with brine, dried over Na$_2$SO$_4$, filtered, and concentrated under reduced pressure. The residue was purified by silica gel column chromatography (hexane/AcOEt = 2/1) to afford (*S*)-**8'** (30.3 mg, 0.112 mmol, 56% yield) as a colorless oil; [α]$_D^{27}$ −17.5 (*c* 1.00, AcOEt) for 94% ee; ^{1}H NMR (400 MHz, CDCl$_3$): δ 7.81 (d, *J* = 8.3 Hz, 2H), 7.38 (d, *J* = 8.3 Hz, 2H), 4.09–3.98 (m, 3H), 3.43 (dd, *J* = 12.9, 4.6 Hz, 1H), 3.38 (dd, *J* = 12.7, 5.4 Hz, 1H), 2.47 (s, 3H), 2.40 (d, *J* = 5.4 Hz, 1H); ^{13}C{^{1}H} NMR (100 MHz, CDCl$_3$): δ 145.4, 132.3, 130.0, 128.0, 70.5, 68.5, 52.7; IR (ATR): 3462, 2100, 1597, 1352, 1173, 1096, 982 cm^{-1}; HRMS (EI) m/z: [M]$^+$ calcd for $C_{10}H_{13}N_3O_4S$ 271.0627, found 271.0622; HPLC analysis: Chiralpak AY-H, hexane/EtOH = 6/1, flow rate 1.0 mL/min, wavelength 254 nm, t_R 28.9 min (minor) and 31.3 min (major).

Supplementary Materials: The following supporting information can be downloaded at: https://www.mdpi.com/article/10.3390/molecules27249025/s1, Figures S1–S10: Copies of ^{1}H and ^{13}C{^{1}H} NMR spectra of compounds **2–12**; Figures S11–S15: Chiral HPLC chromatogram of compounds **2**, **5**, **6**, **7**, and **8'**.

Author Contributions: Conceptualization, O.O.; methodology, O.O.; investigation, K.Y., K.M., M.U. and Y.T.; writing—original draft preparation, K.Y.; writing—review and editing, O.O., M.K. and K.Y.; visualization, K.Y., K.M., M.U. and Y.T.; supervision, O.O.; project administration, K.Y. and M.K.; funding acquisition, O.O., M.K. and K.Y. All authors have read and agreed to the published version of the manuscript.

Funding: This research was funded in part by JSPS Grant-in-Aid for Scientific Research (22K06528, 22K15255, and 19K05459) and a grant from the Japan Soap and Detergent Association.

Institutional Review Board Statement: Not applicable.

Informed Consent Statement: Not applicable.

Data Availability Statement: The data presented in this study are available in Supplementary Material.

Acknowledgments: The spectral data were collected with the research equipment shared in the MEXT Project for promoting public utilization of advanced research infrastructure (Program for Supporting Introduction of the New Sharing System JPMXS0422500320).

Conflicts of Interest: The authors declare no competing financial interests.

Sample Availability: Samples of the compounds are not available from the authors.

References

1. Karmakar, B.; Halder, G. Progress and future of biodiesel synthesis: Advancements in oil extraction and conversion technologies. *Energy Convers. Manag.* **2019**, *182*, 307–339. [CrossRef]
2. Pasha, M.K.; Dai, L.; Liu, D.; Guo, M.; Du, W. An overview to process design, simulation and sustainability evaluation of biodiesel production. *Biotechnol. Biofuels* **2021**, *14*, 129. [CrossRef] [PubMed]
3. Quispe, C.A.G.; Coronado, C.J.R.; Carvalho, J.A. Glycerol: Production, consumption, prices, characterization and new trends in combustion. *Renew. Sustain. Energy Rev.* **2013**, *27*, 475–493. [CrossRef]
4. Pagliaro, M.; Ciriminna, R.; Kimura, H.; Rossi, M.; della Pina, C. From Glycerol to Value-Added Products. *Angew. Chem. Int. Ed.* **2007**, *46*, 4434–4440. [CrossRef]
5. Zhou, C.-H.; Beltramini, J.N.; Fan, Y.-X.; Lu, G.Q. Chemoselective catalytic conversion of glycerol as a biorenewable source to valuable commodity chemicals. *Chem. Soc. Rev.* **2008**, *37*, 527–549. [CrossRef] [PubMed]
6. Bozell, J.J.; Petersen, G.R. Technology development for the production of biobased products from biorefinery carbohydrates—The US Department of Energy's "Top 10" revisited. *Green Chem.* **2010**, *12*, 539–554. [CrossRef]

7. Checa, M.; Nogales-Delgado, S.; Montes, V.; Encinar, J.M. Recent Advances in Glycerol Catalytic Valorization: A Review. *Catalysts* **2020**, *10*, 1279. [CrossRef]
8. Furuta, T.; Sakai, M.; Hayashi, H.; Asakawa, T.; Kataoka, F.; Fujii, S.; Suzuki, T.; Suzuki, Y.; Tanaka, K.; Fishkin, N.; et al. Design and synthesis of artificial phospholipid for selective cleavage of integral membrane protein. *Chem. Commun.* **2005**, 4575–4577. [CrossRef] [PubMed]
9. Andresen, T.L.; Jensen, S.S.; Madsen, R.; Jørgensen, K. Synthesis and Biological Activity of Anticancer Ether Lipids That Are Specifically Released by Phospholipase A_2 in Tumor Tissue. *J. Med. Chem.* **2005**, *48*, 7305–7314. [CrossRef]
10. Zhao, Y.; Zhu, L.; Provencal, D.P.; Miller, T.A.; O'Bryan, C.; Langston, M.; Shen, M.; Bailey, D.; Sha, D.; Palmer, T.; et al. Process Research and Kilogram Synthesis of an Investigational, Potent MEK Inhibitor. *Org. Process Res. Dev.* **2012**, *16*, 1652–1659. [CrossRef]
11. Tangherlini, G.; Torregrossa, T.; Agoglitta, O.; Köhler, J.; Melesina, J.; Sippl, W.; Holl, R. Synthesis and biological evaluation of enantiomerically pure glyceric acid derivatives as LpxC inhibitors. *Bioorg. Med. Chem.* **2016**, *24*, 1032–1044. [CrossRef] [PubMed]
12. Li, Y.; Pasunooti, K.K.; Peng, H.; Li, R.-J.; Shi, W.Q.; Liu, W.; Cheng, Z.; Head, S.A.; Liu, J.O. Design and Synthesis of Tetrazole- and Pyridine-Containing Itraconazole Analogs as Potent Angiogenesis Inhibitors. *ACS Med. Chem. Lett.* **2020**, *11*, 1111–1117. [CrossRef] [PubMed]
13. Tse, B. Total Synthesis of (−)-Galbonolide B and the Determination of Its Absolute Stereochemistry. *J. Am. Chem. Soc.* **1996**, *118*, 7094–7100. [CrossRef]
14. Mukaiyama, T.; Shiina, I.; Iwadare, H.; Saitoh, M.; Nishimura, T.; Ohkawa, N.; Sakoh, H.; Nishimura, K.; Tani, Y.; Hasegawa, M.; et al. Asymmetric Total Synthesis of Taxol®. *Chem. Eur. J.* **1999**, *5*, 121–161. [CrossRef]
15. Byun, H.-S.; Sadlofsky, J.A.; Bittman, R. Enantioselective Synthesis of 3-Deoxy-(R)-sphingomyelin from (S)-1-(4'-Methoxyphenyl) glycerol. *J. Org. Chem.* **1998**, *63*, 2560–2563. [CrossRef]
16. Reymond, S.; Cossy, J. Synthesis of migrastatin and its macrolide core. *Tetrahedron* **2007**, *63*, 5918–5929. [CrossRef]
17. Yoshida, M.; Saito, K.; Kato, H.; Tsukamoto, S.; Doi, T. Total Synthesis and Biological Evaluation of Siladenoserinol A and its Analogues. *Angew. Chem. Int. Ed.* **2018**, *57*, 5147–5150. [CrossRef] [PubMed]
18. Sigurjónsson, S.; Lúthersson, E.; Gudmundsson, H.G.; Haraldsdóttir, H.; Kristinsdóttir, L.; Haraldsson, G.G. Asymmetric Synthesis of Methoxylated Ether Lipids: A Glyceryl Glycidyl Ether Key Building Block Design, Preparation, and Synthetic Application. *J. Org. Chem.* **2022**, *87*, 12306–12314. [CrossRef]
19. Lok, C.M.; Ward, J.P.; van Dorp, D.A. The synthesis of chiral glycerides starting from D- and L-serine. *Chem. Phys. Lipids* **1976**, *16*, 115–122.
20. De Wilde, H.; De Clercq, P.; Vandewalle, M.; Röper, H. L-(S)-erythrulose a novel precursor for L-2,3-O-isopropylidene-C_3 chirons. *Tetrahedron Lett.* **1987**, *28*, 4757–4758.
21. Mikkilineni, A.B.; Kumar, P.; Abushanab, E. The Chemistry of L-Ascorbic and D-Isoascorbic Acids. 2. *R* and *S* Glyceraldehydes from a Common Intermediate. *J. Org. Chem.* **1988**, *53*, 6005–6009.
22. Schmid, C.R.; Bryant, J.D.; Dowlatzedah, M.; Phillips, J.L.; Prather, D.E.; Schantz, R.D.; Sear, N.L.; Vianco, C.S. Synthesis of 2,3-O-Isopropylidene- D-Glyceraldehyde in High Chemical and Optical Purity: Observations on the Development of a Practical Bulk Process. *J. Org. Chem.* **1991**, *56*, 4056–4058.
23. Doboszewski, B.; Herdewijn, P. Simple approach to 1-O-protected (R)- and (S)-glycerols from L- and D-arabinose for glycerol nucleic acids (GNA) monomers research. *Tetrahedron Lett.* **2011**, *52*, 3853–3855. [CrossRef]
24. García-Urdiales, E.; Alfonso, I.; Gotor, V. Update 1 of: Enantioselective Enzymatic Desymmetrizations in Organic Synthesis. *Chem. Rev.* **2011**, *111*, PR110–PR180. [CrossRef] [PubMed]
25. Chenault, H.K.; Chafin, L.F.; Liehr, S. Kinetic Chiral Resolutions of 1,2-Diols and Desymmetrization of Glycerol Catalyzed by Glycerol Kinase. *J. Org. Chem.* **1998**, *63*, 4039–4045. [CrossRef]
26. Batovska, D.I.; Tsubota, S.; Kato, Y.; Asano, Y.; Ubukata, M. Lipase-mediated desymmetrization of glycerol with aromatic and aliphatic anhydrides. *Tetrahedron Asymmetry* **2004**, *15*, 3551–3559. [CrossRef]
27. Caytan, E.; Cherghaoui, Y.; Barril, C.; Jouitteau, C.; Rabiller, C.; Remaud, G.S. Strategy for specific isotope ratio determination by quantitative NMR on symmetrical molecules: Application to glycerol. *Tetrahedron Asymmetry* **2006**, *17*, 1622–1624. [CrossRef]
28. Franke, D.; Machajewski, T.; Hsu, C.-C.; Wong, C.-H. One-Pot Synthesis of L-Fructose Using Coupled Multienzyme Systems Based on Rhamnulose-1-phosphate Aldolase. *J. Org. Chem.* **2003**, *68*, 6828–6831. [CrossRef]
29. Klibanov, A.M.; Alberti, B.N.; Marletta, M.A. Stereospecific Oxidation of Aliphatic Alcohols Catalyzed by Galactose Oxidase. *Biochem. Biophys. Res. Commun.* **1982**, *108*, 804–808. [CrossRef]
30. Enríquez-García, Á.; Kündig, E.P. Desymmetrisation of *meso*-diols mediated by non-enzymatic acyl transfer catalysts. *Chem. Soc. Rev.* **2012**, *41*, 7803–7831. [CrossRef]
31. Nájera, C.; Foubelo, F.; Sansano, J.M.; Yus, M. Enantioselective desymmetrization reactions in asymmetric catalysis. *Tetrahedron* **2022**, *106–107*, 132629. [CrossRef]
32. Mizuta, S.; Sadamori, M.; Fujimoto, T.; Yamamoto, I. Asymmetric Desymmetrization of *meso*-1,2-Diols by Phosphinite Derivatives of Cinchona Alkaloids. *Angew. Chem. Int. Ed.* **2003**, *42*, 3383–3385. [CrossRef] [PubMed]
33. Zhao, Y.; Rodrigo, J.; Hoveyda, A.H.; Snapper, M.L. Enantioselective silyl protection of alcohols catalysed by an amino-acid-based small molecule. *Nature* **2006**, *443*, 67–70. [CrossRef] [PubMed]

34. Zhao, Y.; Mitra, A.W.; Hoveyda, A.H.; Snapper, M.L. Kinetic Resolution of 1,2-Diols through Highly Site- and Enantioselective Catalytic Silylation. *Angew. Chem. Int. Ed.* **2007**, *46*, 8471–8474. [CrossRef] [PubMed]
35. Demizu, Y.; Matsumoto, K.; Onomura, O.; Matsumura, Y. Copper complex catalyzed asymmetric monosulfonylation of *meso*-vic-diols. *Tetrahedron Lett.* **2007**, *48*, 7605–7609. [CrossRef]
36. Sun, X.; Worthy, A.D.; Tan, K.L. Scaffolding Catalysts: Highly Enantioselective Desymmetrization Reactions. *Angew. Chem. Int. Ed.* **2011**, *50*, 8167–8171. [CrossRef]
37. Hamaguchi, N.; Kuriyama, M.; Onomura, O. Chiral copper-catalyzed asymmetric monoarylation of vicinal diols with diaryliodonium salts. *Tetrahedron Asymmetry* **2016**, *27*, 177–181. [CrossRef]
38. Li, R.-Z.; Tang, H.; Yang, K.R.; Wan, L.-Q.; Zhang, X.; Liu, J.; Fu, Z.; Niu, D. Enantioselective Propargylation of Polyols and Desymmetrization of *meso* 1,2-Diols by Copper/Borinic Acid Dual Catalysis. *Angew. Chem. Int. Ed.* **2017**, *56*, 7213–7217. [CrossRef]
39. Hashimoto, Y.; Michimuko, C.; Yamaguchi, K.; Nakajima, M.; Sugiura, M. Selective Monoacylation of Diols and Asymmetric Desymmetrization of Dialkyl *meso*-Tartrates Using 2-Pyridyl Esters as Acylating Agents and Metal Carboxylates as Catalysts. *J. Org. Chem.* **2019**, *84*, 9313–9321. [CrossRef]
40. Trost, B.M.; Mino, T. Desymmetrization of Meso 1,3- and 1,4-Diols with a Dinuclear Zinc Asymmetric Catalyst. *J. Am. Chem. Soc.* **2003**, *125*, 2410–2411. [CrossRef]
41. Honjo, T.; Nakao, M.; Sano, S.; Shiro, M.; Yamaguchi, K.; Sei, Y.; Nagao, Y. Nonenzymatic Enantioselective Monoacetylation of Prochiral 2-Protectedamino-2-Alkyl-1,3-Propanediols Utilizing a Chiral Sulfonamide-Zn Complex Catalyst. *Org. Lett.* **2007**, *9*, 509–512. [CrossRef]
42. Lee, J.Y.; You, Y.S.; Kang, S.H. Asymmetric Synthesis of All-Carbon Quaternary Stereocenters via Desymmetrization of 2,2-Disubstituted 1,3-Propanediols. *J. Am. Chem. Soc.* **2011**, *133*, 1772–1774. [CrossRef] [PubMed]
43. Ke, Z.; Tan, C.K.; Chen, F.; Yeung, Y.-Y. Catalytic Asymmetric Bromoetherification and Desymmetrization of Olefinic 1,3-Diols with C_2-Symmetric Sulfides. *J. Am. Chem. Soc.* **2014**, *136*, 5627–5630. [CrossRef] [PubMed]
44. Zi, W.; Toste, F.D. Gold(I)-Catalyzed Enantioselective Desymmetrization of 1,3-Diols through Intramolecular Hydroalkoxylation of Allenes. *Angew. Chem. Int. Ed.* **2015**, *54*, 14447–14451. [CrossRef] [PubMed]
45. Wu, Z.; Wang, J. Enantioselective Medium-Ring Lactone Synthesis through an NHC-Catalyzed Intramolecular Desymmetrization of Prochiral 1,3-Diols. *ACS Catal.* **2017**, *7*, 7647–7652. [CrossRef]
46. Yamamoto, K.; Tsuda, Y.; Kuriyama, M.; Demizu, Y.; Onomura, O. Copper-Catalyzed Enantioselective Synthesis of Oxazolines from Aminotriols via Asymmetric Desymmetrization. *Chem. Asian. J.* **2020**, *15*, 840–844. [CrossRef]
47. Mandai, H.; Hironaka, T.; Mitsudo, K.; Suga, S. Acylative Desymmetrization of Cyclic *meso*-1,3-Diols by Chiral DMAP Derivatives. *Chem. Lett.* **2021**, *50*, 471–474. [CrossRef]
48. Estrada, C.D.; Ang, H.T.; Vetter, K.-M.; Ponich, A.A.; Hall, D.G. Enantioselective Desymmetrization of 2-Aryl-1,3-Propanediols by Direct O-Alkylation with a Rationally Designed Chiral Hemiboronic Acid Catalyst That Mitigates Substrate Conformational Poisoning. *J. Am. Chem. Soc.* **2021**, *143*, 4162–4167. [CrossRef]
49. Jung, B.; Hong, M.S.; Kang, S.H. Enantioselective Synthesis of Tertiary Alcohols by the Desymmetrizing Benzoylation of 2-Substituted Glycerols. *Angew. Chem. Int. Ed.* **2007**, *46*, 2616–2618. [CrossRef]
50. Jung, B.; Kang, S.H. Chiral imine copper chloride-catalyzed enantioselective desymmetrization of 2-substituted 1,2,3-propanetriols. *Proc. Natl. Acad. Sci. USA* **2007**, *104*, 1471–1475. [CrossRef]
51. You, Z.; Hoveyda, A.H.; Snapper, M.L. Catalytic Enantioselective Silylation of Acyclic and Cyclic Triols: Application to Total Syntheses of Cleroindicins D, F, and C. *Angew. Chem. Int. Ed.* **2009**, *48*, 547–550. [CrossRef] [PubMed]
52. Manville, N.; Alite, H.; Haeffner, F.; Hoveyda, A.H.; Snapper, M.L. Enantioselective silyl protection of alcohols promoted by a combination of chiral and achiral Lewis basic catalysts. *Nat. Chem.* **2013**, *5*, 768–776. [CrossRef] [PubMed]
53. Yamamoto, K.; Suganomata, Y.; Inoue, T.; Kuriyama, M.; Demizu, Y.; Onomura, O. Copper-Catalyzed Asymmetric Oxidative Desymmetrization of 2-Substituted 1,2,3-Triols. *J. Org. Chem.* **2022**, *87*, 6479–6491. [CrossRef] [PubMed]
54. Ichikawa, J.; Asami, M.; Mukaiyama, T. An asymmetric synthesis of glycerol derivatives by the enantioselective acylation of prochiral glycerol. *Chem. Lett.* **1984**, *13*, 949–952. [CrossRef]
55. Lewis, C.A.; Sculimbrene, B.R.; Xu, Y.; Miller, S.J. Desymmetrization of Glycerol Derivatives with Peptide-Based Acylation Catalysts. *Org. Lett.* **2005**, *7*, 3021–3023. [CrossRef]
56. Trost, B.M.; Malhotra, S.; Mino, T.; Rajapaksa, N.S. Dinuclear Zinc-Catalyzed Asymmetric Desymmetrization of Acyclic 2-Substituted-1,3-Propanediols: A Powerful Entry into Chiral Building Blocks. *Chem. Eur. J.* **2008**, *14*, 7648–7657. [CrossRef] [PubMed]
57. Sakakura, A.; Umemura, S.; Ishihara, K. Desymmetrization of *meso*-Glycerol Derivatives Induced by L-Histidine-Derived Acylation Catalysts. *Adv. Synth. Catal.* **2011**, *353*, 1938–1942. [CrossRef]
58. Giustra, Z.X.; Tan, K.L. The efficient desymmetrization of glycerol using scaffolding catalysis. *Chem. Commun.* **2013**, *49*, 4370–4372. [CrossRef]
59. Liu, X.; Cheng, Y.; Tian, Y. Preparation of chiral glycerol sulfonate. Chinese Patent CN114394919A, 26 April 2022.
60. Trost, B.M.; Older, C.M. A Convenient Synthetic Route to [CpRu(CH$_3$CN)$_3$]PF$_6$. *Organometallics* **2002**, *21*, 2544–2546. [CrossRef]
61. Hu, P.; Kan, J.; Su, W.; Hong, M. Pd(O$_2$CCF$_3$)$_2$/Benzoquinone: A Versatile Catalyst System for the Decarboxylative Olefination of Arene Carboxylic Acids. *Org. Lett.* **2009**, *11*, 2341–2344. [CrossRef]

62. Fu, Z.; Huang, S.; Su, W.; Hong, M. Pd-Catalyzed Decarboxylative Heck Coupling with Dioxygen as the Terminal Oxidant. *Org. Lett.* **2010**, *12*, 4992–4995. [CrossRef] [PubMed]
63. Onomura, O.; Takemoto, Y.; Miyamoto, K.; Ito, M. Method for preparing optically active 1,2,3-triol monoesters in the presence of bisoxazoline ligands and copper compounds. World Intellectual Property Organization WO2015/072290A1, 21 May 2015.
64. Yokose, U.; Ishikawa, J.; Morokuma, Y.; Naoe, A.; Inoue, Y.; Yasuda, Y.; Tsujimura, H.; Fujimura, T.; Murase, T.; Hatamochi, A. The ceramide [NP]/[NS] ratio in the stratum corneum is a potential marker for skin properties and epidermal differentiation. *BMC Dermatol.* **2020**, *20*, 6. [CrossRef] [PubMed]
65. Suzuki, T.; Fukasawa, J.; Iwai, H.; Sugai, I.; Yamashita, O.; Kawamata, A. Multilamellar Emulsion of Stratum Corneum Lipid –Formation Mechanism and its Skin Care Effects. *J. Soc. Cosmet. Chem. Japan* **1993**, *27*, 193–205. [CrossRef]
66. Ishida, K. Development and Properties of the Optically Active Ceramides. *Oleoscience* **2004**, *4*, 105–116. [CrossRef]
67. Maki, T.; Ushijima, N.; Matsumura, Y.; Onomura, O. Catalytic monoalkylation of 1,2-diols. *Tetrahedron Lett.* **2009**, *50*, 1466–1468. [CrossRef]
68. Boldwin, J.J.; Raab, A.W.; Mensler, K.; Arison, B.H.; McClure, D.E. Synthesis of (R)- and (S)-Epichlorohydrin. *J. Org. Chem.* **1978**, *43*, 4876–4878. [CrossRef]
69. Tanabe, G.; Sakano, M.; Minematsu, T.; Matusda, H.; Yoshikawa, M.; Muraoka, O. Synthesis and elucidation of absolute stereochemistry of salaprinol, another thiosugar sulfonium sulfate from the ayurvedic traditional medicine *Salacia prinoides*. *Tetrahedron* **2008**, *64*, 10080–10086. [CrossRef]
70. Kurimura, M.; Takemoto, M.; Achiwa, K. Synthesis of Optically Active Lipopeptide Analogs from Outer membrane of *Escherichia coli*. *Chem. Pharm. Bull.* **1991**, *39*, 2590–2596. [CrossRef]
71. Tuin, A.W.; Palachanis, D.K.; Buizert, A.; Grotenbreg, G.M.; Spalburg, E.; de Neeling, A.J.; Mars-Groenendijk, R.H.; Noort, D.; van der Marel, G.A.; Overkleeft, H.S.; et al. Synthesis and Biological Evaluation of Novel Gramicidin S Analogues. *Eur. J. Org. Chem.* **2009**, 4231–4241. [CrossRef]
72. Paterson, I.; Delgado, O.; Florence, G.J.; Lyothier, I.; O'Brien, M.; Scott, J.P.; Sereinig, N. A Second-Generation Total Synthesis of (+)-Discodermolide: The Development of a Practical Route Using Solely Substrate-Based Stereocontrol. *J. Org. Chem.* **2005**, *70*, 150–160. [CrossRef]

Article

Two-Step Synthesis, Structure, and Optical Features of a Double Hetero[7]helicene

Mohamed S. H. Salem [1,2], Ahmed Sabri [1], Md. Imrul Khalid [1], Hiroaki Sasai [1,3] and Shinobu Takizawa [1,*]

1 SANKEN, Osaka University, Ibaraki-shi, Osaka 567-0047, Japan
2 Pharmaceutical Organic Chemistry Department, Faculty of Pharmacy, Suez Canal University, Ismailia 41522, Egypt
3 Graduate School of Pharmaceutical Sciences, Osaka University, Suita-shi, Osaka 565-0871, Japan
* Correspondence: taki@sanken.osaka-u.ac.jp; Tel.: +81-6-6879-8467

Abstract: A novel double aza-oxa[7]helicene was synthesized from the commercially available N^1,N^4-di(naphthalen-2-yl)benzene-1,4-diamine and *p*-benzoquinone in two steps. Combining the acid-mediated annulation with the electrochemical sequential reaction (oxidative coupling and dehydrative cyclization) afforded this double hetero[7]helicene. Moreover, the structural and optical features of this molecule have been studied using X-ray crystallographic analysis, and the absorption and emission behaviors were rationalized based on DFT calculations.

Keywords: polycyclic aromatic hydrocarbon; double hetero[7]helicene; short-step synthesis; electrochemical cross-coupling; nucleus-independent chemical shift

1. Introduction

Helicenes are polycyclic aromatic hydrocarbons (PAHs) in which aromatic rings are annulated in a helical architecture, giving them unique electronic, photophysical and chiroptical properties [1–4]. Over the past couple of decades, the great advances achieved in this chemistry [2–6] promoted a broad spectrum of material-based applications [7–10], transistors [11,12], and semiconductors [13]. Incorporation of one or more heteroatoms in the helicene scaffolds modulate their physical and optical features, and alter the electronic properties in order to expand their applications [14,15]. With these extra features, the trend in helicene chemistry has begun to move towards heterohelicenes after the domination of carbohelicenes [16–21]. Another approach to promote characteristics of helicenes is to induce multihelicity which means combining two or more helical scaffolds in a single molecule [22,23]. Multiple helicenes show a lot of favorable properties due to their amplified non-planarity, diverse conformations, and maximized intermolecular interactions [24,25]. Various smart core scaffolds were used to induce this multihelicity such as perylene diimide (PDI) that afforded valuable twisted structures for different material-based applications [26–29]. Hence, a lot of efforts were dedicated for designing and synthesizing multiple heterohelicenes [30], in particular, double heterohelicenes [14]. After the first report of double helicene reported by Rajca, many examples of these double heterohelicenes were conducted and exhibited clear superiority over their single counterparts, especially in terms of optical properties (Figure 1a) [31–40]. However, during that frantic pursuit to promote the properties of helicenes, another problem, in particular, synthetic difficulty emerged. With the increase in structural complexity, the synthesis of multiple heterohelicenes becomes more challenging and requires many steps. Although few reports succeeded to introduce effective short-step synthetic protocols for double heterohelicene, most of these successes were concentrated in the double hetero[5]helicene derivatives (Figure 1b) [41–43]. In 2016, Narita, Cao, and Müllen introduced an efficient two-step synthesis of a highly strained OBO-fused double hetero[7]helicene **K** via the nucleophilic aromatic substitution reaction of hexabromobenzene, followed by a sequential

step of demethylation and C-H aryl borylation (Figure 1c) [44]. Earlier in the same year, Hatakeyama showed the potential of this synthetic approach to afford their boron-fused double hetero[5]helicene **I** (Figure 1b) [42]. In 2021, Wang and coworkers developed the first examples of B,N-embedded double hetero[7]helicenes **L** that showed excellent chiroptical features in the visible range [45]. With only two steps, they succeeded to prepare this double hetero[7]helicene **L** via the nucleophilic aromatic substitution of dibromotetrafluorobenzene with carbazole, followed by a tandem process of substitution with BBr$_3$ and C-H aryl borylation [45].

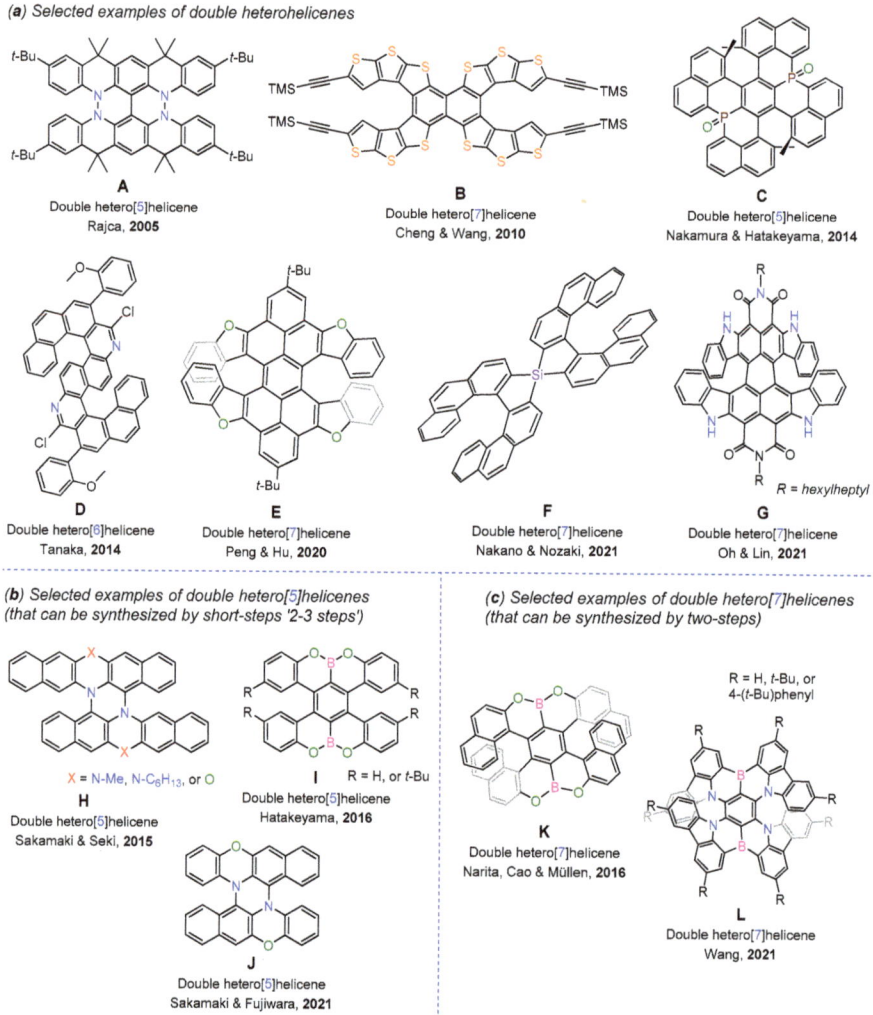

Figure 1. The selected examples of double heterohelicenes in short-step synthesis: (**a**) Double hetero[5–7]helicenes (more than four-step synthesis); (**b**) Double hetero[5]helicenes (two- or three-step synthesis); (**c**) Double hetero[7]helicenes (two-step synthesis).

Notably, these examples (Figure 1c) represent a quantum leap in the short-step synthesis of double hetero[7]helicenes via the tandem process of nucleophilic substitution with BBr$_3$ followed by C-H aryl borylation [44,45]. As part of our effort to explore the electrochemical domino syntheses, we were interested in designing effective sequential reactions to access double helicene motifs [46,47]. Herein, a facile preparation of a double aza-oxa[7]helicene with a phenylene linker has been established through acid-mediated annulation with the electrochemical sequential reaction (oxidative coupling and dehydrative cyclization). We also studied the structural and optical features via x-ray crystallographic analysis, spectrophotometric analysis, and DFT calculations.

2. Results and Discussion

2.1. Synthesis of Double Aza-oxa[7]helicene 3

Recently, Zhang reported a facile acid-mediated synthesis of carbazole in which aniline derivatives react with *p*-benzoquinone to produce 3-hydroxycarbazoles [48]. Combining this approach with our unprecedented electrochemically enabled synthesis of hetero[7]helicenes and dehydro-hetero[7]helicenes [46,47], herein, we achieved the two-step synthesis of double aza-oxa[7]helicenes as depicted in Scheme 1. The acid-mediated annulation of the commercially available substrates; N^1,N^4-di(naphthalen-2-yl)benzene-1,4-diamine **1** and *p*-benzoquinone afforded the corresponding bis-3-hydroxy-benzo[c]carbazole **2** in 54% yield via a tandem process of double Michael addition and subsequent double ring closure. Next, a DCM solution of **2**, β-naphthol, and tetrabutylammonium hexafluorophosphate(V) as an electrolyte, was utilized to a constant current of 1.5 mA in an undivided electrolysis cell with platinum electrodes for 3.5 h at room temperature, affording double aza-oxa[7]helicene **3** in 26% yield. The electrochemical sequential synthesis of **3** proceeds through the oxidative hetero-coupling of arenols to produce a diol intermediate that can readily undergo a subsequent dehydrative cyclization to **3**. All compounds showed good chemical and thermal stabilities and no decomposition was observed upon purification on silica column chromatography and heating at 100 °C in air.

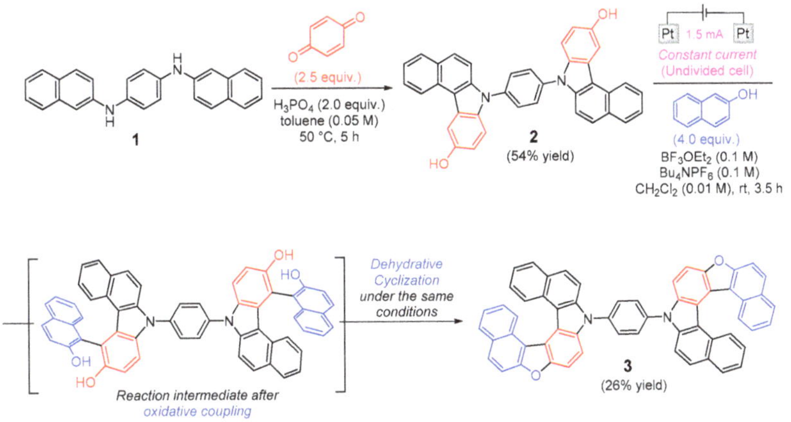

Scheme 1. The synthesis of a double aza-oxa[7]helicenes **3**.

2.2. Structure and Packing Mode of 3

The double aza-oxa[7]helicene structure of **3** was definitely confirmed by X-ray crystallography using a single crystal, grown from its racemic solution. We used the liquid/liquid diffusion technique between ethyl acetate and *n*-hexane to prepare this crystal slowly over three days in a dark environment at −20 °C. As expected, the two helicene moieties are connected via a phenylene linker (Figure 2a,b). The dihedral angles between the phenylene linker's plane and the pyrrole (ring B') are −41.86° for (C_1-C_6-N_7-C_8), and 54.38° for (C_5-C_6-N_7-C_9). Although the experimental values of (C_5-C_6-N_7-C_9) dihedral angle (54.38°) is comparable to that of the optimized structure using DFT calculations at MN15/6-311G(d,p) level of theory (54.72°), (C_1-C_6-N_7-C_8) dihedral angle was smaller than optimized structures at various levels (Table 1). This can be attributed to the intermolecular interactions between the double helicene molecules **3** in the packed structure. Only meso isomer (*P,M*)-**3** was observed in the crystal packing with achiral molecules packed along the b-axis (Figure 2c,d). The packing of **3** shows a herringbone pattern with π-π distance of 4.458 A°. This characteristic arrangement is optimum for many material-based applications, especially semiconductors [49–51]. In addition, it maximizes the optical and electronic properties of the obtained double helicene upon self-assembly [49,52–54]. Most of these larger or multiple helicenes showed significant variations during DFT calculations owing to the long-range conjugation and the effects of charge transfer [55,56]. Among the functions we screened, Minnesota 15 (MN15) function was found to be the most suitable parameters for our molecules (Table 1) [57].

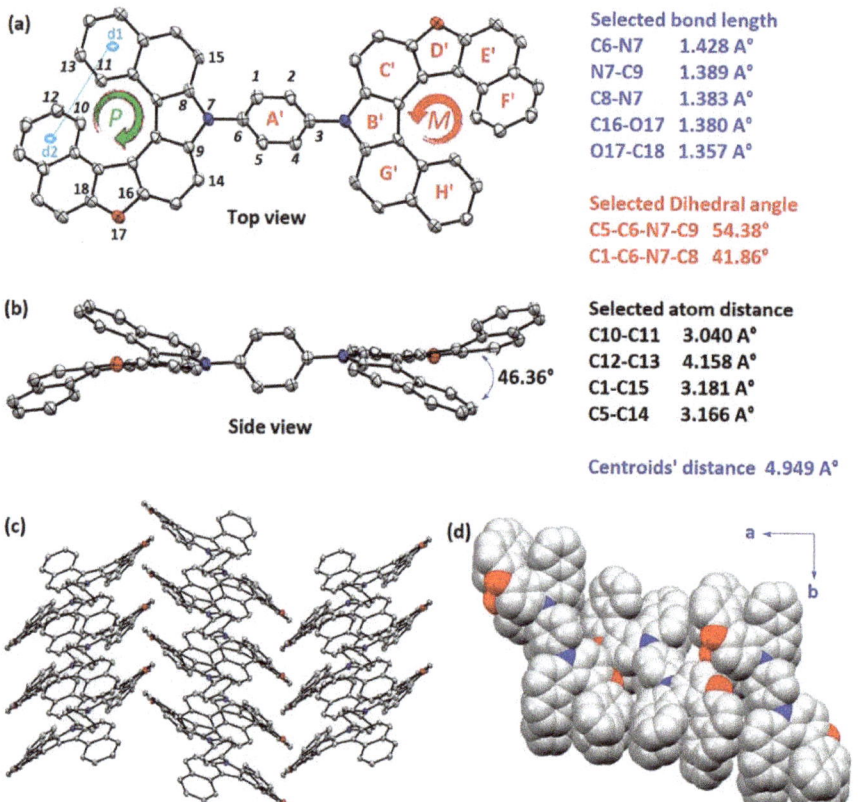

Figure 2. Single crystal structure of the double helicene **3**: (**a**,**b**) ORTEP drawing structure of (*P,M*)-**3** obtained by X-ray crystal analysis with ellipsoids at 50% probability (H atoms were omitted for clarity); (**c**) crystal packing of (*P,M*)-**3** with ellipsoids at 30% probability; (**d**) packing structure of (*P,M*)-**3** is viewed along the c-axis to show the herringbone arrangement.

Table 1. The selected experimental and calculated structural parameters of double aza-oxa[7]helicene 3.

Parameters	Experimental	B3LYP [1]	wB97XD [1]	MN15 [1]
Centroids' distance (rings F'-H')	4.949 A°	4.885 A°	4.721 A°	4.759 A°
d_1-N_7-d_2 Centroid angle	46.36°	45.43°	44.24°	45.51°
C_5-C_6-N_7-C_9 Dihedral angle	54.38°	60.25°	59.03°	54.72°
C_1-C_6-N_7-C_8 Dihedral angle	41.86°	57.08°	55.34°	51.42°
C_1-C_{15} Distance	3.181 A°	3.316 A°	3.281 A°	3.241 A°
C_5-C_{14} Distance	3.166 A°	3.356 A°	3.316 A°	3.266 A°

[1] All calculations are carried out using 6-311G(d,P) basis set at three different functions (B3LYP, wB97XD, and MN15).

Nucleus-independent chemical shift (NICS) calculations revealed the low aromaticity of the central phenylene linker with a NICS (0) value of −5.8 ppm (Figure 3a), much lower than that of benzene −7.6 ppm calculated at the same level of theory. The largest NICS (0) values (between −7.3 ppm and −8.6 ppm) were found on the benzene of 6H-furo[3,2-e]indole (ring C'), pyrrole (ring B') and naphthalene (rings F' and H'). While the lowest NICS (0) values (around −5.8 ppm) were found on the phenylene linker (ring A') and furan rings (D') which is consistent with the aromatic character of this ring. Generally, symmetric double hetero[n]helicenes ($n \geq 4$) have three isomers, those being two chiral enantiomers (P,P) and (M,M), and one meso diasteromer (P,M) [30]. All three isomers of 3 were afforded under our reaction conditions which was confirmed by HPLC separation using DAICEL CHIRALPAK IA column (eluent: n-hexane/i-PrOH = 20/1) (Figure 3b). The experimental ratio among the three isomers was found to be around (1:2:1) with the meso isomer (P,M)-3 as the major formed product (confirmed by the absence of optical rotation). After HPLC chiral resolution, the epimerization rate of 3 was studied at three different temperatures (See SI). Eyring plot (Figure 3c) indicated a low chiral stability of 3 (epimerization barrier ~24.2 kcal mol^{-1}) with an estimated half-life of the epimerization <6.5 h at 25 °C. These observations were matching with our DFT calculations that showed similar epimerization barriers 25.32 kcal mol^{-1} and 25.62 kcal mol^{-1} (Figure 3d).

2.3. Photophysical Properties

Our double aza-oxa[7]helicene 3 shows high luminescence upon photo-irradiation, which can be attributed to the rigid scaffold that hinders the thermal energy loss upon structural changes. The UV/vis absorption of 3 was recorded in different solvents to show its high solubility in most of the organic solvents which increases the potential for some applications that require good solubility such as solution-processed electronics [58–60]. In all measured solvents, compound 3 showed similar UV/vis absorption patterns (Figure 4a). The maximum absorbance exhibited in chloroform was shown at 407 nm (absorption coefficient: 7.5×10^4 M^{-1} cm^{-1}) corresponding to an optical energy gap of (2.18 eV). According to TD-DFT calculations at the MN15/6-311G(d,p) level of theory, this low-energy absorption can be accountable to the HOMO→LUMO transition. The absorption band at 385 nm possibly attributed to the equal contribution of both HOMO−1→LUMO and HOMO→LUMO+1 transitions. The band at 368 nm is estimated to be due to the HOMO−1→LUMO+1 transition, while the higher energy absorption band at 328 nm would be attributed to HOMO−2→LUMO corresponding to an optical energy gap of (2.58 eV). Molecular orbital calculations indicated that the HOMO is spread mainly over the phenylene linker (ring A') and pyrroles (rings B') and LUMO is spread over the whole scaffold rather than the phenylene linker (ring A'), accounting for the substantial stability. Photoluminescence PL spectrum of 3 was recorded in a pure chloroform solution exhibiting emission maxima shifted in a bathochromic way at 415 nm and 440 nm.

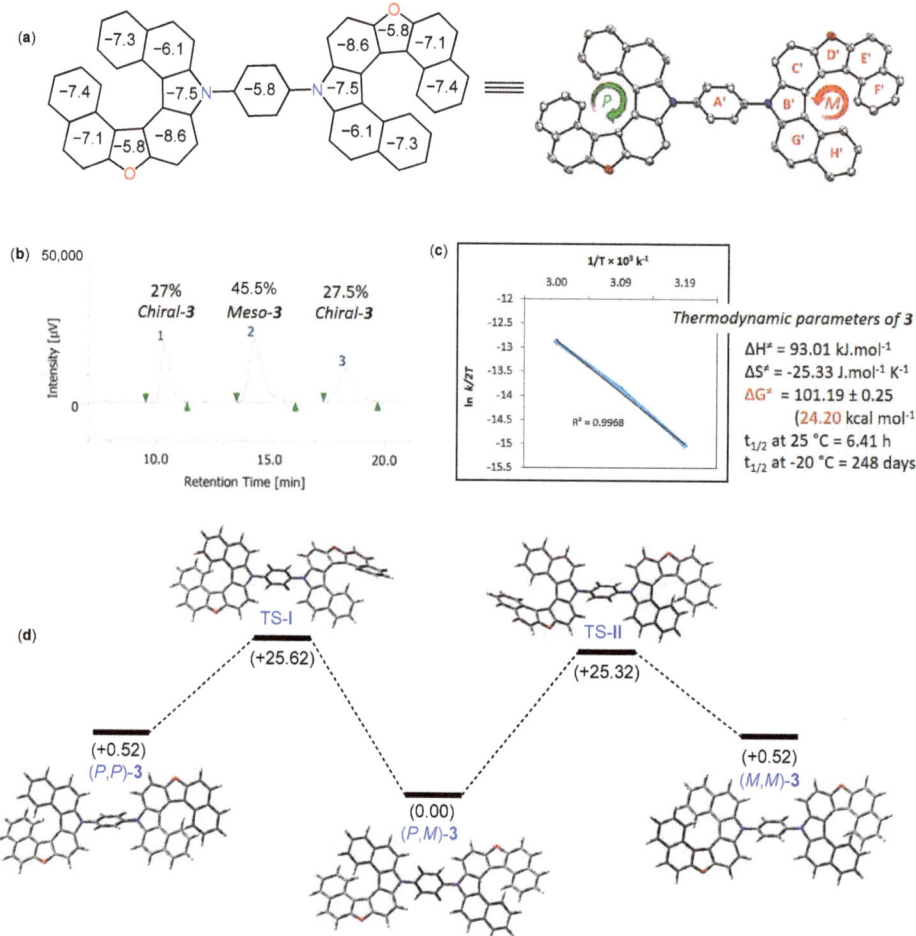

Figure 3. (a) NICS (0) values of (*P,M*)-**3** calculated at the MN15/6-311G+(2d,p) level; (b) HPLC chromatogram determined by (Daicel Chiralpak IA, *n*-hexane/*i*-PrOH = 20/1, flow rate 1.0 mL/min, T = 25 °C, 240 nm): t_1 = 10.36 min, t_2 = 14.30 min, and t_3 = 18.32 min; (c) Eyring plot of compound **3** epimerization and thermodynamic parameters; (d) Epimerization process from (*P,M*)-**3** isomer to (*M,M*)-**3** and (*P,P*)-**3** isomers. The relative Gibbs free energies are calculated in (kcal mol^{-1}) at the MN15/6-311G(d,p) level.

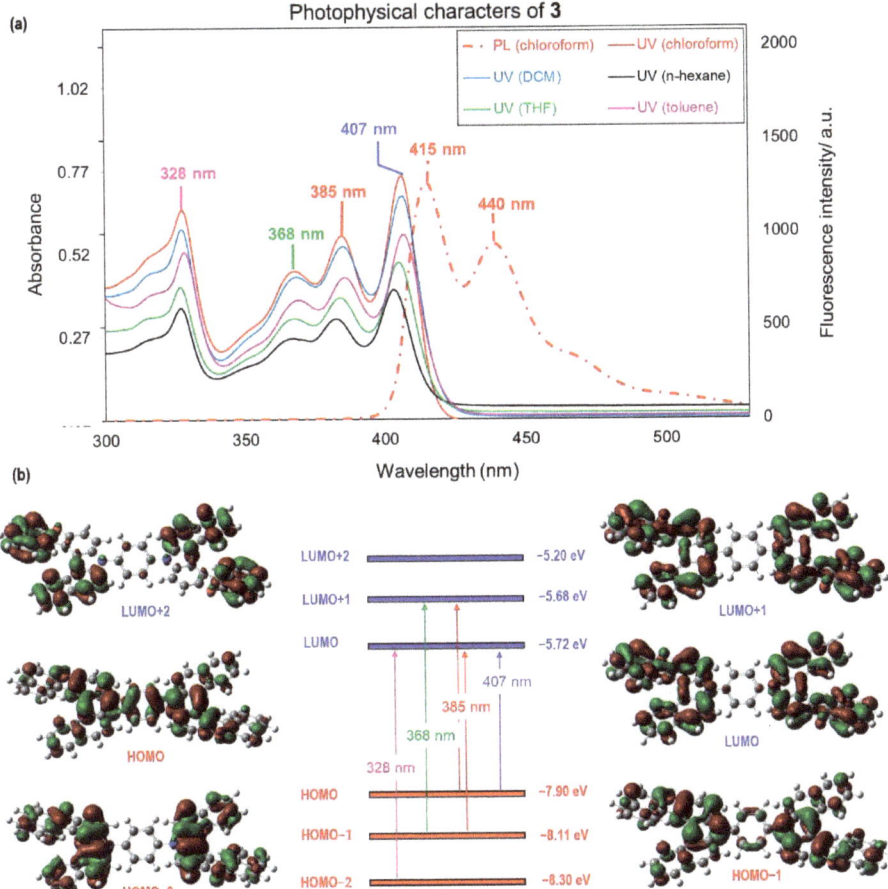

Figure 4. (a) UV/vis absorption and PL spectra of **3** in various solvents (20 µM); (b) Frontier Kohn-Sham molecular orbitals of **3** and TD-DFT calculated electronic transitions at MN15/6-311G (d,p) level of theory.

2.4. Energetic Characterization by Cyclic Voltammetry

Cyclic voltammetry (CV) measurements of our double aza-oxa[7]helicene **3** showed reversible redox peaks in both negative and positive regions indicating the chemical stability of its anion/cation pairs and how they can be reduced or oxidized readily to the neutral form (Figure 5). Using ferrocene and ferrocenium as internal references, the HOMO energy level of **3** was calculated using Bredas empirical equation to be around (−7.83 eV) which is comparable to the DFT-calculated HOMO energy (−7.90 eV) [61]. E_{LUMO} could be estimated after considering the gap between E_{HOMO} and E_{LUMO} (3.04 eV) from the λ_{max} or excitation energy (407 nm) to be around (−4.79 eV) showing little higher energy than the DFT-calculated LUMO (−5.72 eV).

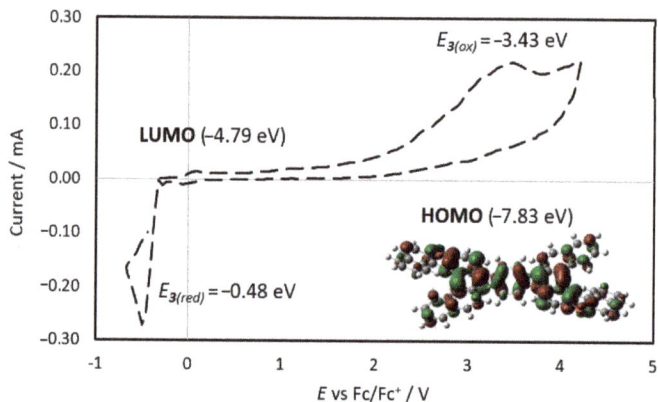

Figure 5. The cyclic voltammetry profile of **3** in (MeCN) with n-Bu$_4$NPF$_6$ (0.1 M) using ferrocene as internal reference.

3. Materials and Methods

3.1. General Experimental Details

^{1}H-, and ^{13}C-NMR were recorded via JNM ECA600 FT NMR (^{1}H-NMR 600 MHz, ^{13}C-NMR 151 MHz). ^{1}H-NMR spectra are reported as follows: a chemical shift in ppm downfield of tetramethylsilane (TMS) and referenced to residual solvent peak (CDCl$_3$) at 7.26 ppm, or ((CD$_3$)$_2$CO) at 2.05 ppm, multiplicities (s = singlet, d = doublet, dd = doublet of doublets, t = triplet, q = quartet, m = multiplet), and coupling constants (Hz). ^{13}C-NMR spectra reported in ppm relative to the central line of triplet for CDCl$_3$ at 77.16 ppm, or the central line of septet for ((CD$_3$)$_2$CO) at 29.84 ppm. APCI-MS spectra were obtained with JMS-T100LC (JEOL). FT-IR spectra were recorded on a JASCO FT-IR system (FT/IR4100). Photoluminescence (PL) spectra were recorded on JASCO FP-8550 Spectrofluorometer. UV spectra were recorded on a JASCO v-770 spectrophotometer. Column chromatography on SiO$_2$ was performed with Kanto Silica Gel 60 (63–210 μm). Commercially available organic and inorganic compounds were used without further purification. The electro-oxidation was carried out using sing ElectraSyn® 2.0 (designed by IKA) at a constant current of 1.5 mA, under air (1 atm.) [62].

3.2. Synthetic Procedures

3.2.1. General Procedure for the Synthesis of Double 3-Hydroxy Benzo[c]carbazole **2**

To a solution of N^1,N^4-di(naphthalen-2-yl)benzene-1,4-diamine **1** (36 mg, 0.1 mmol) and p-benzoquinone (27 mg, 0.25 mmol, 2.5 equiv.) in dry toluene (1.5 mL), orthophosphoric acid (10.6 μL, 2.0 equiv.) dissolved in (0.5 mL) toluene was added dropwise. The reaction mixture was stirred at 50 °C for 5 h under N$_2$ atmosphere until its completion. Next, the reaction was quenched via water, extracted with EtOAc and the combined organic extracts dried over Na$_2$SO$_4$, and evaporated *in vacuo*. The crude mixture was purified on silica column chromatography (eluent: n-hexane/DCM/ethyl acetate = 7/1/1) to give double 3-Hydroxy benzo[c]carbazole **2** as a white solid in 54% yield.

- 7,7'-(1,4-Phenylene)bis(7H-benzo[c]carbazol-10-ol) **2**

^{1}H NMR (600 MHz, (CD$_3$)$_2$CO): δ 8.80 (d, J = 8.2 Hz, 2H), 8.31 (s, 2H), 8.17 (d, J = 2.1 Hz, 2H), 8.07 (d, J = 8.2 Hz, 2H), 7.92–7.96 (m, 6H), 7.75–7.78 (m, 4H), 7.58 (d, J = 8.9 Hz, 2H), 7.50 (dd, J = 8.3, 6.9 Hz, 2H), 7.14 (dd, J = 8.6, 2.4 Hz, 2H); ^{13}C NMR (151 MHz, (CD$_3$)$_2$CO): δ 153.39, 139.70, 137.52, 135.41, 130.82, 130.43, 130.13, 129.80, 128.33, 127.95, 125.47, 123.91, 123.75, 115.93, 115.05, 112.67, 111.86, 107.78; DEPT-135 NMR (151 MHz, (CD$_3$)$_2$CO): δ 130.12, 129.80, 128.32, 127.95, 123.91, 123.74, 115.03, 112.67, 111.86, 107.76; HRMS (APCI): calcd for

$C_{38}H_{24}N_2O_2$: m/z 541.1911 [M + H]$^+$, found 541.1912.; IR (KBr): 3334, 3042, 2977, 2926, 1620, 1517, 1473, 1165, 831, 803 cm^{-1}; mp: 198–199 °C.

3.2.2. General Procedure for the Synthesis of Double Aza-oxa[7]helicene 3

A 10 mL DCM solution of **2** (54 mg, 0.1 mmol), β-naphthol (57.7 mg, 0.4 mmol), tetrabutylammonium hexafluorophosphate(V) (387.4 mg, 1.0 mmol), and BF$_3$OEt$_2$ (0.2 M) was transferred into the undivided electrolysis cell of ElectraSyn® 2.0. This cell is equipped with two Pt electrodes connected to a DC power supply. At room temperature, a constant current of 1.5 mA was applied for 3.5 h. After the completion of reaction, the electrolysis was stopped and crude mixture was purified by column chromatography (SiO$_2$, EtOAc/*n*-hexane) to afford the double aza-oxa[7]helicene **3** as a yellow solid in 26% yield.

- 1,4-Bis(*10H*-benzo[*c*]naphtho[1′,2′:4,5]furo[3,2-*g*]carbazol-10-yl)benzene **3**

^{1}H NMR (600 MHz, CDCl$_3$): δ 8.38 (d, *J* = 8.2 Hz, 2H), 8.31 (d, *J* = 8.2 Hz, 2H), 8.00–8.04 (m, 12H), 7.92 (d, *J* = 8.9 Hz, 2H), 7.85 (d, *J* = 8.9 Hz, 4H), 7.77 (d, *J* = 8.9 Hz, 2H), 7.34–7.39 (m, 4H), 6.95–7.00 (m, 4H); ^{13}C NMR (151 MHz, CDCl$_3$): δ 154.75, 153.31, 138.79, 137.93, 137.32, 130.86, 129.79, 129.73, 129.50, 129.15, 128.59, 128.10, 128.05, 128.03, 125.13, 124.67, 124.37, 123.62, 120.07, 117.98, 117.77, 116.64, 112.65, 111.58, 109.45, 109.06 (Two carbons overlapped); DEPT-135 NMR (151 MHz, CDCl$_3$): δ 129.79, 129.15, 128.59, 128.09, 128.05, 128.03, 125.13, 124.67, 124.37, 123.61, 112.66, 111.58, 109.44, 109.06 (One carbon overlapped); HRMS (APCI): calcd for $C_{58}H_{32}N_2O_2$: *m/z* 789.2537 [M + H]$^+$, found 789.2542; IR (KBr): 3043, 2926, 2856, 1714, 1594, 1500, 1417, 1355, 1209, 805 cm^{-1}; mp: 291–292 °C.

3.3. DFT Calculations

All DFT calculations were performed using the Gaussian 16 package of programs [63]. The geometries of the model compounds were optimized using three DFT functions: B3LYP, wB97XD, and MN15 at 6-311G(d,p) basis set. All stationary points were identified as stable minima by frequency calculations. Geometry optimization was achieved using the normal criteria defined in Gaussian 16. TD-DFT calculations were performed using two different levels of theory B3LYP/6-311G(d,p) and MN15/6-311G(d,p) in both chloroform and gas-phase. All structures were optimized without any symmetry assumptions. For further computational details, see Supplementary Materials.

4. Conclusions

In summary, we introduced a two-step protocol to synthesize double aza-oxa[7]helicene **3** using an electrochemical approach. This novel double hetero[7]helicene shows interesting structural features that were reflected in its excellent optical properties. We have studied the photophysical characteristics of this compound and correlated its absorption and fluorescence behavior based on DFT calculations. Further development for this two-step protocol towards the preparation of other multiple helicenes and PHAs and study their photophysical and chiroptical features are currently under investigation.

Supplementary Materials: The following supporting information can be downloaded at: https://www.mdpi.com/article/10.3390/molecules27249068/s1, Table S1: Optimization of the acid-mediated annulation step; Table S2: Optimization of the electrochemical sequential reaction; Table S3: Selected experimental and calculated structural parameters of double aza-oxa[7]helicene **3**; Tables S4–S6: Summary of the TD-DFT calculation results of **3**; Scheme S1: A plausible mechanism for the electrochemical domino reaction; Figure S1: Crystal measurements; Figure S2: Measurements of the optimized structures; Figure S3: Selected molecular orbitals of **3**; Figure S4: Simulated UV-vis absorption and CD spectra of (*P,M*)-**3**; Figure S5: Further NICS(0) calculations [57,63–68].

Author Contributions: Conceptualization, supervision, and project administration, H.S. and S.T.; methodology, DFT calculations, investigation, data analysis, and writing original draft, M.S.H.S.; methodology, investigation, validation, and formal analysis A.S. and M.I.K. All authors have read and agreed to the published version of the manuscript.

Funding: This work was supported by JSPS KAKENHI Grant Numbers 22K06502 in Grant-in-Aid for Scientific Research (C), Transformative Research Areas (A) 21A204 Digitalization-driven Transformative Organic Synthesis (DigiTOS), 22KK0073 in Fund for the Promotion of Joint International Research (Fostering Joint International Research (B)) from the Ministry of Education, Culture, Sports, Science, and Technology (MEXT), and the Japan Society for the Promotion of Science (JSPS), JST CREST (No. JPMJCR20R1), and Hoansha Foundation.

Informed Consent Statement: The study does not involve humans.

Data Availability Statement: CIF of the crystal of **3** is available as Supplementary. The X-ray crystallographic coordinate for the structure reported in this study has been deposited at the Cambridge Crystallographic Data Centre (CCDC) under deposition numbers CCDC-2156335 (**3**). These data can be obtained free of charge from The Cambridge Crystallographic Data Centre via www.ccdc.cam.ac.uk/data_request/cif.

Acknowledgments: We thank Yoichi Hoshimoto for helping with chiral HPLC resolution. We acknowledge the technical staff of the Comprehensive Analysis Center of SANKEN.

Conflicts of Interest: The authors declare no conflict of interest.

Sample Availability: Samples of compound **3** are available from the corresponding author upon reasonable request.

References

1. Jakubec, M.; Storch, J. Recent advances in functionalizations of helicene backbone. *J. Org. Chem.* **2020**, *85*, 13415–13428. [CrossRef] [PubMed]
2. Gingras, M. One hundred years of helicene chemistry. Part 3: Applications and properties of carbohelicenes. *Chem. Soc. Rev.* **2013**, *42*, 1051–1095. [CrossRef] [PubMed]
3. Shen, Y.; Chen, C.-F. Helicenes: Synthesis and applications. *Chem. Rev.* **2012**, *112*, 1463–1535. [CrossRef] [PubMed]
4. Crassous, J.; Stara, I.G.; Stary, I. *Helicenes: Synthesis, Properties, and Applications*; John Wiley & Sons: Hoboken, NJ, USA, 2022.
5. Gingras, M. One hundred years of helicene chemistry. Part 1: Non-stereoselective syntheses of carbohelicenes. *Chem. Soc. Rev.* **2013**, *42*, 968–1006. [CrossRef]
6. Gingras, M.; Félix, G.; Peresutti, R. One hundred years of helicene chemistry. Part 2: Stereoselective syntheses and chiral separations of carbohelicenes. *Chem. Soc. Rev.* **2013**, *42*, 1007–1050. [CrossRef]
7. Yang, S.Y.; Qu, Y.K.; Liao, L.S.; Jiang, Z.Q.; Lee, S.T. Research progress of intramolecular π-stacked small molecules for device applications. *Adv. Mater.* **2022**, *34*, 2104125. [CrossRef]
8. Han, J.; Guo, S.; Lu, H.; Liu, S.; Zhao, Q.; Huang, W. Recent progress on circularly polarized luminescent materials for organic optoelectronic devices. *Adv. Opt. Mater.* **2018**, *6*, 1800538. [CrossRef]
9. Jeon, S.K.; Lee, H.L.; Yook, K.S.; Lee, J.Y. Recent progress of the lifetime of organic light-emitting diodes based on thermally activated delayed fluorescent material. *Adv. Mater.* **2019**, *31*, 1803524. [CrossRef]
10. Zhang, C.; Wang, X.; Qiu, L. Circularly polarized photodetectors based on chiral materials: A review. *Front. Chem.* **2021**, *9*, 711488. [CrossRef]
11. Brandt, J.R.; Salerno, F.; Fuchter, M.J. The added value of small-molecule chirality in technological applications. *Nat. Rev. Chem.* **2017**, *1*, 0045. [CrossRef]
12. Shang, X.; Wan, L.; Wang, L.; Gao, F.; Li, H. Emerging materials for circularly polarized light detection. *J. Mater. Chem. C* **2022**, *10*, 2400–2410. [CrossRef]
13. Li, Q.; Zhang, Y.; Xie, Z.; Zhen, Y.; Hu, W.; Dong, H. Polycyclic aromatic hydrocarbon-based organic semiconductors: Ring-closing synthesis and optoelectronic properties. *J. Mater. Chem. C* **2022**, *10*, 2411–2430. [CrossRef]
14. Hong, J.; Xiao, X.; Liu, H.; Dmitrieva, E.; Popov, A.A.; Yu, Z.; Li, M.D.; Ohto, T.; Liu, J.; Narita, A.; et al. Controlling the emissive, chiroptical, and electrochemical properties of double [7]helicenes through embedded aromatic rings. *Chem. Eur. J.* **2022**, *28*, e202202243. [CrossRef] [PubMed]
15. Tian, Y.-H.; Park, G.; Kertesz, M. Electronic structure of helicenes, C_2S helicenes, and thiaheterohelicenes. *Chem. Mater.* **2008**, *20*, 3266–3277. [CrossRef]
16. Rajca, A.; Wang, H.; Pink, M.; Rajca, S. Annelated heptathiophene: A fragment of a carbon–sulfur helix. *Angew. Chem. Int. Ed.* **2000**, *39*, 4481–4483. [CrossRef]
17. Pieters, G.; Gaucher, A.; Marque, S.; Maurel, F.o.; Lesot, P.; Prim, D. Regio-defined amino [5]oxa-and thiahelicenes: A dramatic impact of the nature of the heteroatom on the helical shape and racemization barriers. *J. Org. Chem.* **2010**, *75*, 2096–2098. [CrossRef]
18. Nakano, K.; Oyama, H.; Nishimura, Y.; Nakasako, S.; Nozaki, K. λ^5-Phospha[7]helicenes: Synthesis, properties, and columnar aggregation with one-way chirality. *Angew. Chem. Int. Ed.* **2012**, *51*, 695–699. [CrossRef]

19. Žádný, J.; Jančařík, A.; Andronova, A.; Šámal, M.; Vacek Chocholoušová, J.; Vacek, J.; Pohl, R.; Šaman, D.; Císařová, I.; Stará, I.G.; et al. A general approach to optically pure [5]-,[6]-, and [7]heterohelicenes. *Angew. Chem. Int. Ed.* **2012**, *51*, 5857–5861. [CrossRef]
20. Sundar, M.S.; Bedekar, A.V. Synthesis and study of 7,12,17-trioxa[11]helicene. *Org. Lett.* **2015**, *17*, 5808–5811. [CrossRef]
21. Schickedanz, K.; Trageser, T.; Bolte, M.; Lerner, H.-W.; Wagner, M. A boron-doped helicene as a highly soluble, benchtop-stable green emitter. *Chem. Commun.* **2015**, *51*, 15808–15810. [CrossRef]
22. Mori, T. Chiroptical properties of symmetric double, triple, and multiple helicenes. *Chem. Rev.* **2021**, *121*, 2373–2412. [CrossRef] [PubMed]
23. Tsurusaki, A.; Kamikawa, K. Multiple helicenes featuring synthetic approaches and molecular structures. *Chem. Lett.* **2021**, *50*, 1913–1932. [CrossRef]
24. Li, C.; Yang, Y.; Miao, Q. Recent progress in chemistry of multiple helicenes. *Chem. Asian J.* **2018**, *13*, 884–894. [CrossRef]
25. Shen, C.; Gan, F.; Zhang, G.; Ding, Y.; Wang, J.; Wang, R.; Crassous, J.; Qiu, H. Helicene-derived aggregation-induced emission conjugates with highly tunable circularly polarized luminescence. *Mater. Chem. Front.* **2020**, *4*, 837–844. [CrossRef]
26. Meng, D.; Fu, H.; Xiao, C.; Meng, X.; Winands, T.; Ma, W.; Wei, W.; Fan, B.; Huo, L.; Doltsinis, N.L.; et al. Three-bladed rylene propellers with three-dimensional network assembly for organic electronics. *J. Am. Chem. Soc.* **2016**, *138*, 10184–10190. [CrossRef] [PubMed]
27. Meng, D.; Liu, G.; Xiao, C.; Shi, Y.; Zhang, L.; Jiang, L.; Baldridge, K.K.; Li, Y.; Siegel, J.S.; Wang, Z. Corannurylene pentapetalae. *J. Am. Chem. Soc.* **2019**, *141*, 5402–5408. [CrossRef] [PubMed]
28. Lin, Y.-C.; Chen, C.-H.; She, N.-Z.; Juan, C.-Y.; Chang, B.; Li, M.-H.; Wang, H.-C.; Cheng, H.-W.; Yabushita, A.; Yang, Y.; et al. Twisted-graphene-like perylene diimide with dangling functional chromophores as tunable small-molecule acceptors in binary-blend active layers of organic photovoltaics. *J. Mater. Chem. A* **2021**, *9*, 20510–20517. [CrossRef]
29. Lin, Y.-C.; She, N.-Z.; Chen, C.-H.; Yabushita, A.; Lin, H.; Li, M.-H.; Chang, B.; Hsueh, T.-F.; Tsai, B.-S.; Chen, P.-T.; et al. Perylene diimide-fused dithiophenepyrroles with different end groups as acceptors for organic photovoltaics. *ACS Appl. Mater. Interfaces* **2022**, *14*, 37990–38003. [CrossRef]
30. Yang, W.W.; Shen, J.J. Multiple heterohelicenes: Synthesis, properties and applications. *Chem. Eur. J.* **2022**, *28*, e202202069. [CrossRef]
31. Shiraishi, K.; Rajca, A.; Pink, M.; Rajca, S. π-Conjugated conjoined double helicene via a sequence of three oxidative CC-and NN-homocouplings. *J. Am. Chem. Soc.* **2005**, *127*, 9312–9313. [CrossRef]
32. Wang, Z.; Shi, J.; Wang, J.; Li, C.; Tian, X.; Cheng, Y.; Wang, H. Syntheses and crystal structures of benzohexathia[7]helicene and naphthalene cored double helicene. *Org. Lett.* **2010**, *12*, 456–459. [CrossRef] [PubMed]
33. Hashimoto, S.; Nakatsuka, S.; Nakamura, M.; Hatakeyama, T. Construction of a highly distorted benzene ring in a double helicene. *Angew. Chem. Int. Ed.* **2014**, *53*, 14074–14076. [CrossRef] [PubMed]
34. Nakamura, K.; Furumi, S.; Takeuchi, M.; Shibuya, T.; Tanaka, K. Enantioselective synthesis and enhanced circularly polarized luminescence of S-shaped double azahelicenes. *J. Am. Chem. Soc.* **2014**, *136*, 5555–5558. [CrossRef] [PubMed]
35. Chang, H.; Liu, H.; Dmitrieva, E.; Chen, Q.; Ma, J.; He, P.; Liu, P.; Popov, A.A.; Cao, X.-Y.; Wang, X.-Y.; et al. Furan-containing double tetraoxa[7]helicene and its radical cation. *Chem. Commun.* **2020**, *56*, 15181–15184. [CrossRef]
36. Terada, N.; Uematsu, K.; Higuchi, R.; Tokimaru, Y.; Sato, Y.; Nakano, K.; Nozaki, K. Synthesis and properties of spiro-double sila[7]helicene: The LUMO spiro-conjugation. *Chem. Eur. J.* **2021**, *27*, 9342–9349. [CrossRef]
37. Zhang, L.; Song, I.; Ahn, J.; Han, M.; Linares, M.; Surin, M.; Zhang, H.-J.; Oh, J.H.; Lin, J. π-Extended perylene diimide double-heterohelicenes as ambipolar organic semiconductors for broadband circularly polarized light detection. *Nat. Commun.* **2021**, *12*, 142. [CrossRef]
38. Fujikawa, T.; Mitoma, N.; Wakamiya, A.; Saeki, A.; Segawa, Y.; Itami, K. Synthesis, properties, and crystal structures of π-extended double [6]helicenes: Contorted multi-dimensional stacking lattice. *Org. Biomol. Chem.* **2017**, *15*, 4697–4703. [CrossRef]
39. Sun, Z.; Yi, C.; Liang, Q.; Bingi, C.; Zhu, W.; Qiang, P.; Wu, D.; Zhang, F. π-Extended C_2-symmetric double NBN-heterohelicenes with exceptional luminescent properties. *Org. Lett.* **2019**, *22*, 209–213. [CrossRef]
40. Liu, X.; Yu, P.; Xu, L.; Yang, J.; Shi, J.; Wang, Z.; Cheng, Y.; Wang, H. Synthesis for the mesomer and racemate of thiophene-based double helicene under irradiation. *J. Org. Chem.* **2013**, *78*, 6316–6321. [CrossRef]
41. Sakamaki, D.; Kumano, D.; Yashima, E.; Seki, S. A facile and versatile approach to double *N*-heterohelicenes: Tandem oxidative C-N couplings of *N*-heteroacenes via cruciform dimers. *Angew. Chem. Int. Ed.* **2015**, *54*, 5404–5407. [CrossRef]
42. Katayama, T.; Nakatsuka, S.; Hirai, H.; Yasuda, N.; Kumar, J.; Kawai, T.; Hatakeyama, T. Two-step synthesis of boron-fused double helicenes. *J. Am. Chem. Soc.* **2016**, *138*, 5210–5213. [CrossRef] [PubMed]
43. Sakamaki, D.; Tanaka, S.; Tanaka, K.; Takino, H.; Gon, M.; Tanaka, K.; Hirose, T.; Hirobe, D.; Yamamoto, H.M.; Fujiwara, H. Double heterohelicenes composed of benzo [*b*]-and dibenzo [*b*, *i*] phenoxazine: A comprehensive comparison of their electronic and chiroptical properties. *J. Phys. Chem. Lett.* **2021**, *12*, 9283–9292. [CrossRef] [PubMed]
44. Wang, X.-Y.; Wang, X.-C.; Narita, A.; Wagner, M.; Cao, X.-Y.; Feng, X.; Müllen, K. Synthesis, structure, and chiroptical properties of a double [7]heterohelicene. *J. Am. Chem. Soc.* **2016**, *138*, 12783–12786. [CrossRef] [PubMed]
45. Li, J.-K.; Chen, X.-Y.; Guo, Y.-L.; Wang, X.-C.; Sue, A.C.-H.; Cao, X.-Y.; Wang, X.-Y. B,N-embedded double hetero[7]helicenes with strong chiroptical responses in the visible light region. *J. Am. Chem. Soc.* **2021**, *143*, 17958–17963. [CrossRef]

46. Salem, M.S.H.; Khalid, M.I.; Sako, M.; Higashida, K.; Lacroix, C.; Kondo, M.; Takishima, R.; Taniguchi, T.; Miura, M.; Vo-Thanh, G.; et al. Electrochemical synthesis of aza-oxa[7]helicenes via an oxidative heterocoupling and dehydrative cyclization sequence. *Molecules*, 2022; *Submitted*.
47. Khalid, M.I.; Salem, M.S.H.; Sako, M.; Kondo, M.; Sasai, H.; Takizawa, S. Electrochemical synthesis of heterodihydro[7]helicenes. *Commun. Chem.* **2022**, *5*, 166. [CrossRef]
48. Pushkarskaya, E.; Wong, B.; Han, C.; Capomolla, S.; Gu, C.; Stoltz, B.M.; Zhang, H. Single-step synthesis of 3-hydroxycarbazoles by annulation of electron-rich anilines and quinones. *Tetrahedron Lett.* **2016**, *57*, 5653–5657. [CrossRef]
49. Mei, J.; Diao, Y.; Appleton, A.L.; Fang, L.; Bao, Z. Integrated materials design of organic semiconductors for field-effect transistors. *J. Am. Chem. Soc.* **2013**, *135*, 6724–6746. [CrossRef]
50. Rivnay, J.; Jimison, L.H.; Northrup, J.E.; Toney, M.F.; Noriega, R.; Lu, S.; Marks, T.J.; Facchetti, A.; Salleo, A. Large modulation of carrier transport by grain-boundary molecular packing and microstructure in organic thin films. *Nat. Mater.* **2009**, *8*, 952–958. [CrossRef]
51. Minemawari, H.; Tanaka, M.; Tsuzuki, S.; Inoue, S.; Yamada, T.; Kumai, R.; Shimoi, Y.; Hasegawa, T. Enhanced layered-herringbone packing due to long alkyl chain substitution in solution-processable organic semiconductors. *Chem. Mater.* **2017**, *29*, 1245–1254. [CrossRef]
52. Sun, Q.; Ren, J.; Jiang, T.; Peng, Q.; Ou, Q.; Shuai, Z. Intermolecular charge-transfer-induced strong optical emission from herringbone H-Aggregates. *Nano Lett.* **2021**, *21*, 5394–5400. [CrossRef]
53. Gierschner, J.; Ehni, M.; Egelhaaf, H.-J.; Milián Medina, B.; Beljonne, D.; Benmansour, H.; Bazan, G.C. Solid-state optical properties of linear polyconjugated molecules: π-stack contra herringbone. *J. Chem. Phys.* **2005**, *123*, 144914–144922. [CrossRef] [PubMed]
54. Kirstein, S.; Möhwald, H. Herringbone structure in two-dimensional single crystals of cyanine dyes. II. Optical properties. *J. Chem. Phys.* **1995**, *103*, 826–833. [CrossRef]
55. Furche, F.; Ahlrichs, R.; Wachsmann, C.; Weber, E.; Sobanski, A.; Vögtle, F.; Grimme, S. Circular dichroism of helicenes investigated by time-dependent density functional theory. *J. Am. Chem. Soc.* **2000**, *122*, 1717–1724. [CrossRef]
56. Demissie, T.B.; Sundar, M.S.; Thangavel, K.; Andrushchenko, V.; Bedekar, A.V.; Bouř, P. Origins of optical activity in an oxo-helicene: Experimental and computational studies. *ACS Omega* **2021**, *6*, 2420–2428. [CrossRef] [PubMed]
57. Haoyu, S.Y.; He, X.; Li, S.L.; Truhlar, D.G. MN15: A Kohn–Sham global-hybrid exchange–correlation density functional with broad accuracy for multi-reference and single-reference systems and noncovalent interactions. *Chem. Sci.* **2016**, *7*, 5032–5051.
58. Lin, H.A.; Sato, Y.; Segawa, Y.; Nishihara, T.; Sugimoto, N.; Scott, L.T.; Higashiyama, T.; Itami, K. A water-soluble warped nanographene: Synthesis and applications for photoinduced cell death. *Angew. Chem. Int. Ed.* **2018**, *57*, 2874–2878. [CrossRef]
59. Liu, C.; Zhang, S.; Li, J.; Wei, J.; Müllen, K.; Yin, M. A water-soluble, NIR-absorbing quaterrylenediimide chromophore for photoacoustic imaging and efficient photothermal cancer therapy. *Angew. Chem. Int. Ed.* **2019**, *58*, 1638–1642. [CrossRef]
60. Matsuo, Y.; Chen, F.; Kise, K.; Tanaka, T.; Osuka, A. Facile synthesis of fluorescent hetero[8]circulene analogues with tunable solubilities and optical properties. *Chem. Sci.* **2019**, *10*, 11006–11012. [CrossRef]
61. Bredas, J.L.; Silbey, R.; Boudreaux, D.S.; Chance, R.R. Chain-length dependence of electronic and electrochemical properties of conjugated systems: Polyacetylene, polyphenylene, polythiophene, and polypyrrole. *J. Am. Chem. Soc.* **1983**, *105*, 6555–6559. [CrossRef]
62. Yan, M.; Kawamata, Y.; Baran, P.S. Synthetic organic electrochemistry: Calling all engineers. *Angew. Chem. Int. Ed.* **2018**, *57*, 4149–4155. [CrossRef]
63. Frisch, M.E.; Trucks, G.; Schlegel, H.; Scuseria, G.; Robb, M.; Cheeseman, J.; Scalmani, G.; Barone, V.; Petersson, G.; Nakatsuji, H. *Gaussian 16, Rev. C.01*; Gaussian Inc.: Wallingford, CT, USA, 2016.
64. Casida, M.E.; Jamorski, C.; Casida, K.C.; Salahub, D.R. Molecular excitation energies to high-lying bound states from time-dependent density-functional response theory: Characterization and correction of the time-dependent local density approximation ionization threshold. *J. Chem. Phys.* **1998**, *108*, 4439–4449. [CrossRef]
65. Stratmann, R.E.; Scuseria, G.E.; Frisch, M.J. An efficient implementation of time-dependent density-functional theory for the calculation of excitation energies of large molecules. *J. Chem. Phys.* **1998**, *109*, 8218–8224. [CrossRef]
66. Wolinski, K.; Hinton, J.F.; Pulay, P. Efficient implementation of the gauge-independent atomic orbital method for NMR chemical shift calculations. *J. Am. Chem. Soc.* **1990**, *112*, 8251–8260. [CrossRef]
67. Bühl, M.; van Wüllen, C. Computational evidence for a new C84 isomer. *Chem. Phys. Lett.* **1995**, *247*, 63–68. [CrossRef]
68. Schleyer, P.V.R.; Maerker, C.; Dransfeld, A.; Jiao, H.; van Eikema Hommes, N.J. Nucleus-independent chemical shifts: A simple and efficient aromaticity probe. *J. Am. Chem. Soc.* **1996**, *118*, 6317–6318. [CrossRef]

Article

Unexpected Decarbonylation of Acylethynylpyrroles under the Action of Cyanomethyl Carbanion: A Robust Access to Ethynylpyrroles

Denis N. Tomilin [1], Lyubov N. Sobenina [1], Alexandra M. Belogolova [1,2], Alexander B. Trofimov [1,3], Igor A. Ushakov [1] and Boris A. Trofimov [1,*]

[1] A.E. Favorsky Irkutsk Institute of Chemistry, Siberian Branch, Russian Academy of Science, 664033 Irkutsk, Russia
[2] Faculty of Physics, Irkutsk State University, 664003 Irkutsk, Russia
[3] Laboratory of Quantum Chemical Modeling of Molecular Systems, Irkutsk State University, 664003 Irkutsk, Russia
* Correspondence: boris_trofimov@irioch.irk.ru

Abstract: It has been found that the addition of CH_2CN^- anion to the carbonyl group of acylethynylpyrroles, generated from acetonitrile and t-BuOK, results in the formation of acetylenic alcohols, which undergo unexpectedly easy (room temperature) decomposition to ethynylpyrroles and cyanomethylphenylketones (*retro*-Favorsky reaction). This finding allows a robust synthesis of ethynylpyrroles in up to 95% yields to be developed. Since acylethynylpyrroles became available, the strategy thus found makes ethynylpyrroles more accessible than earlier. The quantum-chemical calculations (B2PLYP/6-311G**//B3LYP/6-311G**+C-PCM/acetonitrile) confirm the thermodynamic preference of the decomposition of the intermediate acetylenic alcohols to free ethynylpyrroles rather than their potassium derivatives.

Keywords: acylethynylpyrroles; alkynones; terminal alkynes; deacylation; *retro*-Favorsky reaction

Citation: Tomilin, D.N.; Sobenina, L.N.; Belogolova, A.M.; Trofimov, A.B.; Ushakov, I.A.; Trofimov, B.A. Unexpected Decarbonylation of Acylethynylpyrroles under the Action of Cyanomethyl Carbanion: A Robust Access to Ethynylpyrroles. *Molecules* 2023, 28, 1389. https://doi.org/10.3390/molecules28031389

Academic Editor: Andrea Penoni

Received: 13 December 2022
Revised: 25 January 2023
Accepted: 26 January 2023
Published: 1 February 2023

Copyright: © 2023 by the authors. Licensee MDPI, Basel, Switzerland. This article is an open access article distributed under the terms and conditions of the Creative Commons Attribution (CC BY) license (https://creativecommons.org/licenses/by/4.0/).

1. Introduction

Ethynylpyrroles are valuable building blocks in the synthesis of many natural and synthetic biologically active compounds, such as antibiotic roseophilin, a potent cytotoxic agent against K562 human erythroid leukemia cells [1] and alkaloid quinolactacide with insecticidal activity [2]. They are applied in the syntheses of inhibitors of EGFR tyrosine kinase, an important target for anticancer drug design [3], the HMG-CoA reductase inhibitors for the treatment of hypercholesterolemia, hyperlipoproteinemia, hyperlipidemia and atherosclerosis [4], selective dopamine D4 receptor ligands [5] and foldamers, synthetic receptors, modified for encapsulation of dihydrogenphosphate ions [6]. Pyrroles with terminal acetylenic substituents take part in the syntheses of both lipophilic and highly hydrophilic BODIPY dyes, which fluoresce with high quantum yields and have low cytotoxicity, which makes it possible to visualize cells [7].

These pyrroles are employed in the development of advanced materials capable of detecting various organic and inorganic targets, such as tetrahedral oxoanions ($H_2PO_4^-$ and SO_4^{2-}) [8] and pyrophosphate anions [9].

Also, high-tech materials, including ultrasensitive fluorescent probes for glucopyranoside [10], photoswitchable materials [11–13], components of dye-sensitized solar cells [14], monomers for organic thin-film transistors [15], prospective for energy storage devices, electrochemically active photoluminescence films are based on terminal ethynylpyrroles [16].

In light of the previous, it is clear that the improvement of the synthesis of ethynylpyrroles is a challenge. Indeed, the approaches to the preparation of these functionalized pyrroles are mainly limited to the deprotection of substituted at the triple bond (usually with TMS/TIPS groups) ethynylpyrroles, the products of the reaction of halopyrroles with the corresponding terminal acetylenes (Sonogashira cross-coupling) [2,5,6,8,17–19]. However, in this case, this coupling has limitations, since many halogenated pyrroles, except for representatives with electron-withdrawing substituents, are neither readily available nor stable [20,21]. Variants of the cross-coupling, such as Negishi reaction of halopyrroles with ethynyl magnesium chloride or zinc bromide [22] or cross-coupling of (1-methylpyrrol-2-yl)lithium with fluroacetylene [23], are used albeit less often. It should be especially emphasized that almost all ethynylpyrroles synthesized by the above methods lack the substituents at carbon atoms in the pyrrole ring, i.e., the assortment of accessible ethynylpyrrole remains small and need to be extended.

Among other methods are Corey–Fuchs reaction of pyrrole-2-carbaldehydes with CBr$_4$ with further conversion of dibromoolefins to ethynylpyrroles under the action of bases [1,3,7,24,25] and flash vacuum pyrolysis (FVP) of cyclic and linear 2-alkenylpyrroles (750 °C), limited to a few examples [26–29] due to difficulties in hardware implementation and requirements for substrates. Base-catalyzed elimination of ketones from tertiary acetylenic alcohols (*retro*-Favorsky reaction), affording pyrroles with terminal acetylenic substituents [30], is a rarer approach to such acetylenes because they could decompose or polymerize at high temperatures (up to 180 °C) common for the realization of this synthesis.

The formation of ethynylpyrroles as a result of the deacylation of acylethynylpyrroles was mentioned in only a few cases [31,32], and their yield was insignificant (though alkynones without pyrrole substituents in the presence of alkali metal hydroxides undergo hydrolytic cleavage to form terminal acetylenes [33–35]). For instance, when benzoylethynylpyrrole was treated with NaOH in DMSO (45–50 °C, 4 h), debenzoylation was detected by ^{1}H NMR in negligible extent [31] and 7-days keeping of trifluroacetyl ethynylpyrrole over Al$_2$O$_3$ led to the ethynylpyrrole in 24% yield [32]. Certainly, these results were not suitable for the preparative synthesis of ethynylpyrroles.

Recently [36], we have disclosed the reaction of acylethynylpyrroles **1** with MeCN and metal lithium affording pyrrolyl-cyanopyridines **2** in up 87% yield (Scheme 1).

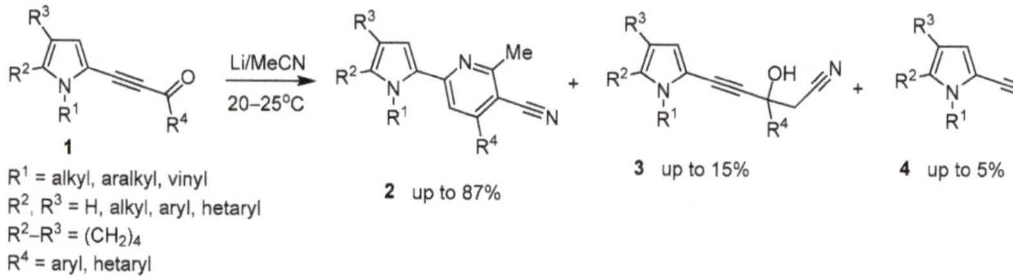

Scheme 1. Previous work. Reaction of acylethynylpyrroles with Li/MeCN system to give pyrrolyl-cyanopyridines.

The synthesis was accompanied by the formation of propargyl alcohols **3** (up to 15%) and small amounts of ethynylpyrroles **4** (up to 5%). The propargyl alcohols **3** were proved to be intermediates in the synthesis of both pyridines and ethynylpyrroles.

These results served as a clue to develop a novel synthesis of ethynylpyrroles, provided we could manage to turn the above side process into a major reaction. Our further successful experiments confirmed this assumption. It appeared that if lithium metal is replaced by *t*-BuOK, the reaction is shifted almost completely to the formation of side ethynylpyrroles. The progress of this synthesis optimization is illustrated in Table 1, wherein the most representative results are presented. As a reference compound, 3-(1-benzyl-4,5,6,7-tetrahydro-1*H*-indol-2-yl)-1-(thiophen-2-yl)prop-2-yn-1-one (**1a**), was chosen believing that the optimal conditions, found for this pyrrolyl acetylenic ketone of higher complexity, will also be valid for the simpler congeners.

In this paper, we report the exceptionally mild decarbonylation of acylethynylpyrroles, readily available from the reaction of pyrroles with electrophilic haloacetylenes in the medium of solid oxides and metal salts [32,37–40], under the action of CH_2CN^- anion generated in situ in the system MeCN/t-BuOK.

2. Results and Discussion

As seen from Table 1, when the reaction was carried out by stirring acylethynylpyrrole **1a** with 2 eq. n-BuLi in MeCN at room temperature under ^{1}H NMR control, the isolated crude product contained 12% of the target ethynylpyrrole **4a** (Table 1, Entry 1), i.e., the expected decarbonylation degree was noticeably increased. The major product, in this case, became tertiary propargylic alcohol **3a** (content in the reaction mixture was 78%). Pyrrolylpyridine **2a**, previously a major product [36], was also present in the reaction mixture but in a much smaller amount (10%). Almost the same results were obtained in the presence of 2 eq. of t-BuONa (Entry 3). But t-BuOLi turned out to be completely inactive in this reaction (Entry 2): the starting acylethynylpyrrole **1a**, in this case, was almost returned from the reaction.

t-BuOK catalyzed the formation of ethynylpyrrole **4a** much more actively: in the crude product obtained with one equivalent of this base, the content of the ethynylpyrrole **4a** in the reaction mixture attained 66% (Entry 4). However, under these conditions, the conversion of the starting acylethynylpyrrole **1a** was only 82%, but the content of pyrrolylpyridine **2a** in the reaction mixture increased to 16%.

When 2 eq. t-BuOK were used, acylethynylpyrrole **1a** reacted completely during the same time, and the content of ethynylpyrrole **4a** in the reaction mixture became 90%. Pyrrolylpyridine **2a** was also present as a by-product (10%) in the reaction mixture (Entry 5).

We found that it was possible to get rid of the pyridine almost completely (Entry 6 and 7) by carrying out the reaction in the mixed solvents (MeCN/THF or MeCN/DMSO in volume ratio 1:1). Thus, under these conditions, the reaction was excellently selective providing ethynylpyrrole **4a** in ~80% isolated yield.

Table 1. Optimization of the ethynylpyrrole **4a** synthesis by decarbonylation of acylethynylpyrrole **1a** [a].

Entry	Base, eq.	Content in the crude, % (^{1}H NMR)			
		1a	2a	3a	4a
1	n-BuLi, 2	traces	10	78	12
2	t-BuOLi, 2	~100	traces	traces	traces
3	t-BuONa, 2	traces	traces	85	15
4	t-BuOK, 1	18	16	traces	66
5	t-BuOK, 2	traces	10	traces	90
6 [b]	t-BuOK, 2	traces	traces	traces	~100 [d]
7 [c]	t-BuOK, 2	traces	traces	traces	~100 [e]

[a]—reaction conditions: 0.5 mmol of **1a**, acetonitrile (2.0 mL), 20–25 °C, nitrogen atmosphere. [b]—the reaction was carried out in the MeCN/THF (1:1) system. [c]—the reaction was carried out in the MeCN/DMSO (1:1) system. [d]—isolated yield 84%. [e]—isolated yield 82%.

Next, with the optimized reaction conditions (2 eq of t-BuOK, THF/MeCN, room temperature, 1 h) in hand, we have evaluated the scope of this reaction using benzoyl-, furoyl-, and thenoylethynylpyrroles with alkyl, aryl and hetaryl substituents at 4(4,5)-

positions and methyl, benzyl, and vinyl moieties at the nitrogen atom of the pyrrole ring. Eventually, the series of earlier unknown ethynylpyrroles **4a–k** were synthesized in good to excellent yields, the exception being pyrrole **4d** (yield 36%) (Scheme 2).

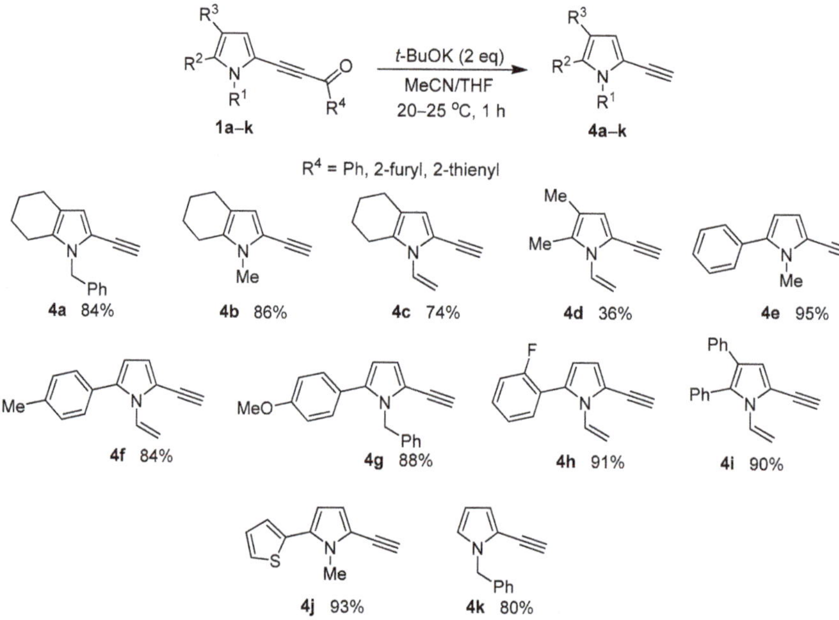

Scheme 2. The scope of the acylethynylpyrroles **1a–k** decarbonylation in the *t*-BuOK/MeCN/THF system.

The method proved to be extendable over indole compounds, as shown in the example of 3-benzoylethynylindole **5**, which was transformed to the expected 1-methyl-3-ethynylindole **6** under the same conditions (Scheme 3).

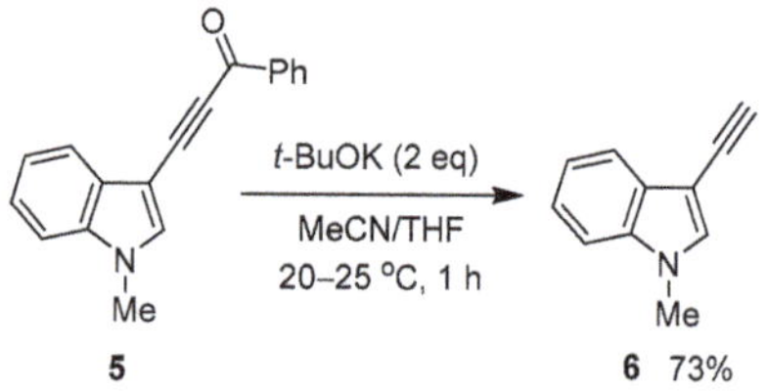

Scheme 3. Reaction of 3-acylethynylindole **5** with *t*-BuOK.

Thus, this result shows that 3-ethynylindoles—valuable synthetic building blocks [41]—could be more accessible than previously due to the above-elaborated strategy. Noteworthy that the starting 3-acylethynylindoles can be easily prepared by the cross-coupling of the corresponding *N*-substituted indoles with acylbromoacetylenes in solid Al_2O_3 media [42].

Also, we have attempted to extend the synthesis of ethynylpyrroles over the furan series. For this, we have chosen menthofuran, a natural antioxidant component of peppermint oil [43]. It turned out that 2-benzoylethynylmenthofuran **7**, which synthesis was previously described in [44], underwent similar decarbonylation under the above conditions to give the expected ethynyl derivative **8** in 80% yield (Scheme 4).

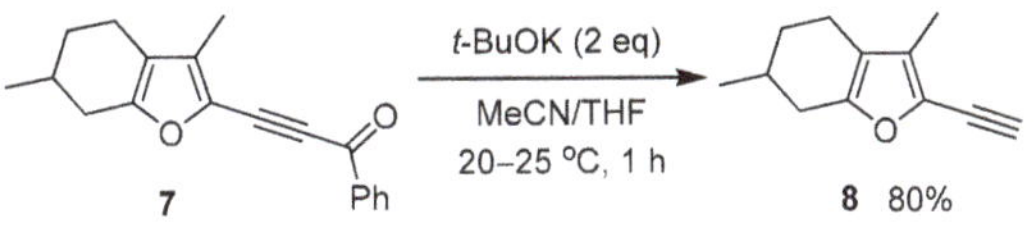

Scheme 4. Reaction of 2-acylethynylmenthofuran **7** with *t*-BuOK.

The narrow range of the yields (74–95%) evidences that the structural effects on the synthesis efficiency are insignificant that are likely to result from the complex character of the process: (i) the formation of intermediate propargyl alcohols **3** and (ii) the decomposition of the latter. Besides, these steps are parallel to the formation of pyridine **2**. Apart from these competing factors, the yields are influenced by the isolation procedure (chromatography on the SiO_2), wherein a noticeable amount of the target products are lost (Table 1, cf. ^{1}H NMR and isolated yields). Nevertheless, the following general trend in yields may be noted: alkyl substituents in the pyrrole ring slightly decrease the reaction efficiency compared to aromatic substituents (74–86% vs. 84–95%). That can be referred to as a higher acidophobicity of the alkyl pyrroles.

It is known that MeCN is easily deprotonated by the action of alkali metals to give acetonitrile dimers via the formation of an intermediate CH_2CN^- anion [45]. Also, it was reported that CH_2CN^- anion was added to ketones to form tertiary cyanomethyl alcohols [46–49]. Correspondingly, in the previous communication, we have shown that the intermediate propargyl alcohol **3** are actually adducts of acylethynylpyrroles and CH_2CN^- anion [36].

Although we failed to isolate propargyl alcohol **3a** in the reaction mixture obtained in the presence of *t*-BuOK, the results produced with *t*-BuONa allowed us to assume that in the first case, the reaction also proceeded with the formation of the intermediate **3a**, which was rapidly decomposed.

To verify this assumption, propargyl alcohols **3a,c,d,f**, prepared from acylethynylpyrroles **1a,c,d,f** and acetonitrile in the presence of *t*-BuONa according to the modified protocol (Scheme 5) [36], were rapidly and quantitatively converted in the presence of *t*-BuOK into the corresponding ethynylpyrroles **4a,c,d,f** (Scheme 5).

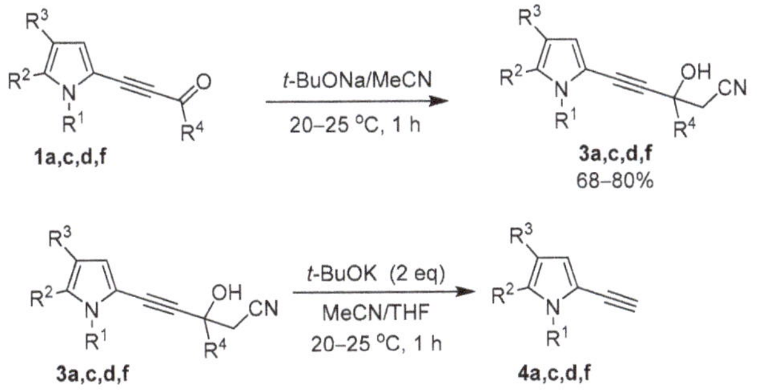

R^1 = CH_2Ph, R^2–R^3 = $(CH_2)_4$, R^4 = 2-thienyl (**3a**)
R^1 = $CH=CH_2$, R^2–R^3 = $(CH_2)_4$, R^4 = Ph (**3c**)
R^1 = $CH=CH_2$, R^2 = R^3 = Me, R^4 = Ph (**3d**)
R^1 = $CH=CH_2$, R^2 = 4-Me-C_6H_4, R^3 = H, R^4 = Ph (**3f**)

Scheme 5. Synthesis and decomposition of propargyl alcohols **3a,c,d,f**.

We performed the reaction in an NMR tube in deuterated acetonitrile. Immediate transformation of the characteristic signals of the protons of the benzoyl group at 8.16 ppm to protons of Ph-substituent occurs after the addition of *t*-BuOK to acylethynylpyrrole **1c** solution, which corresponds to the formation of the intermediate acetylenic alcohol **3c** (Scheme 3). Additionally, two nearly equal singlets at 6.30 (signal of H-3 of pyrrole ring in ethynylpyrrole **4c**) and 6.44 ppm (signal of H-3 of pyrrole ring in intermediate alcohol **3c**) appeared. The singlet at 6.44 ppm decreases rapidly and disappears after about 30 min of reaction. After 1 h reaction mixture contained only terminal alkyne **4c** with a fully deuterated terminal acetylene position. Thus, the results confirm the proposed mechanism of the formation of ethynylpyrroles via intermediate acetylenic alcohol decomposition.

Cyanomethyl ketone (on the example of cyanomethyl-(2-thienyl)ketone **9a**), a second product of the *retro*-Favorsky reaction, was detected (^{1}H NMR) after acidification of the aqueous suspension received during the workup of the reaction mixture (Scheme 6).

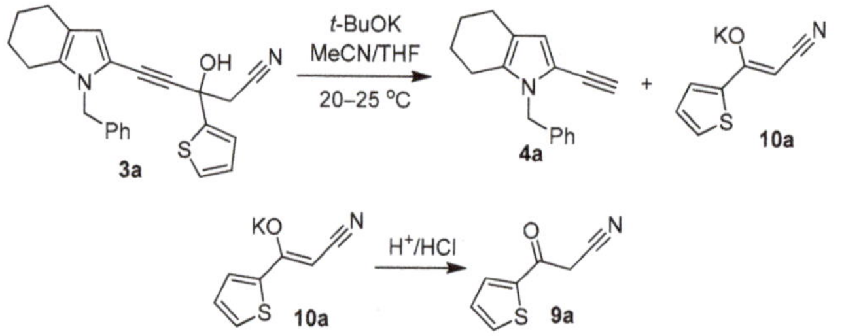

Scheme 6. The decomposition of propargyl alcohol **3a**.

Therefore, it is rigorously confirmed that in this reaction, ethynylpyrroles are the products of tertiary propargyl alcohol **3** decomposition, the *retro*-Favorsky reaction, which in this case occurs under extraordinarily mild (room temperature) conditions. Commonly this reaction requires a considerably higher temperature [120–140 °C (1 mm)] [30].

It could be emphasized that tertiary propargyl alcohols are one of the most attractive synthetic building blocks in organic synthesis [50–58]. This is primarily due to their bifunctionality (acetylene and hydroxyl functions), owing to which they can undergo cascade or multistage reactions with the formation of diverse compounds. In recent years, owing to the development of efficient methods for the synthesis of enantiomerically pure tertiary propargyl alcohols [59–61], interest in this class of compounds has increased significantly.

The tertiary propargyl alcohols here synthesized additionally contain one more synthetically valuable functional group (CN group and active C-H bond adjusted to nitrile function) and a pyrrole ring that significantly expands their potential for the design of novel functionalized compounds.

Despite the experimental evidence highlighting the mechanism of the cascade reaction studied, several mechanistic issues still need a quantum-chemical analysis. These issues mainly relate to the key stage of the synthesis, i.e., the *t*-BuOK-catalyzed decomposition of the intermediate propargyl alcohols **3**. Here the following questions should be clarified: (i) are the intermediates **3** decompose to the corresponding ketones **9** and potassium derivatives **11** of ethynylpyrroles as so far usually considered or free ethynylpyrroles **4** and the corresponding potassium enolates **10** (Scheme 7) are formed? Although the Favorsky *retro*-reaction was synthetically thoroughly studied, this issue was never specially investigated. (ii) Is the experimentally observed formation of enolate from propargyl alcohols kinetically or thermodynamically controlled? (iii) Is the experimentally observed role of alkali metal cation, which fully controls the synthesis direction, an intrinsic (intramolecular) feature of the reaction, or is this influence of intermolecular solvation of the cations? (iv) What is the contribution of the solvent effect to the thermodynamics of this reaction?

To gain a clearer understanding of these mechanistic points, we have performed the quantum chemical calculations of the fundamental characteristics of the above reaction, the Gibbs free energy change, ΔG, using the DFT-based computational approach, which can be briefly referred to as B2PLYP/6-311G**//B3LYP/6-311G**+C-PCM/acetonitrile (see Supplementary Materials for details) and assuming R^1 = Me, $R^2 = R^3$ = H, R^4 = Ph in Scheme 5.

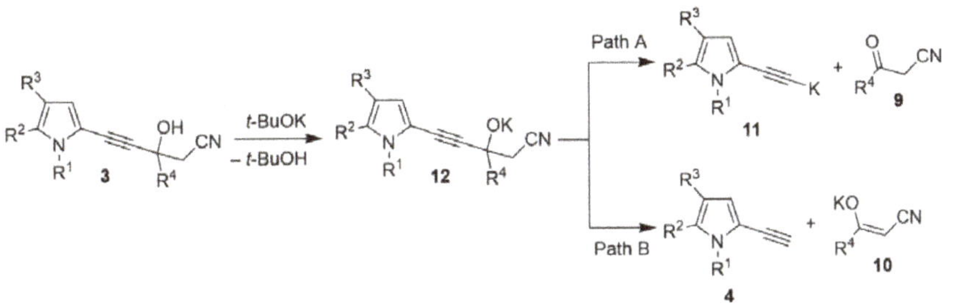

Scheme 7. Possible ways of propargyl alcohols **3** decomposition.

According to the results obtained, path A, i.e., formation of the metallated ethynylpyrroles and ketones (Scheme 5), is thermodynamically closed, whereas path B (Scheme 5), i.e., formation of the ethynylpyrrole and enolate, is thermodynamically opened (see SI for details). The calculations indicate that the decomposition of intermediate **3** proceeds via the formation of free ethynylpyrrole and potassium enolate. Also, these results evidence that path B is thermodynamically controlled. The ΔG values for path B calculated for Li, Na and K derivatives of propargyl alcohols **3** are −66.5, −78.6 and −88.3 kJ/mol, respectively. This explains experimental results according to which with the *t*-BuOLi, no products are formed (Table 1), while *t*-BuONa promotes the formation of sodium enolate, however stable under reaction conditions, and with *t*-BuOK, the decomposition of potassium enolate occurs. Thus, the effect of potassium alkali metal indeed has an intrinsic (intramolecular) character.

The computed O-Li, O-Na and O-K bond lengths in alkali metal derivatives of propargyl alcohols are 1.71, 2.08, and 2.43 Å, respectively, and in the corresponding enolates are 1.83, 2.16, 2.61 Å, respectively. These values correlate with the literature data: 1.95 Å (O-Li), 2.14–2.32 Å (O-Na), 2.60–2.80 Å (O-K) [62], respectively. The reported bond energies are 343, 255 and 238 kJ/mol [63]. From these results, it becomes clear why with *t*-BuOK, the pyridines **2** are not formed: the abstraction of a proton from the CHCN moiety would lead to dianionic-like species that are thermodynamically unfavorable. In the cases of Li- and Na-derivatives of propargyl alcohols, the negative charges on oxygen are smaller since they are tighter ion pairs, especially with lithium cation. Therefore, the reaction takes other directions: with Li cation, expectedly, pyridines are formed, and with *t*-BuONa, the propargyl alcohol decomposition slows down (Table 1, Entry 3).

The mechanism of ethynylpyrroles formation from potassium derivatives of propargyl alcohols (on the example of alcoholate **12**, R^1 = Me, R^2 = R^3 = H, R^4 = Ph) likely represents an intramolecular process (Scheme 8) [36], involving the C_{sp}-CH bond cleavage with simultaneous transfer of a proton from the CH bond.

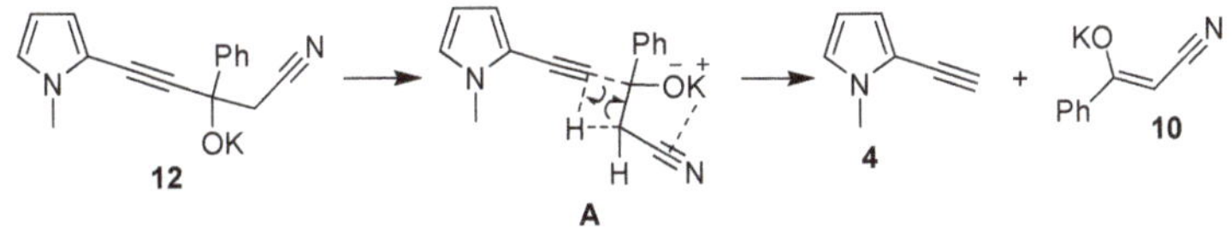

Scheme 8. Ethynylpyrroles formation from potassium derivatives of propargyl alcohols.

This process is probably facilitated by the intramolecular interaction (coordination) between potassium cation and CN-bond (intermediate **A**). This is supported by the fact that the calculated K···N distance in the potassium derivative of propargyl alcohol (3.87 or 4.12 Å, depending on the molecular conformation, see Supplementary Materials) is smaller than the sum of the van der Waals radii of these atoms (4.2 Å). The above-mentioned two conformations are separated only by 0.8 kJ/mol. Since the latter value is well within the error margin of our computational scheme, they both can be considered legitimate propargyl alcohol equilibrium ground-state molecular structures (see Supplementary Materials for more details).

The ΔG values computed for the formation of ethynylpyrroles with the participation of the solvent (MeCN) and then without (gas phase) are close (−88.3 and −84.5 kJ/mol). This means that the contribution of the solvent effect is negligible.

The experiments with MeNO$_2$ showed that in this solvent, the reaction did not proceed at all: the starting acylethynylpyrrole was recovered completely. In our previous work [36], we reported the reaction of benzoylethynylpyrrole **1a** with isobutyronitrile and valeronitrile in the presence of lithium metal. In both cases, respective intermediate alcohols were isolated in 60 and 26% yields. In the presence of *t*-BuOK, both were readily transformed to corresponding ethynylpyrrole **4a**.

3. Experimental Section

3.1. General Information

IR spectra were obtained on a "Bruker IFS-25" spectrometer (Bruker, Billerica, MA, USA) (KBr pellets or films in 400–4000 cm^{-1} region). ^{1}H (400.13 MHz) and ^{13}C (100.6 MHz) NMR spectra were recorded on a "Bruker Avance 400" instrument (Bruker, Billerica, MA, USA) in CDCl$_3$. The assignment of signals in the ^{1}H NMR spectra was made using COSY and NOESY experiments. Resonance signals of carbon atoms were assigned based on ^{1}H-^{13}C HSQC and ^{1}H-^{13}C HMBC experiments. The ^{1}H chemical shifts (δ) were referenced to the residual solvent protons (7.26 ppm, CDCl$_3$), and the ^{13}C chemical shifts were expressed with respect to the deuterated solvent (77.16 ppm). Coupling constants in hertz (Hz) were measured from one-dimensional spectra, and multiplicities were abbreviated as follows: br (broad), s (singlet), d (doublet), t (triplet), and m (multiplet). The chemical shifts were recorded in ppm. The (C, H, N) microanalyses were performed on a Flash EA 1112 CHNS-O/MAS (CHN Analyzer) instrument (Thermo Finnigan, Italy). Sulfur was determined by complexometric titration with Chlorasenazo III. Fluorine content was determined on a SPECOL 11 (Carl Zeiss Jena, Germany) spectrophotometer. Melting points (uncorrected) were determined with SMP50 Stuart Automatic melting point (Cole-Palmer Ltd. Stone, Staffordshire, UK).

3.2. Synthesis of Ethynylpyrroles **4a–k**, Ethynylindole **6**, Ethynylfuran **8**, General Procedure

Acylethynylpyrrole **1a–k**, 3-acylethynylindole **5** or 2-acylethynylfuran **7** (1 mmol) was dissolved in dry THF/MeCN (1:1, 4 mL), and then *t*-BuOK (224 mg, 2 mmol) was added to reaction mixture under nitrogen. Reaction mixture was stirred at room temperature for 1 h while turning into an orange suspension. Then reaction mixture was diluted with cold (0–5 °C) water (30 mL) and extracted by cold (0–5 °C) *n*-hexane (3 × 10 mL). Combined extracts were washed with water (3 × 5 mL) and dried over Na$_2$SO$_4$. The residue, after removing solvent, was purified by flash chromatography (dried SiO$_2$, *n*-hexane) to afford ethynylpyrrole **4a–k**, ethynylindole **6** and ethynylfuran **8**.

1-Benzyl-2-ethynyl-4,5,6,7-tetrahydro-1H-indole (**4a**). Yield: 197 mg (84%), colorless oil; ^{1}H NMR (400.13 MHz, CDCl$_3$): δ 7.37–7.24 (m, 3H, H*m,p*, Ph), 7.13–7.08 (m, 2H, H*o*, Ph), 6.37 (s, 1H, H-3, pyrrole), 5.14 (s, 2H, CH$_2$-Ph), 3.35 (s, 1H, ≡CH), 2.54–2.49 (m, 2H, CH$_2$-7), 2.43–2.38 (m, 2H, CH$_2$-4), 1.82–1.68 (m, 4H, CH$_2$-5, CH$_2$-6); ^{13}C NMR (100.6 MHz, CDCl$_3$): δ 138.3, 130.9, 128.7 (2C), 127.3, 126.7 (2C), 118.0, 114.2, 99.7, 81.2, 77.0, 47.9, 23.5, 23.2, 23.1, 22.5; IR (KBr) 3287, 3087, 3063, 3030, 2928, 2849, 2097, 1495, 1457, 1388, 1357, 1301, 1130, 1077, 1029, 928, 795, 722, 696, 545, 457 cm^{-1}; Anal. Calcd for C$_{17}$H$_{17}$N: C, 86.77; H, 7.28; N, 5.95%. Found: C, 86.47; H, 7.31; N, 6.14%.

2-Ethynyl-1-methyl-4,5,6,7-tetrahydro-1H-indole (**4b**). Yield: 137 mg (86%), white crystals, mp 53–54 °C; ^{1}H NMR (400.13 MHz, CDCl$_3$): δ 6.26 (s, 1H, H-3, pyrrole), 3.50 (s, 3H, NMe), 3.37 (s, 1H, ≡CH), 2.52–2.50 (m, 2H, CH$_2$-7), 2.47–2.45 (m, 2H, CH$_2$-4), 1.83–1.80 (m, 2H, CH$_2$-5), 1.73–1.71 (m, 2H, CH$_2$-6); ^{13}C NMR (100.6 MHz, CDCl$_3$,): δ 130.9, 117.4, 113.6, 112.6, 81.1, 76.9, 30.8, 23.6, 23.2, 23.0, 22.4; IR (film) 3288, 3100, 2929, 2847, 2097, 1570, 1462,

1442, 1386, 1302, 1130, 1055, 790, 667, 536 cm^{-1}; Anal. Calcd for C$_{11}$H$_{13}$N: C, 82.97; H, 8.23; N, 8.80%. Found: C, 82.71; H, 8.44; N, 8.58%.

2-Ethynyl-1-vinyl-4,5,6,7-tetrahydro-1H-indole (**4c**). Yield: 127 mg (74%), colorless oil; ^{1}H NMR (400.13 MHz, CDCl$_3$): δ 6.97 (dd, *J* = 16.1, 9.4 Hz, 1H, H$_x$), 6.34 (s, 1H, H-3, pyrrole), 5.34 (d, *J* = 16.1 Hz, 1H, H$_a$), 4.83 (d, *J* = 9.4 Hz, 1H, H$_b$), 3.39 (s, 1H, ≡CH), 2.66–2.63 (m, 2H, CH$_2$-7), 2.48–2.45 (m, 2H, CH$_2$-4), 1.83–1.80 (m, 2H, CH$_2$-5), 1.71–1.69 (m, 2H, CH$_2$-6); ^{13}C NMR (100.6 MHz, CDCl$_3$): δ 130.5, 119.5, 116.8, 112.3, 102.1, 99.7, 82.1, 76.8, 24.2, 23.4, 23.1, 23.0; IR (film) 3292, 3128, 3049, 2932, 2849, 2099, 1643, 1577, 1483, 1438, 1387, 1324, 1294, 1136, 966, 871, 802, 669, 558 cm^{-1}; Anal. Calcd for C$_{12}$H$_{13}$N: C, 84.17; H, 7.65; N, 8.18%. Found: C, 83.85; H, 7.81; N, 8.36%.

5-Ethynyl-2,3-dimethyl-1-vinyl-1H-pyrrole (**4d**). Yield: 52 mg (36%), colorless oil; ^{1}H NMR (400.13 MHz, CDCl$_3$): δ 6.91 (dd, *J* = 16.0, 9.2 Hz, 1H, H$_x$), 6.36 (s, 1H, H-3, pyrrole), 5.46 (d, *J* = 16.1 Hz, 1H, H$_a$), 4.94 (d, *J* = 9.2 Hz, 1H, H$_b$), 3.37 (s, 1H, ≡CH), 2.21 (s, 3H, Me), 1.99 (s, 3H, Me); ^{13}C NMR (100.6 MHz, CDCl$_3$): δ 130.8, 127.8, 119.0, 116.7, 111.6, 104.5, 81.7, 76.9, 11.4, 11.1; IR (film) 3291, 3106, 2920, 2866, 2099, 1643, 1483, 1432, 1392, 1335, 1310, 1162, 1113, 965, 879, 806, 671, 562 cm^{-1}; Anal. Calcd for C$_{10}$H$_{11}$N: C, 82.72; H, 7.64; N, 9.65%. Found: C, 82.94; H, 7.49; N, 9.80%.

2-Ethynyl-1-methyl-5-phenyl-1H-pyrrole (**4e**). Yield: 172 mg (95%), colorless oil; ^{1}H NMR (400.13 MHz, CDCl$_3$): δ 7.42–7.34 (m, 5H, Ph), 6.55 (d, *J* = 3.8 Hz, 1H, H-3, pyrrole), 6.16 (d, *J* = 3.8 Hz, 1H, H-4, pyrrole), 3.69 (s, 3H, NMe), 3.44 (s, 1H, ≡CH); ^{13}C NMR (100.6 MHz, CDCl$_3$): δ 136.7, 132.9, 128.9 (2C), 128.6 (2C), 127.5, 116.1, 115.6, 108.6, 82.0, 76.5, 33.2; IR (film) 3287, 3106, 3060, 2948, 2102, 1602, 1498, 1457, 1390, 1324, 1234, 1155, 1074, 1028, 758, 698, 568 cm^{-1}; Anal. Calcd for C$_{13}$H$_{11}$N: C, 86.15; H, 6.12; N, 7.73%. Found: C, 85.75; H, 5.86; N, 7.48%.

2-Ethynyl-5-(4-methylphenyl)-1-vinyl-1H-pyrrole (**4f**). Yield: 174 mg (84%), colorless oil; ^{1}H NMR (400.13 MHz, CDCl$_3$): δ 7.34–7.28 (m, 2H, H*o*, Ph), 7.24–7.17 (m, 2H, H*m*, Ph), 6.82 (dd, *J* = 15.9, 9.0 Hz, 1H, H$_x$), 6.63 (d, *J* = 3.8 Hz, 1H, H-3 pyrrole), 6.17 (d, *J* = 3.8 Hz, 1H, H-4, pyrrole), 5.53 (d, *J* = 15.9 Hz, 1H, H$_a$), 4.99 (d, *J* = 9.0 Hz, 1H, H$_b$), 3.43 (s, 1H, ≡CH), 2.38 (s, 3H, Me); ^{13}C NMR (100.6 MHz, CDCl$_3$): δ 137.6, 136.1, 131.1, 129.6, 129.2 (2C), 129.1 (2C), 118.5, 114.5, 109.9, 107.0, 82.5, 76.8, 21.3; IR (KBr) 3287, 3112, 3024, 2921, 2102, 1643, 1547, 1510, 1466, 1419, 1389, 1324, 1297, 1226, 1113, 963, 889, 822, 775, 672, 571, 500 cm^{-1}; Anal. Calcd for C$_{15}$H$_{13}$N: C, 86.92; H, 6.32; N, 6.76%. Found: C, 86.68; H, 6.51; N, 6.85%.

1-Benzyl-2-ethynyl-5-(4-methoxyphenyl)-1H-pyrrole (**4g**). Yield: 253 mg (88%), white crystals; mp 92–93 °C; ^{1}H NMR (400.13 MHz, CDCl$_3$): δ 7.30–7.22 (m, 3H, H*m,p*, Ph), 7.20–7.15 (m, 2H, H*o*, Ph), 6.99–6.93 (m, 2H, H*m*, Ph), 6.87–6.82 (m, 2H, H*o*, Ph), 6.63 (d, *J* = 3.7 Hz, 1H, H-3 pyrrole), 6.16 (d, *J* = 3.7 Hz, 1H, H-4, pyrrole), 5.25 (s, 2H, CH$_2$-Ph), 3.80 (s, 3H, MeO), 3.29 (s, 1H, ≡CH); ^{13}C NMR (CDCl$_3$, 100.6 MHz): δ 159.3, 138.8, 136.7, 130.4 (2C), 128.6 (2C), 127.2, 126.3 (2C), 125.3, 116.1, 115.5, 114.0 (2C), 108.8, 81.8, 76.6, 55.4, 48.9; IR (KBr) 3287, 3087, 3063, 3031, 2955, 2934, 2836, 2100, 1611, 1575, 1547, 1510, 1463, 1442, 1392, 1358, 1321, 1288, 1249, 1178, 1110, 1087, 1031, 977, 909, 836, 767, 731, 695, 575, 524, 459 cm^{-1}; Anal. Calcd for C$_{20}$H$_{17}$NO: C, 83.59; H, 5.96; N, 4.87; O, 5.57%. Found: C, 83.31; H, 6.02; N, 5.02%.

2-Ethynyl-5-(2-fluorophenyl)-1-vinyl-1H-pyrrole (**4h**). Yield: 192 mg (91%), colorless oil; ^{1}H NMR (400.13 MHz, CDCl$_3$): δ 7.40–7.30 (m, 2H, H*m*, Ph), 7.22–7.08 (m, 2H, H*o,p*, Ph), 6.84 (dd, *J* = 15.9, 8.9 Hz, 1H, H$_x$), 6.66 (d, *J* = 3.7 Hz, 1H, H-3 pyrrole), 6.24 (d, *J* = 3.7 Hz, 1H, H-4, pyrrole), 5.34 (d, *J* = 15.9 Hz, 1H, H$_a$), 4.91 (d, *J* = 8.9 Hz, 1H, H$_b$), 3.45 (s, 1H, ≡CH); ^{13}C NMR (100.6 MHz, CDCl$_3$): δ 159.9 (d, *J* = 249.1 Hz, C-2, 2-FC$_6$H$_4$), 132.1 (d, *J* = 2.0 Hz, C-6, 2-FC$_6$H$_4$), 130.9, 130.1 (d, *J* = 8.2 Hz, C-4, 2-FC$_6$H$_4$), 129.2, 124.24 (d, *J* = 3.3 Hz, C-5, 2-FC$_6$H$_4$), 120.6 (d, *J* = 15.5 Hz, C-1, 2-FC$_6$H$_4$), 118.1, 116.1 (d, *J* = 22.0 Hz, C-3, 2-FC$_6$H$_4$), 115.2, 111.8, 106.4, 82.7, 76.4; IR (KBr) 3293, 3115, 3068, 2924, 2104, 1645, 1580, 1547, 1498, 1465, 1397, 1300, 1229, 1109, 963, 890, 817, 780, 759, 672, 577, 471 cm^{-1}; Anal. Calcd for C$_{14}$H$_{10}$FN: C, 79.60; H, 4.77; F, 8.99; N, 6.63%. Found: C, 79.24; H, 4.96; F, 8.75; N, 6.39%.

5-Ethynyl-2,3-diphenyl-1-vinyl-1H-pyrrole (**4i**). Yield: 242 mg (90%), white crystals; mp 93–94 °C; ^{1}H NMR (400.13 MHz, CDCl$_3$): δ 7.39–7.34 (m, 3H, H*o,p*, Ph), 7.31–7.26 (m, 2H, H*o*, Ph), 7.21–7.15 (m, 2H, H*m*, Ph), 7.15–7.09 (m, 3H, H*m,p*, Ph), 6.84 (s, 1H, H-3 pyrrole), 6.71 (dd, *J* = 15.9, 9.2 Hz, 1H, H$_x$), 5.47 (d, *J* = 15.9 Hz, 1H, H$_a$), 4.91 (d, *J* = 9.2 Hz, 1H, H$_b$), 3.46 (s, 1H, ≡CH); ^{13}C NMR (100.6 MHz, CDCl$_3$): δ 135.1, 131.9, 131.8, 131.4 (2C), 130.8, 128.7 (2C), 128.3 (2C), 128.2 (3C), 126.1, 123.8, 118.6, 113.7, 106.5, 82.8, 76.5; IR (KBr) 3274, 3080, 3057, 2923, 2100, 1641, 1601, 1557, 1495, 1446, 1386, 1320, 1305, 1177, 1031, 964, 889, 800, 769, 699, 587, 522 cm^{-1}; Anal. Calcd for C$_{20}$H$_{15}$N: C, 89.19; H, 5.61; N, 5.20%. Found: C, 88.89; H, 5.45; N, 5.34%.

2-Ethynyl-1-methyl-5-(thiophen-2-yl)-1H-pyrrole (**4j**). Yield: 174 mg (93%), colorless oil; ^{1}H NMR (400.13 MHz, CDCl$_3$): δ 7.32–7.28 (m, 1H, H-5, thiophene), 7.10–7.05 (m, 2H, H-3,4, thiophene), 6.51 (d, *J* = 3.9 Hz, 1H, H-3 pyrrole), 6.26 (d, *J* = 3.9 Hz, 1H, H-4, pyrrole), 3.76 (s, 3H, N-CH$_3$), 3.43 (s, 1H, ≡CH); ^{13}C NMR (100.6 MHz, CDCl$_3$): δ 134.4, 129.2, 127.5, 125.8, 125.3, 116.6, 115.6, 109.7, 99.7, 82.2, 33.2; IR (KBr) 3288, 3106, 3074, 2944, 2922, 2101, 1445, 1417, 1395, 1345, 1314, 1201, 1034, 845, 766, 698, 570, 493 cm^{-1}; Anal. Calcd for C$_{11}$H$_9$NS: C, 70.55; H, 4.84; N, 7.48; S, 17.12%. Found: C, 70.26; H, 4.69; N, 7.28; S, 16.82%.

1-Benzyl-2-ethynyl-1H-pyrrole (**4k**). Yield: 145 mg (80%), colorless oil; ^{1}H NMR (400.13 MHz, CDCl$_3$): δ 7.36–7.27 (m, 3H, H*m,p*, Ph), 7.16–7.14 (m, 2H, H*o*, Ph), 6.68–6.65 (m, 1H, H-3, pyrrole), 6.54–6.51 (m, 1H, H-5, pyrrole), 6.13–6.10 (m, 1H, H-4, pyrrole), 5.19 (s, 2H, CH$_2$-Ph), 3.33 (s, 1H, ≡CH); ^{13}C NMR (CDCl$_3$, 100.6 MHz): δ 137.9, 128.8 (2C), 127.7, 127.3 (2C), 123.1, 116.0, 114.7, 108.7, 81.7, 76.0, 51.3; IR (KBr) 3288, 3106, 3064, 3031, 2925, 2853, 2103, 1495, 1466, 1455, 1435, 1300, 1018, 722, 694, 569, 522 cm^{-1}; Anal. Calcd for C$_{13}$H$_{11}$N: C, 86.15; H, 6.12; N, 7.73%. Found: C, 85.84; H, 5.89; N, 7.45%.

3-Ethynyl-1-methyl-1H-indole (**6**). Yield: 113 mg (73%); Spectral characteristics are the same as previously published [64].

2-Ethynyl-3,6-dimethyl-4,5,6,7-tetrahydrobenzofuran (**8**). Yield: 139 mg (80%), colorless oil; ^{1}H NMR (400.13 MHz, CDCl$_3$): δ 3.55 (s, 1H, ≡CH), 2.67–2.62 (m, 1H, CH), 2.33–2.30 (m, 2H, CH$_2$), 2.19–2.12 (m, 1H, CH), 2.00 (s, 3H, Me), 1.93–1.91 (m, 1H, CH), 1.85–1.81 (m, 1H, CH), 1.36–1.30 (m, 1H, CH), 1.07 (d, *J* = 6.7 Hz, 3H, CH*Me*); ^{13}C NMR (100.6 MHz, CDCl$_3$): δ 152.1, 131.5, 127.2, 118.4, 83.8, 74.7, 31.7, 31.2, 29.6, 21.5, 20.0, 9.0; IR (KBr) 3293, 2923, 2849, 2103, 1628, 1558, 1456, 1379, 1295, 1257, 1150, 1107, 1066, 1041, 774, 692 cm^{-1}; Anal. Calcd for C$_{12}$H$_{14}$O: C, 82.72; H, 8.10; O, 9.18%. Found: C, 82.94; H, 7.88%.

3.3. Synthesis of Propargyl Alcohols **3a,c,d,f**

Acylethynylpyrrole **1a,c,d,f** (1 mmol) was dissolved in dry MeCN (4 mL), and then *t*-BuONa (192 mg, 2 mmol) was added to reaction mixture under nitrogen and reaction mixture was stirred at room temperature for 1 h. Then reaction mixture was diluted with water (30 mL) and extracted by diethyl ether (3 × 10 mL). Extracts were washed with water (3 × 5 mL) and dried over Na$_2$SO$_4$. The residue after removing solvents was fractionated by column chromatography (SiO$_2$, *n*-hexane:diethyl ether, 10:1) to afford propargyl alcohol **3a,c,d,f**.

5-(1-Benzyl-4,5,6,7-tetrahydro-1H-indol-2-yl)-3-hydroxy-3-(thiophen-2-yl)pent-4-ynenitrile (**3a**). Spectral characteristics are the same as previously published [36].

3-Hydroxy-3-phenyl-5-(1-vinyl-4,5,6,7-tetrahydro-1H-indol-2-yl)pent-4-ynenitrile (**3c**). Yield: 224 mg (71%), yellow oil; ^{1}H NMR (400.13 MHz, CDCl$_3$): δ 7.72–7.71 (m, 2H, H*o*, Ph), 7.44–7.37 (m, 3H, H*m,p*, Ph), 6.98 (dd, *J* = 15.9, 9.3 Hz, 1H, H$_x$), 6.40 (s, 1H, H-3, pyrrole), 5.34 (d, *J* = 15.9 Hz, 1H, H$_a$), 4.88 (d, *J* = 9.3 Hz, 1H, H$_b$), 3.03 (d, *J* = 4.8 Hz, 2H, CH$_2$CN), 2.85 (s, 1H, OH), 2.67–2.65 (m, 2H, CH$_2$-7), 2.49–2.47 (m, 2H, CH$_2$-4), 1.83–1.81 (m, 2H, CH$_2$-5), 1.74–1.73 (m, 2H, CH$_2$-6); ^{13}C NMR (CDCl$_3$, 100.6 MHz): δ 141.7, 131.5, 130.4, 129.0, 128.8 (2C), 125.4 (2C), 119.9, 117.3, 116.4, 111.3, 103.2, 92.9, 81.4, 71.0, 35.7, 24.1, 23.3, 23.1, 23.0. IR (film) 3422, 3062, 3030, 2931, 2851, 2215, 1643, 1492, 1447, 1383, 1295, 1241, 1143,

1102, 1053, 968, 910, 805, 765, 733, 700, 646 cm^{-1}; Anal. Calcd for $C_{21}H_{20}N_2O$: C, 79.72; H, 6.37; N, 8.85; O, 5.06%. Found: C, 79.44; H, 6.20; N, 8.59%.

5-(4,5-Dimethyl-1-vinyl-1H-pyrrol-2-yl)-3-hydroxy-3-phenylpent-4-ynenitrile (**3d**). Yield: 197 mg (68%), yellow crystals, mp 101–102 °C; ^{1}H NMR (400.13 MHz, CDCl$_3$): δ 7.72–7.70 (m, 2H, H*o*, Ph), 7.42–7.40 (m, 2H, H*m,p*, Ph), 6.91 (dd, *J* = 15.9, 9.1 Hz, 1H, H$_x$), 6.41 (s, 1H, H-3, pyrrole), 5.45 (d, *J* = 15.9 Hz, 1H, H$_a$), 4.99 (d, *J* = 9.1 Hz, 1H, H$_b$), 3.02 (d, *J* = 5.1 Hz, 2H, CH$_2$CN), 2.86 (s, 1H, OH), 2.22 (s, 3H, Me), 2.00 (s, 3H, Me); ^{13}C NMR (100.6 MHz, CDCl$_3$): δ 141.7, 130.6, 128.9, 128.7 (2C), 128.6, 125.4 (2C), 119.5, 117.0, 116.4, 110.6, 105.6, 92.6, 81.4, 70.9, 35.6, 11.3, 11.1; IR (KBr) 3422, 3062, 3030, 2921, 2215, 1643, 1493, 1449, 1392, 1357, 1304, 1172, 1100, 1049, 967, 910, 809, 765, 733, 700, 634 cm^{-1}; Anal. Calcd for $C_{19}H_{18}N_2O$: C, 78.59; H, 6.25; N, 9.65; O, 5.51%. Found: C, 78.22; H, 6.02; N, 9.42%.

3-Hydroxy-3-phenyl-5-(5-(4-methylphenyl)-1-vinyl-1H-pyrrol-2-yl)pent-4-ynenitrile (**3f**). Yield: 281 mg (80%), yellow oil; ^{1}H NMR (CDCl$_3$, 400 MHz): δ 7.74–7.72 (m, 2H, H*o*, Ph), 7.45–7.39 (m, 2H, H*m,p*, Ph), 7.31 (d, *J* = 7.9 Hz, 2H, H*o*, C$_6$H$_4$), 7.21 (d, *J* = 7.9 Hz, 2H, H*m*, C$_6$H$_4$), 6.83 (dd, *J* = 15.8, 8.9 Hz, 1H, H$_x$), 6.67 (d, *J* = 3.8 Hz, 1H, H-4, pyrrole), 6.22 (d, *J* = 3.8 Hz, 1H, H-3, pyrrole), 5.52 (d, *J* = 15.8 Hz, 1H, H$_a$), 5.05 (d, *J* = 8.9 Hz, 1H, H$_b$), 3.06 (d, *J* = 5.0 Hz, 2H, CH$_2$CN), 2.85 (s, 1H, OH), 2.39 (s, 3H, Me); ^{13}C NMR (CDCl$_3$, 100.6 MHz): δ 141.6, 137.8, 136.8, 131.2, 129.4, 129.3 (2C), 129.1 (3C), 128.8 (2C), 125.4 (2C), 118.9, 116.3, 113.7, 110.1, 108.0, 93.1, 81.4, 71.0, 35.6, 21.4; IR (KBr) 3416, 3061, 3028, 2922, 2218, 1643, 1515, 1472, 1449, 1418, 1389, 1324, 1301, 1224, 1112, 1042, 964, 909, 823, 773, 733,701, 622, 503 cm^{-1}, Anal. Calcd for $C_{24}H_{20}N_2O$: C, 81.79; H, 5.72; N, 7.95; O, 4.54%. Found: C, 81.35; H, 5.60; N, 7.68%.

4. Conclusions

In conclusion, we have found efficient and extraordinarily easy (room temperature) access to ethynylpyrroles via decarbonylation of available acylethynylpyrroles. The reaction proceeds in the MeCN-THF/*t*-BuOK system via the addition of CH$_2$CN$^-$ anion to the carbonyl group of acylethynylpyrroles followed by *retro*-Favorsky reaction of the intermediated propargylic alcohols. Thermodynamic aspects of the intermediate alcohol decomposition have been considered in the framework of B2PLYP/6-311G**//B3LYP/6-311G**+C-PCM/acetonitrile methodology. The substrate scope of the reaction includes benzoyl-, furoyl-, thenoylethynylpyrroles with alkyl, vinyl, aryl and hetaryl substituents at 1(4,5)-positions of the pyrrole ring, and methyl, benzyl, and vinyl moieties at the nitrogen atom, as well as acylethynyl derivatives of 1-methylindole and menthofuran.

Supplementary Materials: The following supporting information can be downloaded at: https://www.mdpi.com/article/10.3390/molecules28031389/s1. Synthesis of ethynylpyrroles **4a–k**, ethynylindole **6**, ethynylfuran **8** (S3–S7); Synthesis of propargyl alcohols **3a,c,d,f** (S7–S8); Quantum chemical calculations details (S9–S12); ^{1}H and ^{13}C NMR spectra of synthesized compounds **3a,c,d,f**, **4a–k**, **6**, **8** (S13–S44). References [36,64–74] are cited in the Supplementary Materials.

Author Contributions: Conceptualization, B.A.T.; investigation, D.N.T.; writing—review and editing, D.N.T., L.N.S. and B.A.T.; writing—original draft preparation, D.N.T. and L.N.S.; formal analysis, I.A.U.; calculation part, A.M.B. and A.B.T. All authors have read and agreed to the published version of the manuscript.

Funding: This research was funded by the Ministry of Education and Science of Russian Federation (State Registration No. 121021000199-6).

Institutional Review Board Statement: Not applicable.

Informed Consent Statement: Not applicable.

Data Availability Statement: The data presented in this study are available in the Supplementary Material.

Acknowledgments: This work was supported by the research project plans in the State Register of the IPC RAS no. 121021000199-6. The authors acknowledge Baikal Analytical Center for the collective use of SB RAS for the equipment. A.B.T. gratefully acknowledges Grant No. FZZE-2020-0025 from the Ministry of Science and Higher Education of the Russian Federation. A.M.B. thanks the Irkutsk Supercomputer Center of SB RAS for providing computational resources of the HPC-cluster "Akademik V. M. Matrosov".

Conflicts of Interest: The authors declare no conflict of interest.

Sample Availability: Samples of compounds **1a–k**, **5**, and **7** are available from the authors.

References

1. Bitar, A.Y.; Frontier, A.J. Formal Synthesis of (±)-Roseophilin. *Org. Lett.* **2009**, *11*, 49–52. [CrossRef]
2. Saito, K.; Yoshida, M.; Uekusa, H.; Doi, T. Facile Synthesis of Pyrrolyl 4-Quinolinone Alkaloid Quinolactacide by 9-AJ-Catalyzed Tandem Acyl Transfer–Cyclization of o-Alkynoylaniline Derivatives. *ACS Omega* **2017**, *2*, 4370–4381. [CrossRef]
3. Kitano, Y.; Suzuki, T.; Kawahara, E.; Yamazaki, T. Synthesis and inhibitory activity of 4-alkynyl and 4-alkenylquinazolines: Identification of new scaffolds for potent EGFR tyrosine kinase inhibitors. *Bioorg. Med. Chem. Lett.* **2007**, *17*, 5863–5867. [CrossRef]
4. Thottathil, J.K.; Li, W.S. Process for the Preparation of 4-phosphinyl-3-Keto-Carboxylate and 4-Phosphonyl-3-Keto-Carboxylate Intermediates Useful in the Preparation of Phosphorus Containing HMG-CoA Reductase Inhibitors. U.S. Patent US5298625A, 29 March 1994.
5. Haubmann, C.; Hübner, H.; Gmeiner, P. Piperidinylpyrroles: Design, synthesis and binding properties of novel and selective dopamine D4 receptor ligands. *Bioorg. Med. Chem. Lett.* **1999**, *9*, 3143–3146. [CrossRef] [PubMed]
6. Lee, C.; Lee, H.; Lee, S.; Jeon, H.-G.; Jeong, K.-S. Encapsulation of dihydrogenphosphate ions as a cyclic dimer to the cavities of site-specifically modified indolocarbazole-pyridine foldamers. *Org. Chem. Front.* **2019**, *6*, 299–303. [CrossRef]
7. Guérin, C.; Jean-Gérard, L.; Octobre, G.; Pascal, S.; Maury, O.; Pilet, G.; Ledoux, A.; Andrioletti, B. Bis-triazolyl BODIPYs: A simple dye with strong red-light emission. *RSC Adv.* **2015**, *5*, 76342–76345. [CrossRef]
8. Lim, J.Y.C.; Beer, P.D. A pyrrole-containing cleft-type halogen bonding receptor for oxoanion recognition and sensing in aqueous solvent media. *New J. Chem.* **2018**, *42*, 10472–10475. [CrossRef]
9. Sessler, J.L.; Cai, J.; Gong, H.-Y.; Yang, X.; Arambula, J.F.; Hay, B.P. A Pyrrolyl-Based Triazolophane: A Macrocyclic Receptor with CH and NH Donor Groups That Exhibits a Preference for Pyrophosphate Anions. *J. Am. Chem. Soc.* **2010**, *132*, 14058–14060. [CrossRef]
10. Liao, J.-H.; Chen, C.-T.; Chou, H.-C.; Cheng, C.-C.; Chou, P.-T.; Fang, J.-M.; Slanina, Z.; Chow, T.J. 2,7-Bis(1H-pyrrol-2-yl)ethynyl-1,8naphthyridine: An Ultrasensitive Fluorescent Probe for Glucopyranoside. *Org. Lett.* **2002**, *4*, 3107–3110. [CrossRef]
11. Li, Y.; Zuo, Z.; Liu, H.; Li, Y. Highly-conductive carbon material and low-temperature preparation method thereof. Chinese Patent CN108298516A, 20 July 2018.
12. Heynderickx, A.; Mohamed Kaou, A.; Moustrou, C.; Samat, A.; Guglielmetti, R. Synthesis and photochromic behaviour of new dipyrrolylperfluorocyclopentenes. *New J. Chem.* **2003**, *27*, 1425–1432. [CrossRef]
13. Tanaka, Y.; Ishisaka, T.; Koike, T.; Akita, M. Synthesis and properties of diiron complexes with heteroaromatic linkers: An approach for modulation of organometallic molecular wire. *Polyhedron* **2015**, *86*, 105–110. [CrossRef]
14. Cheema, H.; Baumann, A.; Loya, E.K.; Brogdon, P.; McNamara, L.E.; Carpenter, C.A.; Hammer, N.I.; Mathew, S.; Risko, C.; Delcamp, J.H. Near-Infrared-Absorbing Indolizine-Porphyrin Push–Pull Dye for Dye-Sensitized Solar Cells. *ACS Appl. Mater. Interfaces* **2019**, *11*, 16474–16489. [CrossRef] [PubMed]
15. Debnath, S.; Singh, S.; Bedi, A.; Krishnamoorthy, K.; Zade, S.S. Site-selective synthesis and characterization of BODIPY–acetylene copolymers and their transistor properties. *J. Polym. Sci. A Polym. Chem.* **2016**, *54*, 1978–1986. [CrossRef]
16. Yasuda, T.; Imase, T.; Nakamura, Y.; Yamamoto, T. New Alternative Donor−Acceptor Arranged Poly(Aryleneethynylene)s and Their Related Compounds Composed of Five-Membered Electron-Accepting 1,3,4-Thiadiazole, 1,2,4-Triazole, or 3,4-Dinitrothiophene Units: Synthesis, Packing Structure, and Optical Properties. *Macromolecules* **2005**, *38*, 4687–4697. [CrossRef]
17. Martire, D.O.; Jux, N.; Aramendia, P.F.; Martin Negri, R.; Lex, J.; Braslavsky, S.E.; Schaffner, K.; Vogel, E. Photophysics and photochemistry of 22π and 26π acetylene-cumulene porphyrinoids. *J. Am. Chem. Soc.* **1992**, *114*, 9969–9978. [CrossRef]
18. Tu, B.; Ghosh, B.; Lightner, D.A. Novel Linear Tetrapyrroles: Hydrogen Bonding in Diacetylenic Bilirubins. *Mon. Für Chem. Chem. Mon.* **2004**, *135*, 519–541. [CrossRef]
19. Rana, A.; Lee, S.; Kim, D.; Panda, P.K. β-Octamethoxy-Substituted 22π and 26π Stretched Porphycenes: Synthesis, Characterization, Photodynamics, and Nonlinear Optical Studies. *Chem. Eur. J.* **2015**, *21*, 12129–12135. [CrossRef]
20. Blomquist, A.T.; Wasserman, H.H. Organic Chemistry: A Series of Monographs. In *Organic Chemistry: A Series of Monographs*; Jones, R.A., Bean, G.P., Eds.; Academic Press: Cambridge, MA, USA, 1977; Volume 34, pp. 129–140.
21. Gossauer, A. *Die Chemie der Pyrrole*; Springer: Berlin/Heidelberg, Germany, 2013; Volume 15.
22. Negishi, E.; Xu, C.; Tan, Z.; Kotora, M. Direct Synthesis of Heteroarylethynes via Palladium-catalyzed Coupling of Heteroaryl Halides with Ethynylzinc Halides. Its Application to an Efficient Synthesis of a Thiophenelactone from Chamaemelum nobile L. *Heterocycles* **1997**, *46*, 209–214. [CrossRef]

23. Sauvêtre, R.; Normant, J.F. Une nouvelle preparation du fluoroacetylene—Sa reaction avec les organometalliques. Synthese d'alcynes et d'enynes divers. *Tetrahedron Lett.* **1982**, *23*, 4325–4328. [CrossRef]
24. Tietze, L.F.; Kettschau, G.; Heitmann, K. Synthesis of N-Protected 2-Hydroxymethylpyrroles and Transformation into Acyclic Oligomers. *Synthesis* **1996**, *1996*, 851–857. [CrossRef]
25. Morri, A.K.; Thummala, Y.; Doddi, V.R. The Dual Role of 1,8-Diazabicyclo [5.4.0]undec-7-ene (DBU) in the Synthesis of Terminal Aryl- and Styryl-Acetylenes via Umpolung Reactivity. *Org. Lett.* **2015**, *17*, 4640–4643. [CrossRef] [PubMed]
26. Wentrup, C.; Winter, H.-W. A General and Facile Synthesis of Aryl- and Hetero-arylacetylenes. *Angew. Chem. Int. Ed.* **1978**, *17*, 609–610. [CrossRef]
27. Benzies, D.W.M.; Fresneda, P.M.; Jones, R.A.; McNab, H. Flash vacuum pyrolysis of 5-(indol-2- and -3-ylmethylene)-2,2-dimethyl-1,3-dioxane-4,6-diones. *J. Am. Chem. Soc. Perkin Trans.* **1986**, *1*, 1651–1654. [CrossRef]
28. Comer, M.C.; Despinoy, X.L.M.; Gould, R.O.; McNab, H.; Parsons, S. Synthesis and unexpectedly facile dimerisation of 1-methoxycarbonylpyrrolizin-3-one. *Chem. Commun.* **1996**, *9*, 1083–1084. [CrossRef]
29. Despinoy, X.L.M.; McNab, H. 1-Methoxycarbonylpyrrolizin-3-one and related compounds. *Org. Biomol. Chem.* **2009**, *7*, 2187–2194. [CrossRef]
30. Vasilevskii, S.F.; Sundukova, T.A.; Shvartsberg, M.S.; Kotlyarevskii, I.L. Synthesis of acetylenyl-N-methylpyrroles. *Izv. Akad. Nauk. SSSR Seriya Khimicheskaya* **1980**, *8*, 1346–1350. [CrossRef]
31. Sobenina, L.N.; Tomilin, D.N.; Gotsko, M.D.; Ushakov, I.A.; Mikhaleva, A.I.; Trofimov, B.A. From 4,5,6,7-tetrahydroindoles to 3- or 5-(4,5,6,7-tetrahydroindol-2-yl)isoxazoles in two steps: A regioselective switch between 3- and 5-isomers. *Tetrahedron* **2014**, *70*, 5168–5174. [CrossRef]
32. Tomilin, D.N.; Gotsko, M.D.; Sobenina, L.N.; Ushakov, I.A.; Afonin, A.V.; Soshnikov, D.Y.; Trofimov, A.B.; Koldobsky, A.B.; Trofimov, B.A. N-Vinyl-2-(trifluoroacetylethynyl)pyrroles and E-2-(1-bromo-2-trifluoroacetylethenyl)pyrroles: Cross-coupling vs. addition during CH-functionalization of pyrroles with bromotrifluoroacetylacetylene in solid Al$_2$O$_3$ medium. H-bonding control. *J. Fluorine Chem.* **2016**, *186*, 1–6. [CrossRef]
33. Vereshchagin, L.I.; Kirillova, L.P.; Buzilova, S.R. Unsaturated carbonyl-containing compounds. 18. Alkaline cleavage of alpha-acetylenic ketones. *Zh. Org. Khim.* **1975**, *11*, 292.
34. Fedenok, L.G.; Shvartsberg, M.S. A method for the preparation of terminal acetylenes. *Bull. Acad. Sci. USSR Div. Chem. Sci.* **1990**, *39*, 2376–2377. [CrossRef]
35. Trofimov, B.A.; Sobenina, L.N.; Mikhaleva, A.I.; Ushakov, I.A.; Vakul'skaya, T.I.; Stepanova, Z.V.; Toryashinova, D.-S.D.; Mal'kina, A.G.; Elokhina, V.N. N- and C-Vinylation of Pyrroles with Disubstituted Activated Acetylenes. *Synthesis* **2003**, *2003*, 1272–1278. [CrossRef]
36. Tomilin, D.N.; Sobenina, L.N.; Saliy, I.V.; Ushakov, I.A.; Belogolova, A.M.; Trofimov, B.A. Substituted pyrrolyl-cyanopyridines on the platform of acylethynylpyrroles via their 1:2 annulation with acetonitrile under the action of lithium metal. *New J. Chem.* **2022**, *46*, 13149–13155. [CrossRef]
37. Trofimov, B.A.; Stepanova, Z.V.; Sobenina, L.N.; Mikhaleva, A.I.; Ushakov, I.A. Ethynylation of pyrroles with 1-acyl-2-bromoacetylenes on alumina: A formal 'inverse Sonogashira coupling'. *Tetrahedron Lett.* **2004**, *45*, 6513–6516. [CrossRef]
38. Trofimov, B.A.; Sobenina, L.N. *Targets in Heterocyclic Systems*.; Attanasi, O.A., Spinelli, D., Eds.; Società Chimica Italiana: Roma, Italy, 2009; Volume 13, pp. 92–119.
39. Sobenina, L.N.; Tomilin, D.N.; Petrova, O.V.; Gulia, N.; Osowska, K.; Szafert, S.; Mikhaleva, A.I.; Trofimov, B.A. Cross-coupling of 4,5,6,7-tetrahydroindole with functionalized haloacetylenes on active surfaces of metal oxides and salts. *Russ. J. Org. Chem.* **2010**, *46*, 1373–1377. [CrossRef]
40. Sobenina, L.N.; Trofimov, B.A. Recent Strides in the Transition Metal-Free Cross-Coupling of Haloacetylenes with Electron-Rich Heterocycles in Solid Media. *Molecules* **2020**, *25*, 2490. [CrossRef]
41. Tarshits, D.L.; Przhiyalgovskaya, N.M.; Buyanov, V.N.; Tarasov, S.Y. Ethynylindoles and their derivatives. Methods of synthesis and chemical transformations (review). *Chem. Heterocycl. Compd.* **2009**, *45*, 501–523. [CrossRef]
42. Sobenina, L.N.; Demenev, A.P.; Mikhaleva, A.I.; Ushakov, I.A.; Vasil'tsov, A.M.; Ivanov, A.V.; Trofimov, B.A. Ethynylation of indoles with 1-benzoyl-2-bromoacetylene on Al$_2$O$_3$. *Tetrahedron Lett.* **2006**, *47*, 7139–7141. [CrossRef]
43. Tisserand, R.; Young, R. 12—Cancer and the immune system. In *Essential Oil Safety*, 2nd ed.; Tisserand, R., Young, R., Eds.; Churchill Livingstone: St. Louis, MO, USA, 2014; pp. 165–186.
44. Sobenina, L.N.; Tomilin, D.N.; Gotsko, M.D.; Ushakov, I.A.; Trofimov, B.A. Transition metal-free cross-coupling of furan ring with haloacetylenes. *Tetrahedron* **2018**, *74*, 1565–1570. [CrossRef]
45. Uchida, A.; Doyama, A.; Matsuda, S. Reaction of Acetonitrile with Carboxylic Esters. *Bull. Chem. Soc. Jpn.* **1970**, *43*, 963–965. [CrossRef]
46. Ko, E.-Y.; Lim, C.-H.; Chung, K.-H. Additions of Acetonitrile and Chloroform to Aromatic Aldehydes in the Presence of Tetrabutylammonium Fluoride. *Bull. Korean Chem. Soc.* **2006**, *27*, 432–434. [CrossRef]
47. Xiao, S.; Chen, C.; Li, H.; Lin, K.; Zhou, W. A Novel and Practical Synthesis of Rameltoen. *Org. Process Res. Dev.* **2015**, *19*, 373–377. [CrossRef]
48. Yu, Y.; Li, G.; Jiang, L.; Zu, L. An Indoxyl-Based Strategy for the Synthesis of Indolines and Indolenines. *Angew. Chem. Int. Ed.* **2015**, *54*, 12627–12631. [CrossRef] [PubMed]
49. Hoff, B.H. Acetonitrile as a Building Block and Reactant. *Synthesis* **2018**, *50*, 2824–2852. [CrossRef]

50. Engel, D.A.; Dudley, G.B. The Meyer–Schuster rearrangement for the synthesis of α,β-unsaturated carbonyl compounds. *Org. Biomol. Chem.* **2009**, *7*, 4149–4158. [CrossRef]
51. Wang, L.-X.; Tang, Y.-L. Cycloisomerization of Pyridine-Substituted Propargylic Alcohols or Esters To Construct Indolizines and Indolizinones. *Eur. J. Org. Chem.* **2017**, *2017*, 2207–2213. [CrossRef]
52. Roy, R.; Saha, S. Scope and advances in the catalytic propargylic substitution reaction. *RSC Adv.* **2018**, *8*, 31129–31193. [CrossRef]
53. Qian, H.; Huang, D.; Bi, Y.; Yan, G. 2-Propargyl Alcohols in Organic Synthesis. *Adv. Synth. Catal.* **2019**, *361*, 3240–3280. [CrossRef]
54. Kumar, G.R.; Rajesh, M.; Lin, S.; Liu, S. Propargylic Alcohols as Coupling Partners in Transition-Metal-Catalyzed Arene C−H Activation. *Adv. Synth. Catal.* **2020**, *362*, 5238–5256. [CrossRef]
55. Liu, X.-Y.; Liu, Y.-L.; Chen, L. Tandem Annulations of Propargylic Alcohols to Indole Derivatives. *Adv. Synth. Catal.* **2020**, *362*, 5170–5195. [CrossRef]
56. Du, S.; Zhou, A.-X.; Yang, R.; Song, X.-R.; Xiao, Q. Recent advances in the direct transformation of propargylic alcohols to allenes. *Org. Chem. Front.* **2021**, *8*, 6760–6782. [CrossRef]
57. Song, X.-R.; Yang, R.; Xiao, Q. Recent Advances in the Synthesis of Heterocyclics via Cascade Cyclization of Propargylic Alcohols. *Adv. Synth. Catal.* **2021**, *363*, 852–876. [CrossRef]
58. Bai, J.-F.; Tang, J.; Gao, X.; Jiang, Z.-J.; Tang, B.; Chen, J.; Gao, Z. Regioselective Cycloaddition and Substitution Reaction of Tertiary Propargylic Alcohols and Heteroareneboronic Acids via Acid Catalysis. *Org. Lett.* **2022**, *24*, 4507–4512. [CrossRef] [PubMed]
59. Noda, H.; Kumagai, N.; Shibasaki, M. Catalytic Asymmetric Synthesis of α-Trifluoromethylated Carbinols: A Case Study of Tertiary Propargylic Alcohols. *Asian J. Org. Chem.* **2018**, *7*, 599–612. [CrossRef]
60. Kobayashi, D.; Miura, M.; Toriyama, M.; Motohashi, S. Stereoselective synthesis of secondary and tertiary propargylic alcohols induced by a chiral sulfoxide auxiliary. *Tetrahedron Lett.* **2019**, *60*, 120–123. [CrossRef]
61. Wang, J.; Zhang, W.; Wu, P.; Huang, C.; Zheng, Y.; Zheng, W.-F.; Qian, H.; Ma, S. Chiral tertiary propargylic alcohols via Pd-catalyzed carboxylative kinetic resolution. *Org. Chem. Front.* **2020**, *7*, 3907–3911. [CrossRef]
62. Bradley, D.C.; Mehrotra, R.C.; Rothwell, I.P.; Singh, A. 4—X-Ray Crystal Structures of Alkoxo Metal Compounds. In *Alkoxo and Aryloxo Derivatives of Metals*; Bradley, D.C., Mehrotra, R.C., Rothwell, I.P., Singh, A., Eds.; Academic Press: London, UK, 2001; pp. 229–382.
63. Dean, J.A.; Lange, N.A. *Lange's Handbook of Chemistry*; McGraw-Hill: New York, NY, USA, 1999.
64. Gupton, J.T.; Telang, N.; Gazzo, D.F.; Barelli, P.J.; Lescalleet, K.E.; Fagan, J.W.; Mills, B.J.; Finzel, K.L.; Kanters, R.P.F.; Crocker, K.R.; et al. Preparation of indole containing building blocks for the regiospecific construction of indole appended pyrazoles and pyrroles. *Tetrahedron* **2013**, *69*, 5829–5840. [CrossRef]
65. Grimme, S. Semiempirical hybrid density functional with perturbative second-order correlation. *J. Chem. Phys.* **2006**, *124*, 034108. [CrossRef]
66. Becke, A.D. Density-functional thermochemistry. III. The role of exact exchange. *J. Chem. Phys.* **1993**, *98*, 5648–5652. [CrossRef]
67. Lee, C.; Yang, W.; Parr, R.G. Development of the Colle-Salvetti correlation-energy formula into a functional of the electron density. *Phys. Rev. B* **1988**, *37*, 785–789. [CrossRef]
68. Krishnan, R.; Binkley, J.S.; Seeger, R.; Pople, J.A. Self-consistent molecular orbital methods. XX. A basis set for correlated wave functions. *J. Chem. Phys.* **1980**, *72*, 650–654. [CrossRef]
69. Tomasi, J.; Mennucci, B.; Cammi, R. Quantum Mechanical Continuum Solvation Models. *Chem. Rev.* **2005**, *105*, 2999–3094. [CrossRef]
70. Cossi, M.; Rega, N.; Scalmani, G.; Barone, V. Energies, structures, and electronic properties of molecules in solution with the C-PCM solvation model. *J. Comput. Chem.* **2003**, *24*, 669–681. [CrossRef] [PubMed]
71. Vitkovskaya, N.M.; Orel, V.B.; Absalyamov, D.Z.; Trofimov, B.A. Self-Assembly of N-Phenyl-2,5-dimethylpyrrole from Acetylene and Aniline in KOH/DMSO and KOBu^t/DMSO Superbase Systems: A Quantum-Chemical Insight. *J. Org. Chem.* **2020**, *85*, 10617–10627. [CrossRef] [PubMed]
72. Frisch, M.J.; Trucks, G.W.; Schlegel, H.B.; Scuseria, G.E.; Robb, M.A.; Cheeseman, J.R.; Scalmani, G.; Barone, V.; Mennucci, B.; Petersson, G.A.; et al. *Gaussian 09*; Gaussian, Inc.: Wallingford, CT, USA, 2009.
73. Petersson, G.A. Complete Basis Set Models for Chemical Reactivity: From the Helium Atom to Enzyme Kinetics. In *Quantum-Mechanical Prediction of Thermochemical Data*; Cioslowski, J., Ed.; Springer: Dordrecht, The Netherlands, 2001; pp. 99–130.
74. Montgomery, J.A., Jr.; Frisch, M.J.; Ochterski, J.W.; Petersson, G.A. A complete basis set model chemistry. VII. Use of the minimum population localization method. *J. Chem. Phys.* **2000**, *112*, 6532–6542. [CrossRef]

Disclaimer/Publisher's Note: The statements, opinions and data contained in all publications are solely those of the individual author(s) and contributor(s) and not of MDPI and/or the editor(s). MDPI and/or the editor(s) disclaim responsibility for any injury to people or property resulting from any ideas, methods, instructions or products referred to in the content.

Article

Synthesis and Catalytic Activity of Bifunctional Phase-Transfer Organocatalysts Based on Camphor

Luka Ciber [1], Franc Požgan [1], Helena Brodnik [1], Bogdan Štefane [1], Jurij Svete [1], Mario Waser [2,*] and Uroš Grošelj [1,*]

[1] Chair of Organic Chemistry, Faculty of Chemistry and Chemical Technology, University of Ljubljana, Večna pot 113, 1000 Ljubljana, Slovenia
[2] Institute of Organic Chemistry, Johannes Kepler University Linz, Altenbergerstrasse 69, 4040 Linz, Austria
* Correspondence: mario.waser@jku.at (M.W.); uros.groselj@fkkt.uni-lj.si (U.G.); Tel.: +386-1-479-8565 (U.G.)

Abstract: Ten novel bifunctional quaternary ammonium salt phase-transfer organocatalysts were synthesized in four steps from (+)-camphor-derived 1,3-diamines. These quaternary ammonium salts contained either (thio)urea or squaramide hydrogen bond donor groups in combination with either trifluoroacetate or iodide as the counteranion. Their organocatalytic activity was evaluated in electrophilic heterofunctionalizations of β-keto esters and in the Michael addition of a glycine Schiff base with methyl acrylate. α-Fluorination and chlorination of β-keto esters proceeded with full conversion and low enantioselectivities (up to 29% ee). Similarly, the Michael addition of a glycine Schiff base with methyl acrylate proceeded with full conversion and up to 11% ee. The new catalysts have been fully characterized; the stereochemistry at the C-2 chiral center was unambiguously determined.

Keywords: asymmetric organocatalysis; quaternary ammonium salts; phase-transfer catalysts (PTCs); camphor; camphor-derived diamines; β-keto esters; enantioselective α-fluorination; electrophilic α-chlorination; asymmetric Michael addition

1. Introduction

Since the seminal contributions of Wynberg [1], Dolling [2], and O'Donnell [3] in the 1970s and 1980s, in which chiral quaternary ammonium salts based on Cinchona alkaloids were used as catalysts for enantioselective epoxidations and α-alkylations of prochiral substrates, the use of chiral quaternary ammonium salts as phase-transfer catalysts (PTCs) has been successfully demonstrated in a multitude of asymmetric organic transformations and now represents an established fundamental catalysis principle in asymmetric organocatalysis [4–8]. In addition to Cinchona alkaloid-based PTCs, other chiral backbones have also been successfully used to access high-performance catalysts. A group of highly efficient binaphthyl-based ammonium salts was introduced by Maruoka (the so-called Maruoka catalysts) [9,10], which have since established themselves as the second most privileged class of chiral ammonium salt PTCs, alongside Cinchona alkaloids (Figure 1). Over the years, efficient chiral quaternary ammonium salts based on tartaric acid [11,12], α-amino acids [13,14], *trans*-cyclohexane-1,2-diamine [15], and others [16] have been developed. Many of the developed quaternary ammonium salts, especially catalysts based on Cinchona alkaloids and some Maruoka-type catalysts, possess a hydrogen-bonding donor in the form of a OH group, which leads to improved catalytic properties [17]. The incorporation of (thio)urea-containing hydrogen bond donors in catalysts based on Cinchona alkaloids, and, in particular, in catalysts based on amino acids and cyclohexane-1,2-diamine, contributed significantly to the diversification of the available catalysts and extended the scope of catalyzed asymmetric transformations [13,15,18,19].

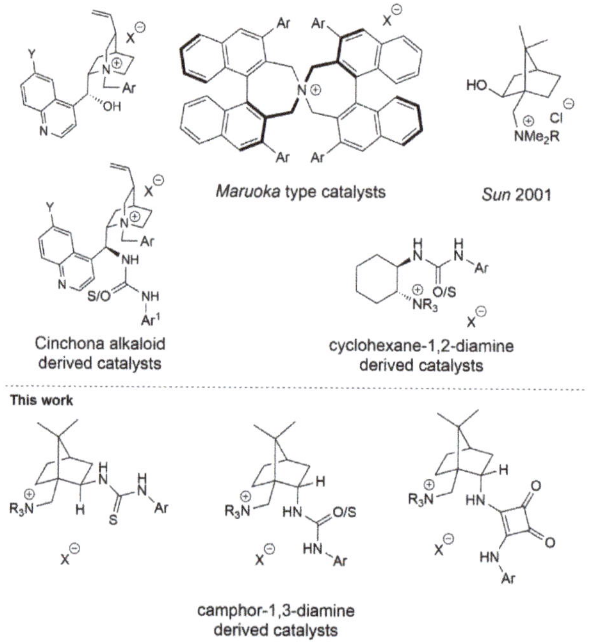

Figure 1. Selected efficient chiral quaternary ammonium salt phase-transfer catalysts (PTCs) and novel bifunctional camphor-based PTCs reported herein [20].

Camphor is one of nature's most privileged scaffolds, readily available in both enantiomeric forms. In addition, camphor undergoes a variety of interesting chemical transformations that functionalize, at first sight, inactive positions [21,22], allowing the synthesis of structurally and functionally very different products [23–27], thus making camphor a desirable starting material. The first reports on the application of camphor-derived organocatalysts date back to 2001. Camphor-derived phase-transfer organocatalysts were employed to catalyze the α-alkylation of a glycine Schiff base with enantioselectivities up to 39% ee (Figure 1) [20]. Since then, several types of camphor-based organocatalysts have been reported, exhibiting covalent or noncovalent activation modes, both those with a camphor backbone as the sole chiral fragment and those in which the camphor backbone is covalently linked to a chiral amino acid, usually proline, via a suitable spacer [28].

As part of our ongoing study of camphor-based diamines as potential organocatalyst scaffolds [29], we reported the synthesis of 1,3-diamine-based bifunctional squaramide organocatalysts prepared from camphor and their application as efficient catalysts in Michael additions of 1,3-dicarbonyl compounds and pyrrolones as nucleophiles to *trans*-β-nitrostyrene derivatives [30,31]. Extending this work, we report here the synthesis of a new type of 1,3-diamine-based bifunctional quaternary ammonium salt phase-transfer organocatalyst (Figure 1) and its evaluation in the electrophilic α-functionalization of β-keto ester and the alkylation of a glycine-derived Schiff base with methyl acrylate.

2. Results and Discussion
2.1. Synthesis

Camphor-derived *endo*-diamines **1a,b** and *exo*-diamines **2a** were prepared in four steps from commercially available (1*S*)-(+)-10-camphorsulfonic acid (Scheme 1) [29,30]. Camphorsulfonic acid was transformed into 10-iodocamphor in an Apple-type reaction, followed by nucleophilic substitution with pyrrolidine or dimethylamine in dimethyl sulfoxide. The thus formed tertiary amines were transformed into the corresponding oximes. The final oxime reduction with sodium in isopropanol gave a mixture of the corresponding major *endo*-diamines **1a,b** and minor *exo*-diamines **2a**, separable by column chromatography.

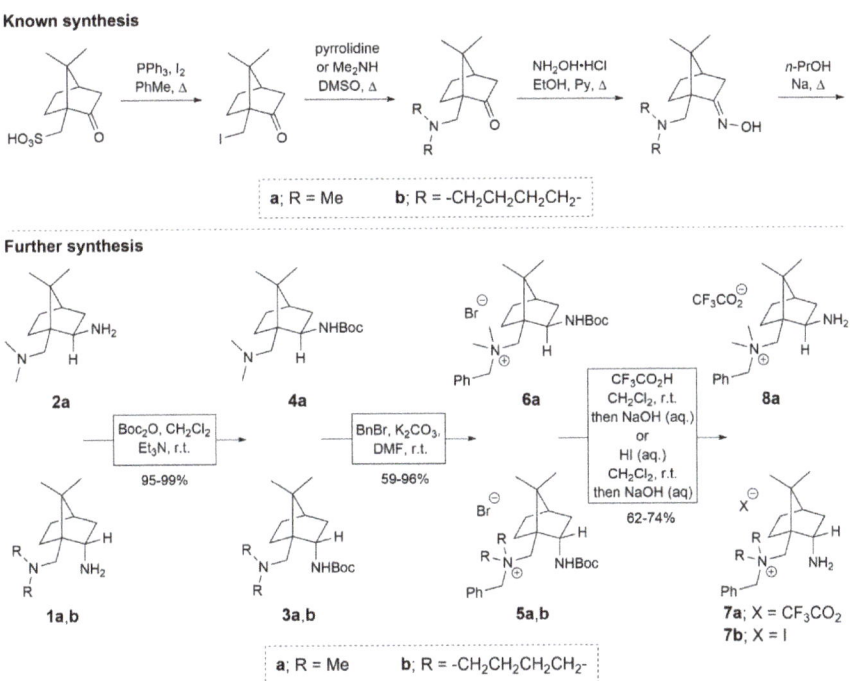

Scheme 1. Synthesis of camphor-derived *endo*- **7a,b** and *exo*-quaternary ammonium salts **8a**.

Next, the primary amino group of diamines **1a,b** and **2a** was Boc-protected, yielding **3a,b** and **4a**, respectively. In the following step, we introduced the benzyl group to the tertiary amine (benzyl groups have been very successfully established as useful motives for numerous quaternary ammonium salt phase-transfer catalysts [4–8]). Alkylation with benzyl bromide thus gave the quaternary ammonium salts **5a,b** and **6a**. Potassium carbonate was added to ensure complete conversion. They were subsequently Boc-deprotected with trifluoroacetic acid or aqueous hydrogen iodide, furnishing ammonium salts **7a,b** and **8a**, respectively (Scheme 1).

Finally, the ammonium salts **7a,b** and **8a** were reacted with aromatic iso(thio)cyanates **9** and squaramate **10** to give the quaternary trifluoroacetate ammonium salts **I, IV, VI, VII,** and **IX**. The quaternary iodide ammonium salts **II, V,** and **VIII** were formed from the corresponding trifluoroacetates **I, IV,** and **VII**, respectively, via anion metathesis with excess NaI in dichloromethane (Scheme 2). The quaternary iodide ammonium salts **III** and **X** were formed directly from the iodide ammonium salt **7b** and the corresponding isothiocyanate. The catalysts thus formed have either (thio)urea or squaramide hydrogen bond donors (Supplementary Materials).

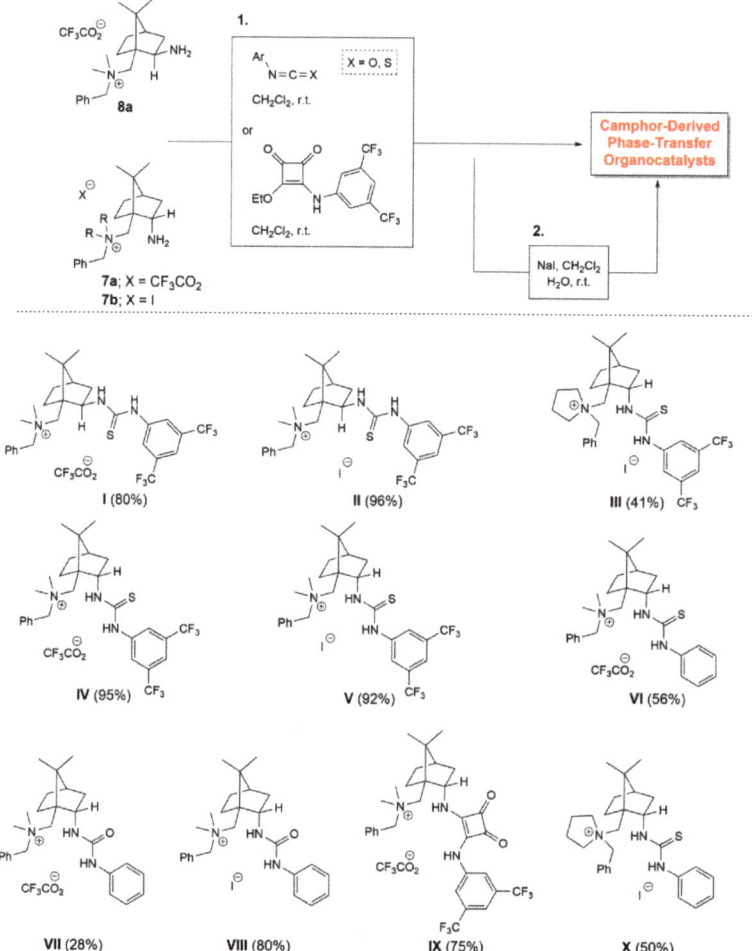

Scheme 2. Synthesis of camphor-derived phase-transfer organocatalysts.

2.2. Structure Determination

The intermediates **3–8** were characterized by ^{1}H- and ^{13}C-NMR, IR, and HRMS. Compounds **1a** and **2a** were characterized by ^{1}H-NMR. Phase-transfer catalysts **I–X** have been fully characterized. The structures of the thioureas **III** and **VI-Br** (the bromide analog of the compound **VI**) were determined by single-crystal X-ray analysis (Figure 2). In both structures, the *endo*-stereochemistry was confirmed at the C-2 chiral center. The conformational differences in the two structures in the solid state are shown in Figure 3. The main differences are due to the conformation of the benzyl group and the arylthiourea structural elements.

The *endo*-stereochemistry at the C-2 chiral center of compounds **III–X** was further confirmed by NOESY measurements based on the cross-peak between the methyl group and the *exo*-H(2) proton (Figure 4). Similarly, the *exo*-stereochemistry at the C-2 chiral center of compounds **I** and **II** was in line with the cross-peak between the methyl group and the *exo*-H–N proton observed in the NOESY spectra (Supplementary Materials).

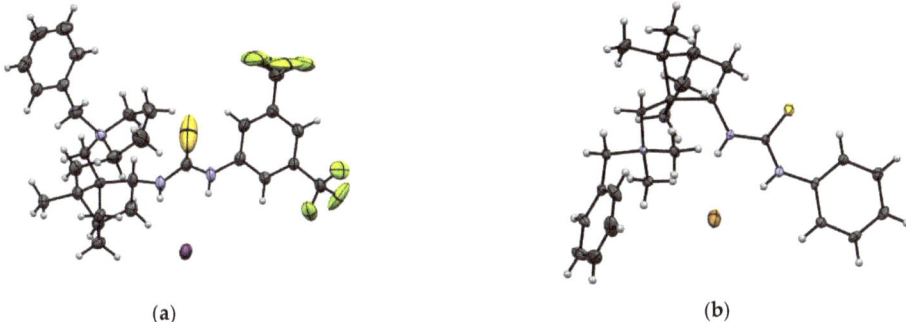

(a) (b)

Figure 2. Molecular structures of products **III** (a) and **VI-Br** (b). Thermal ellipsoids are shown at 50% probability.

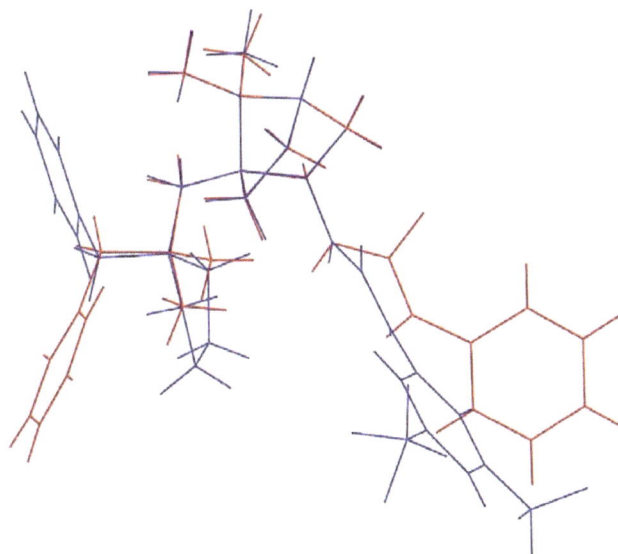

Figure 3. Stick presentation of the superimposed molecular structures of products **III** (blue) and **VI-Br** (red).

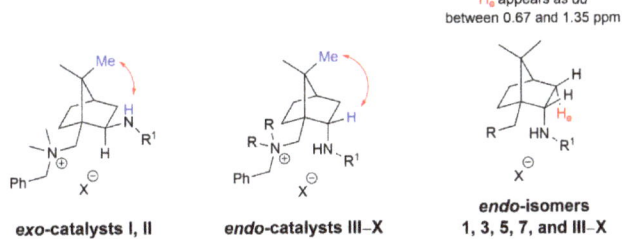

Figure 4. The determination of stereochemistry at the C-2 chiral center of compounds **I–X** by NOESY experiments and the correlation between the multiplicity of the H–C(3)-*endo* proton (H$_e$) and the *endo* absolute configuration at the C-2 chiral center of compounds **1, 3, 5, 7**, and **III–X** (Supplementary Materials).

Finally, the stereochemistry at the C-2 chiral center can be correlated on the basis of the multiplicity of the H–C(3)-*endo* proton (H_e) (Figure 4). Exclusively in the *endo*-isomers of compounds **1**, **3**, **5**, **7**, and **III–X**, the H–C(3)-*endo* proton appears as a doublet of doublet between 0.67 and 1.35 ppm (Table S7 in Supplementary Materials).

2.3. Organocatalytic Activity

First, the organocatalytic activity of camphor-derived phase-transfer organocatalysts **I–IX** was tested in electrophilic functionalizations of β-keto ester **9** (Scheme 3). Details of the optimization reactions can be found in the Supporting Information. The asymmetric α-fluorination of β-keto ester **9** with *N*-fluorobenzenesulfonimide (NFSI) proceeded under complete conversion and gave the product **10** with low enantioselectivity (87% yield and up to 29% ee). The best result was obtained with the catalyst **VIII** in toluene in the presence of K_3PO_4. In contrast, up to 86% ee has been reported in the literature for the α-fluorination of β-keto ester **9** with NFSI [15,32]. Similarly, the α-chlorination of **9** with *N*-chlorosuccinimide (NCI) gave product **11** in complete conversion and a meager 7% ee when squaramide PTC **IX** was used (for comparison, up to 80% ee has been reported in the literature when using alternative catalyst scaffolds [33]). Disappointingly, both the α-hydroxylation [34] of **9** with tosylimine **12**/H_2O_2 and the ring opening of arylaziridine **14** [35] with **9** did not give the expected products **13** and **16**, respectively. In both cases, no conversion was observed. Finally, the asymmetric Michael addition of glycine Schiff base **16** to methyl acrylate (**17**) was investigated, with up to 90% ee reported in the literature [36]. Catalyst **V** gave the expected product **18** with full conversion but low enantioselectivity (11% ee).

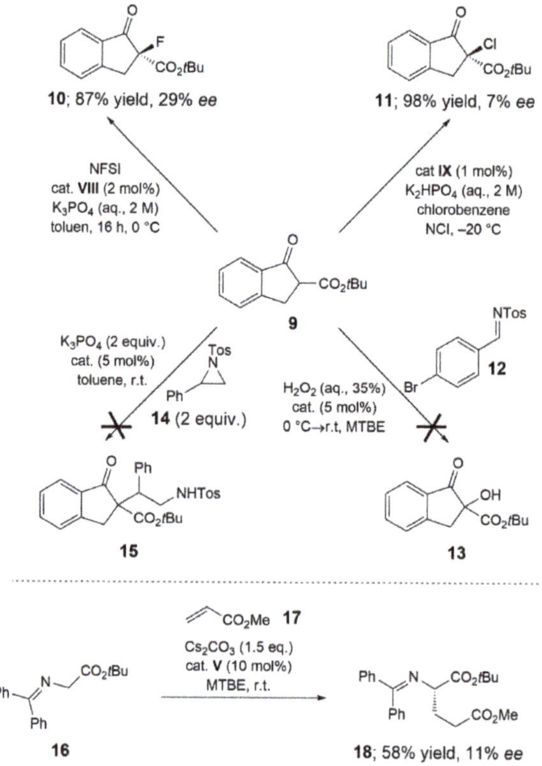

Scheme 3. Organocatalytic activity of camphor-derived phase-transfer organocatalysts; MTBE: methyl *tert*-butyl ether.

3. Materials and Methods

Solvents for extractions and chromatography were of technical grade and were distilled prior to use. Extracts were dried over technical-grade anhydrous Na_2SO_4. Melting points were determined on a Kofler micro hot stage. The NMR spectra were obtained on a Bruker Avance DPX 300 and Bruker Avance III 300 at 300 MHz for ^1H nucleus, Bruker UltraShield 500 plus (Bruker, Billerica, MA, USA) at 500 MHz for ^1H and 126 MHz for ^{13}C nucleus, and Bruker Ascend 600 (Bruker, Billerica, MA, USA) at 600 MHz for ^1H and 151 MHz for ^{13}C nucleus, using DMSO-d_6 and $CDCl_3$, with TMS as the internal standard, as solvents. Mass spectra were recorded on an Agilent 6224 Accurate Mass TOF LC/MS (Agilent Technologies, Santa Clara, CA, USA), and IR spectra on a Perkin-Elmer Spectrum BX FTIR spectrophotometer (PerkinElmer, Waltham, MA, USA). CD spectra were recorded on a J-1500 Circular Dichroism Spectrophotometer (JASCO corporation, Tokyo, Japan). Column chromatography (CC) was performed on silica gel (Silica gel 60, particle size: 0.035–0.070 mm (Sigma-Aldrich, St. Louis, MI, USA)). HPLC analyses were performed on an Agilent 1260 Infinity LC (Agilent Technologies, Santa Clara, CA, USA) and Dionex Summit HPLC system (Dionex Corporation, Sunnyvale, CA, USA) using CHIRALPAK AD-H (0.46 cm ø × 25 cm) and CHIRALPAK OJ-H (0.46 cm ø × 25 cm), as the chiral columns (Chiral Technologies, Inc., West Chester, PA, United States). All the commercially available chemicals used were purchased from Sigma-Aldrich (St. Louis, MI, USA). In addition, (1S,2S,4R)-7,7-dimethyl-1-(pyrrolidin-1-ylmethyl)bicyclo[2.2.1]heptan-2-amine (**1b**) was prepared following the literature procedure [30].

3.1. Reduction of (1S,4R,E)-1-[(Dimethylamino)methyl]-7,7-dimethylbicyclo[2.2.1]heptan-2-one oxime

Oxime (7.6 mmol, 1.6 g) was dissolved in propan-1-ol (86 mL) and heated to 95 °C. Then, small pieces of sodium (approximately 50 mg) were added continuously for 1 h at 95 °C; care was taken to ensure that the unreacted sodium (excess sodium) remained present in the reaction mixture at all times during the reaction. After completion of the reaction, the volatiles were evaporated in vacuo. The residue was dissolved in a mixture of water (20 mL) and Et_2O (80 mL). The organic phase was washed with water (2 × 20 mL) and NaCl (aq. sat., 1 × 20 mL), dried over anhydrous Na_2SO_4, and the volatiles were evaporated in vacuo. Diastereomers **1a** and **2a** were formed in a ratio of 2.6:1. The diastereomers were separated by column chromatography (Silica gel 60, EtOAc/MeOH/Et_3N = 4:1:1).

3.1.1. (1S,2R,4R)-1-[(Dimethylamino)methyl]-7,7-dimethylbicyclo[2.2.1]heptan-2-amine (**2a**)

Elutes first from the column. Yield: 175 mg (0.89 mmol, 12%) of colorless oil. ^{1}H-NMR (500 MHz, $CDCl_3$): δ 0.79 (s, 3H), 1.05 (s, 3H), 1.06–1.13 (m, 1H), 1.34 (ddd, J = 12.8, 9.4, 3.9, 1H), 1.54–1.61 (m, 3H), 1.63 (t, J = 4.3, 1H), 1.64–1.75 (m, 2H), 1.93 (d, J = 11.4, 1H), 2.02 (d, J = 13.0, 1H), 2.27 (s, 6H), 2.74 (d, J = 13.0, 1H), 3.11 (dd, J = 8.7, 5.1, 1H).

3.1.2. (1S,2S,4R)-1-[(Dimethylamino)methyl]-7,7-dimethylbicyclo[2.2.1]heptan-2-amine (**1a**)

Elutes second from the column. Yield: 850 mg (4.33 mmol, 57%) of colorless oil. ^{1}H-NMR (500 MHz, $CDCl_3$): δ 0.67 (dd, J = 12.9, 4.3, 1H), 0.86 (s, 3H), 0.89 (s, 3H), 1.22 (ddd, J = 12.3, 9.5, 4.4, 1H), 1.38 (ddd, J = 12.3, 4.5, 2.0, 1H), 1.49 (t, J = 4.6, 1H), 1.70–1.79 (m, 1H), 1.80 (br s, 2H), 2.10 (d, J = 13.1, 1H), 2.13–2.17 (m, 1H), 2.20 (s, 6H), 2.21–2.26 (m, 1H), 2.45 (d, J = 13.0, 1H), 3.36 (ddd, J = 10.6, 4.3, 2.0, 1H).

3.2. Boc Protection of Chiral Amines—General Procedure 1 (GP1)

To a solution of amine **1** or **2** and triethylamine (1.4 equivalents) in anhydrous CH_2Cl_2 was added di-*tert*-butyl dicarbonate (1.4 equivalents). The resulting reaction mixture was stirred at 25 °C for 24 h. Dichloromethane was evaporated in vacuo and the residue was purified by column chromatography (CC). The fractions containing product **3** or **4** were combined and the volatiles were evaporated in vacuo.

3.2.1. tert-Butyl {(1S,2S,4R)-1-[(dimethylamino)methyl]-7,7-dimethylbicyclo[2.2.1]heptan-2-yl}carbamate (3a)

Following *GP1*. Prepared from *endo*-amine **1a** (4.69 mmol, 920 mg) and di-*tert*-butyl dicarbonate (6.56 mmol, 1.431 g), Et$_3$N (6.56 mmol, 915 µL), CH$_2$Cl$_2$ (20 mL), 25 °C, 24 h. Isolation by column chromatography (Silica gel 60, EtOAc/petroleum ether = 1:5). Yield: 1.39 g (4.69 mmol, 99%) of colorless oil. [α]$^{r.t.}_D$ = +11.2 (0.15, MeOH). EI-HRMS: m/z = 297.2646 (MH)$^+$; C$_{17}$H$_{33}$N$_2$O$_2{}^+$ requires: m/z = 297.2536 (MH)$^+$; ν$_{max}$ 3346, 2935, 2819, 2765, 1698, 1483, 1454, 1389, 1364, 1297, 1242, 1167, 1114, 1065, 1040, 1014, 946, 874, 837, 780 cm^{-1}. ^{1}H-NMR (500 MHz, CDCl$_3$): δ 0.86 (s, 3H), 0.90 (s, 3H), 1.04 (dd, *J* = 13.4, 4.3, 1H), 1.21 (ddd, *J* = 12.2, 9.5, 4.4, 1H), 1.43 (s, 9H), 1.45–1.51 (m, 1H), 1.56 (t, *J* = 4.6, 1H), 1.72 (tq, *J* = 12.1, 4.1, 1H), 1.86 (br t, 1H), 2.21 (s, 6H), 2.24 (d, *J* = 13.6, 1H), 2.28–2.33 (m, 1H), 2.36 (d, *J* = 13.8, 1H), 3.75 (s, 1H), 6.00 (s, 1H). ^{13}C-NMR (126 MHz, CDCl$_3$): δ 19.19, 20.31, 25.40, 28.39, 28.64, 37.92, 45.07, 48.10, 48.33, 50.98, 56.25, 61.97, 78.72, 157.52.

3.2.2. tert-Butyl {(1S,2R,4R)-1-[(dimethylamino)methyl]-7,7-dimethylbicyclo[2.2.1]heptan-2-yl}carbamate (4a)

Following *GP1*. Prepared from *exo*-amine **2a** (0.81 mmol, 160 mg) and di-*tert*-butyl dicarbonate (1.134 mmol, 247 mg), Et$_3$N (1.19 mmol, 166 µL), CH$_2$Cl$_2$ (4 mL), 25 °C, 24 h. Isolation by column chromatography (Silica gel 60, EtOAc/petroleum ether = 1:5). Yield: 230 mg (0.78 mmol, 95%) of colorless oil. [α]$^{r.t.}_D$ = +25.7 (0.175, MeOH). EI-HRMS: m/z = 297.2536 (MH)$^+$; C$_{17}$H$_{33}$N$_2$O$_2{}^+$ requires: m/z = 297.2537 (MH)$^+$; ν$_{max}$ 3344, 2935, 2819, 2765, 1698, 1484, 1453, 1389, 1364, 1297, 1243, 1167, 1113, 1065, 1040, 1004, 943, 874, 837, 780 cm^{-1}. ^{1}H-NMR (500 MHz, CDCl$_3$): δ 0.87 (s, 3H), 0.99 (s, 3H), 1.09–1.17 (m, 1H), 1.34 (t, *J* = 9.4, 1H), 1.42–1.45 (m, 1H), 1.43 (s, 9H), 1.67 (d, *J* = 3.5, 2H), 1.69–1.75 (m, 1H), 1.86 (d, *J* = 8.4, 1H), 2.24 (s, 6H), 2.25 (d, *J* = 13.9, 1H), 2.40 (d, *J* = 13.9, 1H), 3.71 (br s, 1H), 5.58 (br s, 1H). ^{13}C-NMR (126 MHz, CDCl$_3$): δ 20.95, 27.15, 28.50, 28.67, 30.48, 33.79, 40.55, 45.67, 48.03, 50.94, 57.44, 58.86, 78.90, 155.72.

3.2.3. tert-Butyl [(1S,2S,4R)-7,7-dimethyl-1-(pyrrolidin-1-ylmethyl)bicyclo[2.2.1]heptan-2-yl}carbamate (3b)

Following *GP1*. Prepared from *endo*-amine **1b** (3.91 mmol, 869 mg) and di-*tert*-butyl dicarbonate (5.474 mmol, 1.194 g), Et$_3$N (5.474 mmol, 763 µL), CH$_2$Cl$_2$ (20 mL), 25 °C, 24 h. Isolation by column chromatography (Silica gel 60, EtOAc/petroleum ether = 1:5). Yield: 1.251 g (3.88 mmol, 99%) of brownish oil. [α]$^{r.t.}_D$ = +1.1 (0.295, MeOH). EI-HRMS: m/z = 323.2688 (MH)$^+$; C$_{19}$H$_{35}$N$_2$O$_2{}^+$ requires: m/z = 323.2693 (MH)$^+$; ν$_{max}$ 3300, 2979, 2937, 2879, 2794, 1808, 1757, 1715, 1460, 1395, 1371, 1306, 1250, 1211, 1168, 1113, 1062, 950, 844, 775, 664 cm^{-1}. ^{1}H-NMR (600 MHz, CDCl$_3$): δ 0.85 (s, 3H), 0.90 (s, 3H), 1.07 (dd, *J* = 13.4, 4.4, 1H), 1.21 (ddd, *J* = 12.8, 9.5, 4.5, 1H), 1.4–1.45 (m, 1H), 1.41 (s, 9H), 1.56 (t, *J* = 4.6, 1H), 1.65–1.73 (m, 6H), 1.90 (ddd, *J* = 13.6, 8.9, 4.1, 1H), 2.31 (s, 1H), 2.37 (d, *J* = 13.4, 1H), 2.41–2.46 (m, 2H), 2.56–2.60 (m, 2H), 2.66 (d, *J* = 13.4, 1H), 3.72 (br s, 1H), 6.31 (br s, 1H). ^{13}C-NMR (151 MHz, CDCl$_3$): δ 19.18, 20.23, 24.16, 26.03, 28.43, 28.61, 37.58, 45.22, 47.88, 50.81, 56.47, 56.82, 58.03, 78.48, 157.80.

3.3. Benzylation of Tertiary Amines—General Procedure 2 (GP2)

To a stirred mixture of tertiary amine **3** or **4** and K$_2$CO$_3$ (1.1 equivalents) in anhydrous DMF was added benzyl bromide (1.1 equivalents). The resulting reaction mixture was stirred at 25 °C for 24 h. DMF was evaporated in vacuo and the residue was purified by column chromatography (CC). The fractions containing product **5** or **6** were combined and the volatiles were evaporated in vacuo.

3.3.1. N-Benzyl-1-{(1S,2S,4R)-2-[(tert-butoxycarbonyl)amino]-7,7-dimethylbicyclo[2.2.1]heptan-1-yl}-N,N-dimethylmethanaminium Bromide (5a)

Following *GP2*. Prepared from compound **3a** (1.06 mmol, 315 mg) and benzyl bromide (1.16 mmol, 139 µL), K$_2$CO$_3$ (1.16 mmol, 160 mg), DMF (5 mL), 25 °C, 24 h. Isolation by column chromatography (Silica gel 60, EtOAc/MeOH = 4:1). Yield: 340 mg (0.73 mmol,

69%) of colorless oil. $[\alpha]^{r.t.}_D$ = +14.0 (0.087, MeOH). EI-HRMS: m/z = 387.3003 (M)$^+$; $C_{24}H_{39}N_2O_2$ requires: m/z = 387.3006 (M)$^+$; ν_{max} 3369, 3197, 2951, 2199, 2163, 2098, 1989, 1685, 1540, 1490, 1477, 1454, 1392, 1379, 1366, 1299, 1284, 1271, 1252, 1217, 1158, 1125, 1065, 1042, 1012, 947, 917, 882, 868, 854, 839, 783, 752, 732, 706 cm^{-1}. ^{1}H-NMR (500 MHz, CDCl$_3$): δ 0.87–0.93 (m, 1H), 0.94 (s, 3H), 0.98 (s, 3H), 1.29–1.34 (m, 1H), 1.36 (s, 9H), 1.58 (t, J = 4.4, 1H), 1.86 (br t, J = 11.7, 1H), 1.93–2.03 (m, 1H), 2.21 (br t, J = 13.1, 1H), 2.47 (d, J = 11.7, 1H), 3.17 (s, 3H), 3.25 (s, 3H), 3.44 (br d, J = 13.6, 1H), 4.11 (br d, J = 13.5, 1H), 4.18 (br t, J = 9.9, 1H), 4.97 (d, J = 12.3, 1H), 5.03 (br s, 1H), 5.20 (d, J = 12.2, 1H), 7.32–7.43 (m, 3H), 7.61 (d, J = 7.4, 2H). ^{13}C-NMR (126 MHz, CDCl$_3$): δ 19.48, 20.76, 27.34, 28.28, 28.75, 40.19, 43.67, 50.40, 51.55, 53.72, 54.15, 69.24, 70.42, 80.67, 127.78, 128.94, 130.51, 133.47, 156.14.

3.3.2. N-Benzyl-1-{(1S,2R,4R)-2-[(tert-butoxycarbonyl)amino]-7,7-dimethylbicyclo[2.2.1]heptan-1-yl}-N,N-dimethylmethanaminium Bromide (6a)

Following GP2. Prepared from compound 4a (2.53 mmol, 748 mg) and benzyl bromide (3.795 mmol, 453 µL), K$_2$CO$_3$ (2.78 mmol, 385 mg), DMF (13 mL), 25 °C, 24 h. Isolation by column chromatography (Silica gel 60, EtOAc/MeOH = 4:1). Yield: 904 mg (1.93 mmol, 76%) of colorless oil. $[\alpha]^{r.t.}_D$ = −4.3 (0.26, MeOH). EI-HRMS: m/z = 387.3000 (M)$^+$; $C_{24}H_{39}N_2O_2^+$ requires: m/z = 387.3006 (M)$^+$; ν_{max} 3341, 2965, 2885, 2156, 1698, 1606, 1508, 1475, 1456, 1365, 1278, 1247, 1168, 1060, 1019, 953, 860, 782, 732, 706 cm^{-1}. ^{1}H-NMR (500 MHz, CDCl$_3$): δ 0.92 (dd, J = 13.4, 3.5, 1H), 0.97 (s, 3H), 1.03 (s, 3H), 1.35 (t, J = 4.8, 1H), 1.39 (s, 9H), 1.62 (t, J = 4.5, 1H), 1.88–1.99 (m, 2H), 2.29 (br t, J = 12.9, 1H), 2.46–2.56 (m, 1H), 3.19 (s, 3H), 3.27 (s, 3H), 3.43 (br d, J = 14.3, 1H), 4.16–4.26 (m, 2H), 4.92 (d, J = 10.9, 1H), 4.97 (d, J = 12.4, 1H), 5.21 (d, J = 12.4, 1H), 7.37–7.46 (m, 3H), 7.63 (d, J = 6.9, 2H). ^{13}C-NMR (126 MHz, CDCl$_3$): δ 19.60, 20.95, 27.47, 28.40, 28.91, 40.49, 43.82, 50.54, 51.73, 53.85, 54.30, 69.39, 70.63, 80.94, 127.79, 129.12, 130.72, 133.57, 156.18.

3.3.3. 1-Benzyl-1-({(1S,2S,4R)-2-[(tert-butoxycarbonyl)amino]-7,7-dimethylbicyclo[2.2.1]heptan-1-yl}methyl)pyrrolidin-1-ium Bromide (5b)

Following GP2. Prepared from compound 3b (2.48 mmol, 828 mg) and benzyl bromide (2.73 mmol, 324 µL), K$_2$CO$_3$ (2.73 mmol, 377 mg), DMF (13 mL), 25 °C, 24 h. Isolation by column chromatography (Silica gel 60, EtOAc/MeOH = 4:1). Yield: 469 mg (1.45 mmol, 59%) of brownish semisolid. $[\alpha]^{r.t.}_D$ = +17.7 (0.12, MeOH). EI-HRMS: m/z = 413.3161 (M)$^+$; $C_{26}H_{41}N_2O_2^+$ requires: m/z = 413.3162 (M)$^+$; ν_{max} 3323, 3270, 2965, 2923, 1708, 1639, 1531, 1452, 1388, 1363, 1307, 1247, 1159, 1121, 1066, 1028, 1002, 923, 901, 855, 839, 780, 710 cm^{-1}. ^{1}H-NMR (500 MHz, CDCl$_3$): δ 0.92 (s, 3H), 0.91–0.96 (m, 1H), 0.96 (s, 3H), 1.32–1.35 (br t, 1H), 1.37 (s, 9H), 1.55 (t, J = 4.5, 1H), 1.62–1.73 (m, 1H), 1.74–1.86 (m, 2H), 1.88–2.04 (m, 2H), 2.04–2.13 (m, 1H), 2.17–2.27 (m, 1H), 2.39–2.48 (m, 1H), 3.39 (d, J = 14.0, 1H), 3.41–3.50 (m, 1H), 3.58–3.74 (m, 2H), 3.96 (ddd, J = 12.3, 8.1, 6.3, 1H), 4.13 (d, J = 14.0, 1H), 4.24 (tt, J = 10.8, 3.1, 1H), 4.59 (d, J = 12.6, 1H), 5.22 (br s, 1H), 5.26 (d, J = 10.8, 1H), 7.28–7.39 (m, 3H), 7.57 (d, J = 7.0, 2H). ^{13}C-NMR (126 MHz, CDCl$_3$): δ 19.33, 20.70, 21.60, 22.07, 27.88, 28.29, 28.81, 39.70, 43.63, 51.48, 53.44, 53.74, 59.55, 62.11, 63.58, 67.05, 80.48, 128.21, 128.99, 130.40, 133.30, 156.03.

3.4. Boc Deprotection of Amines—General Procedure 3 (GP3)

To a solution of amine 5 or 6 in anhydrous CH$_2$Cl$_2$ (2.5 mL/mmol) was added trifluoroacetic acid (2.5 mL/mmol). The resulting reaction mixture was stirred at 25 °C for 2 h. Dichloromethane and trifluoroacetic acid were evaporated in vacuo and the residue was dissolved in CH$_2$Cl$_2$ (2.5 mL/mmol). The organic phase was washed with NaOH (aq., 2 M, 2 × 2.5 mL/mmol) and NaCl (aq. sat., 1 × 2.5 mL/mmol). The volatiles were evaporated in vacuo to give product 7 or 8.

3.4.1. (1S,2S,4R)-1-[(Benzyldimethylammonio)methyl]-7,7-dimethylbicyclo[2.2.1]heptan-2-aminium 2,2,2-Trifluoroacetate (7a)

Following GP3. Prepared from compound 5a (2.1 mmol, 1 g), trifluoroacetic acid (5 mL), CH$_2$Cl$_2$ (5 mL), 25 °C, 2 h. Volatile components were evaporated in vacuo, and the residue was dissolved in dichloromethane and washed with NaOH (aq., 2 M) and NaCl (aq. sat.). Yield: 605 mg (1.51 mmol, 72%) of colorless solid, mp = 179.9–182.1 °C. $[\alpha]^{r.t.}{}_D$ = +16.2 (0.125, MeOH). EI-HRMS: m/z = 287.2483 (M)$^+$; $C_{19}H_{31}N_2{}^+$ requires: m/z = 287.2482 (M)$^+$; ν_{max} 3377, 3292, 3042, 2943, 2881, 1685, 1585, 1479, 1457, 1401, 1372, 1302, 1196, 1157, 1113, 1048, 1025, 1010, 989, 935, 917, 881, 854, 819, 780, 785, 753, 733, 716, 707, 632, 607 cm^{-1}. ^{1}H-NMR (500 MHz, CDCl$_3$): δ 0.89 (dd, J = 3.4, 13.1, 1H); 0.93 (s, 3H); 0.97 (s, 3H); 1.31–1.38 (m, 1H); 1.61 (t, J = 4.6, 1H); 1.77–1.84 (m, 1H); 1.86–1.96 (m, 1H); 2.05–2.13 (m, 1H); 2.44–2.53 (m, 1H); 3.28 (s, 3H); 3.37 (s, 3H); 3.43–3.49 (m, 1H); 3.68 (d, J = 13.9, 1H); 3.72 (d, J = 13.9, 1H); 4.97 (d, J = 12.4, 1H); 5.11 (d, J = 12.3, 1H); 7.40–7.48 (m, 3H); 7.56–7.60 (m, 2H), signal for NH$_2$ is missing. ^{13}C-NMR (126 MHz, CDCl$_3$): δ 19.41, 20.51, 26.70, 29.29, 44.19, 44.63, 49.97, 50.49, 52.44, 53.20, 53.69, 69.11, 71.98, 117.64 (q, J = 297.3), 128.29, 129.14, 130.57, 133.49, 161.16 (q, J = 32.7).

3.4.2. (1S,2R,4R)-1-[(Benzyldimethylammonio)methyl]-7,7-dimethylbicyclo[2.2.1]heptan-2-aminium 2,2,2-Trifluoroacetate (**8a**)

Following *GP3*. Prepared from compound **6a** (1.92 mmol, 900 mg), trifluoroacetic acid (5 mL), CH$_2$Cl$_2$ (5 mL), 25 °C, 2 h. Volatile components were evaporated in vacuo, and the residue was dissolved in dichloromethane and washed with NaOH (aq., 2 M) and NaCl (aq. sat.). Yield: 567 mg (1.41 mmol, 74%) of colorless solid, mp = 157.1–158.8 °C. $[\alpha]^{r.t.}{}_D$ = −9.8 (0.11, MeOH). EI-HRMS: m/z = 287.2477 (M)$^+$; $C_{19}H_{31}N_2{}^+$ requires: m/z = 287.2482 (M)$^+$; ν_{max} 2953, 2883, 1684, 1476, 1456, 1393, 1371, 1311, 1196, 1153, 1113, 1035, 1009, 936, 911, 851, 820, 798, 784, 735, 714, 631 cm^{-1}. ^{1}H-NMR (600 MHz, CDCl$_3$): δ 0.86 (s, 3H), 0.90 (s, 3H), 1.15–1.22 (m, 1H), 1.34–1.41 (m, 1H), 1.50–1.55 (m, 1H), 1.67 (s, 1H), 1.77 (d, J = 13.9, 3H), 1.92 (dd, J = 13.0, 7.3, 2H), 3.01 (dd, J = 9.0, 4.9, 1H), 3.19 (s, 6H), 3.33 (d, J = 13.5, 1H), 4.14 (d, J = 13.5, 1H), 4.75 (d, J = 12.5, 1H), 4.86 (d, J = 12.8, 1H), 7.35–7.41 (m, 3H), 7.56 (d, J = 7.4, 2H). ^{13}C-NMR (126 MHz, CDCl$_3$): δ 20.58, 21.40, 27.73, 32.27, 44.01, 44.29, 50.60, 51.35, 51.53, 52.44, 57.11, 64.64, 71.26, 117.56 (q, J = 296.8), 127.79, 129.26, 130.81, 133.53, 161.25 (q, J = 32.7).

3.4.3. Synthesis of 1-{[(1S,2S,4R)-2-Ammonio-7,7-dimethylbicyclo[2.2.1]heptan-1-yl]methyl}-1-benzylpyrrolidin-1-ium Iodide (**7b**)

Compound **5b** (0.55 mmol, 270 mg) was dissolved in anhydrous CH$_2$Cl$_2$ (8 mL), and then HI (aq., 48%, 5 equivalents, 2.75 mmol, 495 μL) was added. The reaction mixture was stirred for 4 h at 25 °C. Volatile components were evaporated in vacuo, and the residue was dissolved in dichloromethane (5 mL) and washed with NaOH (aq., 2 M, 2 × 5mL) and NaCl (aq. sat., 1 × 5mL). Yield: 150 mg (0.34 mmol, 62%) of yellowish semisolid. $[\alpha]^{r.t.}{}_D$ = +18.8 (0.15, MeOH). EI-HRMS: m/z = 313.2635 (M)$^+$; $C_{21}H_{33}N_2{}^+$ requires: m/z = 313.2635 (M)$^+$; ν_{max} 3273, 2951, 2881, 2188, 2152, 1969, 1594, 1458, 1372, 1303, 1217, 1142, 1077, 1033, 1004, 917, 822, 764, 725, 641 cm^{-1}. ^{1}H-NMR (600 MHz, CDCl$_3$): δ 1.03–1.06 (m, 1H), 1.06 (s, 3H), 1.07 (s, 3H), 1.41–1.50 (m, 1H), 1.67 (t, J = 4.6, 1H), 1.77 (td, J = 10.7, 9.7, 6.5, 1H), 1.81–1.88 (m, 1H), 1.92–1.99 (m, 1H), 2.05–2.11 (m, 1H), 2.12–2.21 (m, 2H), 2.26 (ddd, J = 13.4, 9.4, 4.1, 1H), 2.36–2.75 (m, 3H), 3.68 (dt, J = 10.5, 3.1, 1H), 3.75 (ddd, J = 12.3, 8.5, 5.9, 1H), 3.81–3.87 (m, 1H), 3.83 (d, J = 14.1, 1H), 3.89 (d, J = 14.2, 1H), 4.07 (ddd, J = 12.2, 8.3, 6.3, 1H), 4.16 (ddd, J = 11.9, 8.2, 6.2, 1H), 5.06 (d, J = 12.6, 1H), 5.25 (d, J = 12.6, 1H), 7.42–7.50 (m, 3H), 7.66 (d, J = 6.6, 2H). ^{13}C-NMR (151 MHz, CDCl$_3$): δ 19.88, 21.16, 21.96, 22.08, 26.80, 29.23, 42.76, 44.69, 52.63, 53.69, 54.48, 60.85, 61.96, 63.75, 67.23, 128.66, 129.39, 130.71, 133.42.

3.5. Synthesis of Phase-Transfer Bifunctional Catalysts—General Procedure 4 (GP4)

Amine **7** or **8** was dissolved in anhydrous CH$_2$Cl$_2$, the appropriate electrophile was added (1.2–1.4 equivalents), and the reaction mixture was stirred for 16 h at room temperature. The volatiles were evaporated in vacuo. The residue was purified by column chromatography (CC). The fractions containing product **I–X** were combined and the volatiles were evaporated in vacuo.

3.6. *Trifluoroacetate Anion Exchange—General Procedure 5 (GP5)*

The column was packed with NaI (5 g) and conditioned with ethyl acetate. The trifluoroacetate phase-transfer catalyst was dissolved in ethyl acetate and applied to the NaI column. The fractions containing the product were combined and the volatiles were evaporated in vacuo. Based on the ^{19}F NMR spectra (presence of a signal for fluorine from trifluoroacetate anion), the procedure was repeated as necessary.

3.6.1. *N*-Benzyl-1-((1*R*,2*R*,4*R*)-2-{3-[3,5-bis(trifluoromethyl)phenyl]thioureido}-7,7-dimethylbicyclo[2.2.1]heptan-1-yl)-*N*,*N*-dimethylmethanaminium 2,2,2-Trifluoroacetate (**I**)

Following *GP4*. Prepared from compound **8a** (0.585 mmol, 300 mg) and 3,5-bis(trifluoromethyl)phenyl isothiocyanate (1.05 mmol, 192 µL), CH$_2$Cl$_2$ (4 mL), 25 °C, 16 h. Isolation by evaporation followed by column chromatography (Silica gel 60, EtOAc/MeOH = 4:1). Yield: 316 mg (0.47 mmol, 80%) of yellowish solid, mp = 87.5–89.0 °C. $[\alpha]^{r.t.}_D$ = +6.8 (0.13, MeOH). EI-HRMS: m/z = 558.2374 (M)$^+$; C$_{28}$H$_{34}$F$_6$N$_3$S$^+$ requires: m/z = 558.2372 (M)$^+$; ν_{max} 3260, 2962, 2885, 2091, 1679, 1622, 1523, 1472, 1382, 1333, 1274, 1218, 1171, 1125, 999, 970, 883, 848, 828, 780, 760, 727, 700, 680 cm^{-1}. ^{1}H-NMR (500 MHz, CDCl$_3$): δ 0.89 (s, 3H); 1.18 (s, 3H); 1.36–1.45 (m, 1H); 1.66–1.74 (m, 1H); 1.81–1.94 (m, 4H); 2.22 (dd, *J* = 8.6, 13.4, 1H); 3.00 (s, 3H); 3.08 (s, 3H); 3.29 (d, *J* = 14.5, 1H); 4.65 (d, *J* = 12.6, 1H); 4.74–4.80 (m, 1H); 4.84–4.93 (m, 2H); 7.40–7.54 (m, 6H); 8.36 (s, 2H); 8.79 (d, *J* = 7.9, 1H); 10.97 (s, 1H). ^{13}C NMR (126 MHz, CDCl$_3$): δ 20.76, 20.93, 27.84, 34.28, 41.11, 43.92, 50.38, 50.44, 51.25, 53.08, 58.99, 66.90, 71.59, 113.57, 116.74–117.00 (m), 117.08 (q, *J* = 294.0), 122.13 (d, *J* = 4.0), 123.41 (q, *J* = 272.6), 124.49, 126.58, 129.53, 131.34 (q, *J* = 33.2), 131.37, 133.09, 141.70, 162.07 (q, *J* = 34.3), 180.02.

3.6.2. *N*-Benzyl-1-((1*R*,2*R*,4*R*)-2-{3-[3,5-bis(trifluoromethyl)phenyl]thioureido}-7,7-dimethylbicyclo[2.2.1]heptan-1-yl)-*N*,*N*-dimethylmethanaminium Iodide (**II**)

Following *GP5*. Prepared from catalyst **I** (0.21 mmol, 140 mg), dissolved in ethyl acetate (4 mL) and filtered through a pad of NaI. Volatile components were evaporated in vacuo. Yield: 140 mg (0.20 mmol, 96%) of colorless solid, mp = 61.4–62.9 °C. $[\alpha]^{r.t.}_D$ = +5 (0.19, MeOH). EI-HRMS: m/z = 558.2365 (M)$^+$; C$_{28}$H$_{34}$F$_6$N$_3$S$^+$ requires: m/z = 558.2372 (M)$^+$; ν_{max} 3305, 2964, 1674, 1536, 1473, 1384, 1336, 1275, 1172, 1123, 1000, 971, 884, 846, 801, 780, 759, 725, 701, 679 cm^{-1}. ^{1}H-NMR (500 MHz, CDCl$_3$): δ 0.96 (s, 3H); 1.28 (s, 3H); 1.39–1.46 (m, 1H); 1.64–1.73 (m, 1H); 1.81–2.01 (m, 4H); 2.23 (dd, *J* = 8.8, 13.4, 1H); 3.08 (s, 3H); 3.14 (s, 3H); 3.51 (d, *J* = 14.3, 1H); 4.81–4.89 (m, 1H); 4.96 (d, *J* = 12.5, 1H); 5.07 (d, *J* = 13.9, 1H); 5.17 (d, *J* = 12.7, 1H); 7.38–7.45 (m, 2H); 7.46–7.51 (m, 1H); 7.54 (s, 1H); 7.55–7.60 (m, 2H); 8.08 (d, *J* = 8.1, 1H); 8.43 (s, 2H); 10.52 (s, 1H). ^{13}C-NMR (126 MHz, CDCl$_3$): δ 21.37, 21.94, 27.90, 34.23, 41.04, 43.83, 50.02, 50.70, 51.55, 53.44, 59.26, 66.70, 70.85, 117.35, 122.27, 123.36 (q, *J* = 272.7), 126.56, 129.49, 131.36, 131.36 (q, *J* = 33.3), 133.31, 141.23, 179.91.

3.6.3. 1-Benzyl-1-[((1*S*,2*S*,4*R*)-2-{3-[3,5-bis(trifluoromethyl)phenyl]thioureido}-7,7-dimethylbicyclo[2.2.1]heptan-1-yl)methyl]yrrolidine-1-ium Iodide (**III**)

Following *GP4*. Prepared from compound **7b** (0.34 mmol, 150 mg) and 3,5-bis(trifluoromethyl)phenyl isothiocyanate (0.68 mmol, 124 µL), CH$_2$Cl$_2$ (3 mL), 25 °C, 16 h. Isolation by column chromatography (Silica gel 60, EtOAc/MeOH = 10:1). Yield: 99 mg (0.14 mmol, 41%) of brownish semisolid. $[\alpha]^{r.t.}_D$ = +20 (0.047, MeOH). EI-HRMS: m/z = 584.2519 (M)$^+$; C$_{30}$H$_{36}$F$_6$N$_3$S$^+$ requires: m/z = 584.2529 (M)$^+$; ν_{max} 3194, 3126, 2968, 2149, 1625, 1589, 1542, 1492, 1472, 1381, 1324, 1271, 1249, 1222, 1166, 1136, 1108, 1094, 1061, 1025, 999, 967, 909, 885, 847, 756, 721, 701, 679, 612 cm^{-1}. ^{1}H-NMR (500 MHz, CDCl$_3$): δ 1.03 (s, 3H); 1.06 (s, 3H); 1.15 (dd, *J* = 13.4, 3.9, 1H); 1.51–1.93 (m, 5H); 1.98–2.06 (m, 1H); 2.16–2.31 (m, 2H); 2.61–2.70 (m, 1H); 2.99–3.07 (m, 1H); 3.41–3.51 (m, 2H); 3.67–3.83 (m, 3H); 3.83–3.93 (m, 1H); 4.63 (d, *J* = 12.9, 1H); 4.96 (d, *J* = 12.8, 1H); 5.34–5.45 (m, 1H); 7.39–7.45 (m, 2H); 7.47–7.52 (m, 1H); 7.54–7.58 (m, 2H); 7.60 (s, 1H); 8.34–8.43 (m, 3H); 10.67 (s, 1H). ^{13}C-NMR (126 MHz, CDCl$_3$): δ 19.98, 20.65, 21.49, 21.93, 28.65, 29.57, 38.39, 43.68, 51.83, 54.54, 57.00, 60.89,

61.91, 64.34, 67.34, 117.85–118.26 (m), 123.08–123.29 (m), 123.38 (q, J = 209.2), 127.38, 129.67, 131.28, 131.57 (q, J = 33.4), 133.39, 140.92, 182.20.

3.6.4. N-Benzyl-1-((1R,2S,4R)-2-{3-[3,5-bis(trifluoromethyl)phenyl]thioureido}-7,7-dimethylbicyclo[2.2.1]heptan-1-yl)-N,N-dimethylmethanaminium 2,2,2-Trifluoroacetate (IV)

Following *GP4*. Prepared from compound **7a** (0.39 mmol, 200 mg) and 3,5-bis(trifluoromethyl)phenyl isothiocyanate (0.70 mmol, 128 µL), CH$_2$Cl$_2$ (4 mL), 25 °C, 16 h. Isolation by column chromatography (Silica gel 60, EtOAc/MeOH = 4:1). Yield: 250 mg (0.37 mmol, 95%) of colorless solid, mp = 153–155 °C. [α]$^{r.t.}_D$ = +2.1 (0.11, MeOH). EI-HRMS: m/z = 558.2363 (M)$^+$; C$_{28}$H$_{34}$F$_6$N$_3$S$^+$ requires: m/z = 558.2372 (M)$^+$; ν$_{max}$ 3275, 3247, 3047, 2961, 2890, 1682, 1542, 1473, 1385, 1278, 1177, 1132, 966, 887, 848, 801, 719, 702, 680 cm^{-1}. ^{1}H-NMR (500 MHz, CDCl$_3$): δ 0.97 (s, 3H), 1.05 (s, 3H), 1.10 (dd, J = 13.4, 3.7, 1H), 1.51 (ddd, J = 13.4, 9.2, 4.7, 1H), 1.73 (t, J = 4.6, 1H), 1.78–1.89 (m, 1H), 1.95–2.12 (m, 1H), 2.59–2.67 (m, 1H), 2.68–2.76 (m, 1H), 3.03 (s, 3H), 3.04 (s, 3H), 3.47 (d, J = 13.7, 1H), 3.72 (d, J = 13.7, 1H), 4.55 (d, J = 12.6, 1H), 4.71 (d, J = 12.6, 1H), 5.22 (tt, J = 10.2, 3.0, 1H), 7.34–7.40 (m, 4H), 7.43–7.48 (m, 1H), 7.53 (s, 1H), 8.29 (s, 2H), 8.91 (d, J = 9.8, 1H), 11.09 (s, 1H). ^{13}C-NMR (126 MHz, CDCl$_3$): δ 19.81, 20.25, 28.05, 28.57, 38.68, 43.75, 50.78, 51.03, 51.76, 54.43, 56.70, 69.58, 73.03, 117.02 (q, J = 294.4), 117.24–117.53 (m), 122.72 (q, J = 3.4), 123.39 (q, J = 272.7), 126.96, 129.47, 131.24, 131.48 (q, J = 33.4), 133.03, 141.44, 161.62 (q, J = 34.1), 181.96.

3.6.5. N-Benzyl-1-((1R,2S,4R)-2-{3-[3,5-bis(trifluoromethyl)phenyl]thioureido}-7,7-dimethylbicyclo[2.2.1]heptan-1-yl)-N,N-dimethylmethanaminium Iodide (V)

Following *GP5*. Prepared from catalyst **IV** (0.17 mmol, 116 mg), dissolved in ethyl acetate (3 mL) and filtered through a pad of NaI. Volatile components were evaporated in vacuo. Yield: 109 mg (0.16 mmol, 92%) of white solid, mp = decomposition above 350 °C. [α]$^{r.t.}_D$ = +69.2 (0.013, MeOH). EI-HRMS: m/z = 558.2368 (M)$^+$; C$_{28}$H$_{34}$F$_6$N$_3$S$^+$ requires: m/z = 558.2372 (M)$^+$; ν$_{max}$ 3247, 2960, 2928, 2857, 2175, 2163, 2135, 2034, 1996, 1954, 1722, 1595, 1534, 1473, 1385, 1277, 1177, 1135, 965, 887, 730, 701, 680 cm^{-1}. ^{1}H-NMR (500 MHz, DMSO-d_6): δ 0.98 (s, 3H), 1.02 (dd, J = 13.0, 3.7, 1H), 1.05 (s, 3H), 1.43–1.52 (m, 1H), 1.72 (t, J = 4.4, 1H), 1.90–2.06 (m, 2H), 2.16–2.24 (m, 1H), 2.45–2.49 (m, 1H), 2.95 (s, 3H), 2.99 (s, 3H), 3.60 (d, J = 14.1, 1H), 3.71 (d, J = 14.0, 1H), 4.52–4.63 (m, 2H), 5.04–5.12 (m, 1H), 7.43–7.59 (m, 5H), 7.81 (s, 1H), 8.28 (s, 2H), 8.38 (d, J = 9.8, 1H), 10.32 (s, 1H). ^{13}C-NMR (126 MHz, DMSO-d_6): δ 19.25, 20.04, 26.97, 28.11, 38.10, 42.97, 49.08, 50.46, 51.23, 53.88, 56.13, 67.81, 70.46, 116.76–117.00 (m), 122.27–122.44 (m), 123.17 (q, J = 272.8), 128.03, 128.86, 130.17 (q, J = 32.8), 130.38, 133.06, 141.46, 181.00.

3.6.6. N-Benzyl-1-[(1R,2S,4R)-7,7-dimethyl-2-(3-phenylthioureido)bicyclo[2.2.1]heptan-1-yl]-N,N-dimethylmethanaminium 2,2,2-Trifluoroacetate (VI)

Following *GP4*. Prepared from compound **7a** (0.39 mmol, 200 mg) and phenyl isothiocyanate (0.70 mmol, 84 µL), CH$_2$Cl$_2$ (4 mL), 25 °C, 16 h. Isolation by evaporation followed by column chromatography (Silica gel 60, EtOAc/MeOH = 4:1). Yield: 119 mg (0.22 mmol, 56%) of colorless solid, mp = 180–183 °C. [α]$^{r.t.}_D$ = +6.7 (0.06, MeOH). EI-HRMS: m/z = 422.2618 (M)$^+$; C$_{26}$H$_{36}$N$_3$S$^+$ requires: m/z = 422.2624 (M)$^+$; ν$_{max}$ 3244, 2959, 2884, 1683, 1540, 1507, 1489, 1473, 1457, 1362, 1317, 1202, 1148, 1056, 1033, 851, 801, 727 cm^{-1}. ^{1}H-NMR (500 MHz, CDCl$_3$): δ 0.96 (s, 3H); 1.04 (s, 3H); 1.08 (dd, J = 13.3, 3.7, 1H); 1.45–1.53 (m, 1H); 1.67–1.79 (m, 2H); 1.96–2.06 (m, 1H); 2.57–2.65 (m, 1H); 2.67–2.76 (m, 1H); 2.98 (s, 3H); 3.03 (s, 3H); 3.41 (d, J = 13.6, 1H); 3.73 (d, J = 13.8, 1H); 4.54 (d, J = 12.5, 1H); 4.79 (d, J = 12.4, 1H); 5.24–5.31 (m, 1H); 7.04–7.10 (m, 1H); 7.22–7.29 (m, 2H); 7.33–7.47 (m, 5H); 7.60–7.68 (m, 2H); 8.64 (d, J = 9.9, 1H); 10.45 (s, 1H). ^{13}C-NMR (126 MHz, CDCl$_3$): δ 19.95, 20.50, 28.53, 28.65, 38.67, 43.85, 51.11, 51.22, 51.72, 54.34, 56.50, 69.08, 72.83, 123.82, 124.84, 127.30, 128.46, 129.38, 130.98, 133.35, 139.65, 182.21 (two signals missing).

3.6.7. N-Benzyl-1-[(1R,2S,4R)-7,7-dimethyl-2-(3-phenylureido)bicyclo[2.2.1]heptan-1-yl]-N,N-dimethylmethanaminium 2,2,2-Trifluoroacetate (**VII**)

Following *GP4*. Prepared from compound **7a** (0.39 mmol, 200 mg) and phenyl isocyanate (0.69 mmol, 76 µL), CH_2Cl_2 (4 mL), 25 °C, 16 h. Isolation by evaporation followed by column chromatography (Silica gel 60, EtOAc/MeOH = 5:1). Yield: 57 mg (0.11 mmol, 28%) of colorless solid, mp = 120.0–123.8 °C. $[\alpha]^{r.t.}_D$ = +5 (0.08, MeOH). EI-HRMS: m/z = 406.2850 (M)$^+$; $C_{26}H_{36}N_3O^+$ requires: m/z = 406.2853 (M)$^+$; ν_{max} 3261, 2960, 2886, 2150, 1683, 1598, 1550, 1489, 1457, 1313, 1202, 1139, 846, 801, 727, 702 cm^{-1}. ^{1}H-NMR (500 MHz, CDCl$_3$): δ 0.95 (s, 3H), 1.00 (s, 3H), 1.12 (dd, *J* = 13.3, 3.6, 1H), 1.44–1.53 (m, 1H), 1.55–1.68 (m, 2H), 2.03–2.11 (m, 1H), 2.47–2.55 (m, 1H), 2.60–2.67 (m, 1H), 3.04 (s, 3H), 3.08 (s, 3H), 3.36 (d, *J* = 13.7, 1H), 3.88 (d, *J* = 13.7, 1H), 4.52–4.59 (m, 1H), 4.73 (d, *J* = 12.4, 1H), 4.97 (d, *J* = 12.3, 1H), 6.91–6.98 (m, 1H), 7.18–7.24 (m, 2H), 7.25–7.30 (m, 3H), 7.35–7.40 (m, 1H), 7.42–7.47 (m, 2H), 7.51–7.57 (m, 2H), 9.31 (s, 1H). ^{13}C-NMR (126 MHz, CDCl$_3$): δ 19.71, 20.51, 28.46, 28.68, 39.90, 44.00, 50.46, 51.29, 51.76, 52.02, 54.06, 68.41, 72.90, 117.34 (d, *J* = 295.3), 118.79, 122.19, 127.49, 128.82, 129.24, 130.75, 133.32, 139.93, 156.51, 161.52 (q, *J* = 33.5).

3.6.8. N-Benzyl-1-[(1R,2S,4R)-7,7-dimethyl-2-(3-phenylureido)bicyclo[2.2.1]heptan-1-yl]-N,N-dimethylmethanaminium Iodide (**VIII**)

Following *GP5*. Prepared from catalyst **VII** (0.26 mmol, 139 mg), dissolved in ethyl acetate (5 mL), and filtered through a pad of NaI. All volatile components were evaporated in vacuo. Yield: 111 mg (0.21 mmol, 80%) of colorless solid, mp = 153–155 °C. $[\alpha]^{r.t.}_D$ = +78 (0.073, MeOH). EI-HRMS: m/z = 406.2850 (M)$^+$; $C_{26}H_{36}N_3O^+$ requires: m/z = 406.2853 (M)$^+$; ν_{max} 3277, 2967, 2881, 1678, 1597, 1543, 1487, 1442, 1377, 1311, 1217, 1158, 1128, 1030, 949, 852, 816, 753, 729, 694 cm^{-1}. ^{1}H-NMR (500 MHz, CDCl$_3$): δ 1.01 (s, 6H), 1.14 (dd, *J* = 13.3; 3.6, 1H), 1.46–1.60 (m, 2H), 1.62 (t, *J* = 4.5, 1H), 2.23–2.31 (m, 1H), 2.48–2.56 (m, 1H), 2.67–2.76 (m, 1H), 3.08 (s, 3H), 3.11 (s, 3H), 3.34 (d, *J* = 13.7, 1H), 4.07 (d, *J* = 13.6, 1H), 4.54–4.65 (m, 1H), 4.80 (d, *J* = 12.3, 1H), 5.05 (d, *J* = 12.3, 1H), 6.75 (d, *J* = 10.8, 1H), 6.94–7.00 (m, 1H), 7.19–7.25 (m, 2H), 7.31 (t, *J* = 7.6, 2H), 7.38–7.43 (m, 1H), 7.48–7.53 (m, 2H), 7.56–7.62 (m, 2H), 8.93 (s, 1H). ^{13}C-NMR (126 MHz, CDCl$_3$): δ 19.88, 20.81, 28.41, 29.13, 39.96, 44.02, 50.51, 51.57, 51.95, 52.34, 54.45, 68.01, 72.56, 118.82, 122.49, 127.33, 128.83, 129.29, 130.86, 133.38, 139.60, 156.34.

3.6.9. N-Benzyl-1-{(1R,2S,4R)-2-[(2-{[3,5-bis(trifluoromethyl)phenyl]amino}-3,4-dioxocyclobut-1-en-1-yl)amino]-7,7-dimethylbicyclo[2.2.1]heptan-1-yl}-N,N-dimethylmethanaminium 2,2,2-Trifluoroacetate (**IX**)

Following *GP4*. Prepared from compound **7a** (0.19 mmol, 100 mg) and 3-((3,5-bis(trifluoromethyl)phenyl)amino)-4-ethoxycyclobut-3-ene-1,2-dione (0.30 mmol, 106.4 mg), CH_2Cl_2 (2 mL), 25 °C, 16 h. Isolation by evaporation followed by column chromatography (Silica gel 60, EtOAc/MeOH = 4:1). Yield: 106 mg (0.15 mmol, 75%) of colorless solid, mp = 148.9–150.1 °C. $[\alpha]^{r.t.}_D$ = +65 (0.006, MeOH). EI-HRMS: m/z = 594.2545 (M)$^+$; $C_{31}H_{34}F_6N_3O_2$ requires: m/z = 594.2550 (M)$^+$; ν_{max} 3420, 3153, 3034, 2967, 2888, 1791, 1686, 1603, 1551, 1475, 1427, 1377, 1276, 1176, 1127, 948, 931, 880, 848, 831, 730, 701, 684, 666 cm^{-1}. ^{1}H-NMR (500 MHz, CDCl$_3$): δ 1.02 (s, 3H), 1.17 (s, 3H), 1.35 (dd, *J* = 13.2, 3.6, 1H), 1.60–1.66 (m, 1H), 1.77 (t, *J* = 4.5, 1H), 1.89 (br t, *J* = 13.3, 1H), 1.93–2.05 (m, 1H), 2.52–2.62 (m, 1H), 3.01 (s, 1H), 3.14 (s, 3H), 3.16 (s, 3H), 3.42 (d, *J* = 13.8, 1H), 4.33 (d, *J* = 13.9, 1H), 4.63 (d, *J* = 12.5, 1H), 4.76 (d, *J* = 12.5, 1H), 5.27 (t, *J* = 9.9, 1H), 7.41 (t, *J* = 7.4, 2H), 7.45–7.50 (m, 2H), 7.55–7.66 (m, 2H), 8.21 (s, 2H), 9.13 (d, *J* = 9.2, 1H), 11.34 (s, 1H). ^{13}C-NMR (126 MHz, CDCl$_3$): δ 19.80, 20.66, 26.61, 28.93, 41.29, 44.05, 50.80, 51.47, 52.79, 55.04, 58.56, 70.65, 73.17, 116.24, 119.01, 123.33 (q, *J* = 272.9), 126.85, 129.59, 131.38, 132.65 (q, *J* = 33.4), 133.40, 140.74, 165.81, 169.02, 181.07, 185.00 (two carbons missing).

3.6.10. 1-Benzyl-1-{[(1S,2S,4R)-7,7-dimethyl-2-(3-phenylthioureido)bicyclo[2.2.1]heptan-1-yl]methyl}pyrrolidin-1-ium Iodide (X)

Following *GP4*. Prepared from compound (**7b**) (0.25 mmol, 109 mg) and phenyl isothiocyanate (0.38 mmol, 45 µL), CH_2Cl_2 (2 mL), 25 °C, 16 h. Isolation by column chromatography (Silica gel 60, EtOAc/MeOH = 10:1). Yield: 72 mg (0.13 mmol, 50%) of colorless solid, mp = 178–180 °C. $[\alpha]^{r.t.}_D$ = +7.4 (0.14, MeOH). EI-HRMS: m/z = 448.2776 (M)$^+$; $C_{28}H_{38}N_3S$ requires: m/z = 448.2781 (M)$^+$; ν_{max} 3209, 3030, 2953, 1685, 1597, 1528, 1495, 1450, 1360, 1308, 1243, 1144, 1089, 1027, 1002, 915, 758, 716, 698, 607 cm^{-1}. ^{1}H-NMR (500 MHz, CDCl$_3$): δ 0.99 (s, 3H), 1.02 (s, 3H), 1.12 (dd, *J* = 13.4, 3.9, 1H), 1.42–1.52 (m, 2H), 1.65 (t, *J* = 4.3, 1H), 1.71–1.89 (m, 2H), 1.96–2.07 (m, 1H), 2.15–2.23 (m, 2H), 2.53–2.64 (br t, *J* = 11.8, 1H), 2.97 (br s, 1H), 3.40 (d, *J* = 13.9, 1H), 3.42–3.48 (m, 1H), 3.70 (dt, *J* = 12.3, 7.4, 1H), 3.81 (d, *J* = 13.8, 1H), 3.83–3.99 (m, 2H), 4.64 (d, *J* = 12.7, 1H), 5.12 (d, *J* = 12.7, 1H), 5.40 (br t, *J* = 10.6, 1H), 7.14 (t, *J* = 7.3, 1H), 7.31 (t, *J* = 7.6, 2H), 7.41 (t, *J* = 7.4, 2H), 7.46 (t, *J* = 7.3, 1H), 7.60 (d, *J* = 6.8, 2H), 7.76 (d, *J* = 7.2, 2H), 8.10 (d, *J* = 10.4, 1H), 10.10 (s, 1H). ^{13}C-NMR (126 MHz, CDCl$_3$): δ 19.98, 20.70, 21.34, 22.00, 28.48, 29.80, 38.39, 43.70, 51.66, 54.42, 56.54, 60.24, 62.02, 63.82, 67.18, 123.99, 125.16, 127.84, 128.55, 129.49, 130.93, 133.61, 139.32, 182.27.

3.7. General Procedure for the α-Fluorination of β-Keto Ester 9

Aqueous K_3PO_4 or other base (2 M, 2 equivalents, 0.1 mL) was added to a mixture of β-keto ester **9** (0.1 mmol, 24.4 mg, ω = 95%) and organocatalyst **I–IX** (2 mol%) in toluene or in CH_2Cl_2 (2 mL) under argon atmosphere. The mixture was cooled to −10°C and NFSI (1.1 equivalents, 34.7 mg) was added in two portions over 2 h. The reaction mixture was stirred for another 12 h at −10 °C. After completion, the reaction was quenched by addition of NH_4Cl (aq. sat, 4 mL) and extracted with CH_2Cl_2 (10 mL). The organic phase was dried over anhydrous Na_2SO_4, filtered, and the volatiles were evaporated in vacuo. The residue was purified by column chromatography (Silica gel 60, EtOAc/*n*-Heptane = 1:15). Enantiomeric excess (ee) was determined by HPLC (Chiralpak AD-H, *n*-Hexane/*i*-PrOH = 200:1, flow rate 0.75 mL/min, λ = 250 nm, 10 °C) after isolation by column chromatography.

3.8. General Procedure for the α-Chlorination of β-Keto Ester 9

To a mixture of β-keto ester **9** (0.1 mmol, 24.4 mg, ω = 95%), organocatalyst **II, III, VI–VIII**, or **IX** (1 mol%), and K_2HPO_4 (solid, 1 equivalent, 17.4 mg) in chlorobenzene (2 mL), at −20 °C under argon atmosphere, was added *N*-chlorosuccinimide (NCS, 1.2 equivalents, 16 mg), and the reaction mixture was stirred for 2 h at −20 °C. After completion, the reaction was quenched by addition of NH_4Cl (aq. sat, 4 mL) and extracted with CH_2Cl_2 (10 mL). The organic phase was dried over anhydrous Na_2SO_4, filtered, and the volatiles were evaporated in vacuo. The residue was purified by column chromatography (Silica gel 60, EtOAc/*n*-Heptane = 1:12). Enantiomeric excess (ee) was determined by HPLC (Chiralpak OJ-H, *n*-Hexane/*i*-PrOH = 70:30, flow rate 0.7 mL/min, λ = 250 nm, 10 °C) isolation with column chromatography (Silica gel 60, EtOAc/*n*-Heptane = 1:12).

3.9. General Procedure for the α-Hydroxylation of β-Keto Ester 9

Into a flame-dried Schlenk flask under argon atmosphere at 0 °C, a mixture of β-keto ester **9** (0.1 mmol, 24.4 mg, ω = 95%) and *N*-(4-bromobenzylidene)-4-methylbenzenesulfonamide (**12**) (1 equivalent, 33.8 mg) was added. Catalyst **III, IV, VII, VIII**, or **IX** (5 mol%) was dissolved in anhydrous methyl *tert*-butyl ether (MTBE, 5 mL) and slowly added via syringe into the reaction mixture. After addition of H_2O_2 (1 equivalent, 35% in water, 8.6 µL), the reaction mixture was stirred for 20 h at room temperature. After 24 h at 25 °C, the reaction mixture was filtrated trough a plug of anhydrous Na_2SO_4 and washed with dichloromethane.

3.10. General Procedure for the Ring-Opening of Aryl-Aziridine 14 with β-Keto Ester 9

To a mixture of β-keto ester **9** (0.1 mmol, 24.4 mg, ω = 95%), catalyst **IV, VIII**, or **IX** (5 mol%), and K_3PO_4 (2 equivalents, 42 mg) in toluene (2.5 mL) under argon atmosphere,

2-phenyl-1-tosylaziridine (**14**) (2 equivalent, 54.6 mg) was added and stirred at room temperature for 24 h. After 24 h at 25 °C, the reaction mixture was filtrated through a plug of anhydrous Na_2SO_4 and washed with dichloromethane.

3.11. General Procedure for the Michael Addition of Glycine Schiff Base **16** *with Methyl Acrylate (***17***)*

Degassed solvent (2.5 mL) was added to a mixture of *tert*-butyl 2-((diphenylmethylene)amino)acetate (**16**) (0.05 mmol, 14.8 mg), catalyst **I**, **III–V**, **VII**, **VIII**, or **IX** (10 mol%), and Cs_2CO_3 (1.5–10.0 equivalents) in a Schlenk tube at 25 °C or 0 °C; then, methyl acrylate (**17**) (1.5 equivalents, 6.8 μL) was added. After 24 h at 25 °C or 0 °C, the reaction mixture was filtrated through a plug of anhydrous Na_2SO_4 and washed with ethyl acetate. The volatiles were evaporated in vacuo. The crude product **18** was purified by column chromatography (Silica gel 60, EtOAc/Heptane = 1:15). Enantiomeric excess (ee) was determined by HPLC (Chiralpak AD-H, *n*-Hexane/*i*-PrOH = 95:5, flow rate 0.5 mL/min, λ = 250 nm, 10 °C) after filtration of the reaction mixture through a plug of Na_2SO_4.

3.12. X-ray Crystallography

Single-crystal X-ray diffraction data were collected on an Agilent Technologies SuperNova Dual diffractometer with an Atlas detector using monochromated Mo-K$_α$ radiation (λ = 0.71073 Å) at 150 K. The data were processed using CrysAlis PRO [37]. Using Olex2.1.2. [38], the structures were solved by direct methods implemented in SHELXS [39] or SHELXT [40] and refined by a full-matrix least-squares procedure based on F^2 with SHELXT-2014/7 [41]. All nonhydrogen atoms were refined anisotropically. Hydrogen atoms were placed in geometrically calculated positions and were refined using a riding model. The drawings and the analysis of bond lengths, angles, and intermolecular interactions were carried out using Mercury [42] and Platon [43]. Structural and other crystallographic details on data collection and refinement for compounds **VI-Br** and **III** have been deposited with the Cambridge Crystallographic Data Centre as a supplementary publication under CCDC Deposition Numbers 2204647 and 2204648, respectively. These data can be obtained free of charge via www.ccdc.cam.ac.uk/conts/retrieving.html (or from the CCDC, 12 Union Road, Cambridge CB2 1EZ, UK; fax: +44 1223 336033; e-mail: deposit@ccdc.cam.ac.uk).

4. Conclusions

Ten novel quaternary ammonium salt bifunctional phase-transfer organocatalysts based on a chiral (+)-camphor framework were prepared. Starting from camphor-derived 1,3-diamines, catalysts **I–X** were synthesized in a four-step sequence: Boc protection–benzylation–Boc deprotection–reaction with iso(thio)cyanate/squaramate. The catalysts prepared bear either a (thio)urea or squaramide hydrogen bond donor and possess either a trifluoroacetate or iodide counteranion. The quaternary iodide ammonium salts **II**, **V**, and **VIII** were formed from the corresponding trifluoroacetates **I**, **IV**, and **VII**, respectively, via anion methathesis with an excess of NaI. The phase-transfer catalysts have been fully characterized; the stereochemistry at the C-2 chiral center was unambiguously determined. Their organocatalytic activity was investigated in the electrophilic functionalization of the β-keto ester **9**. The α-fluorination and chlorination of β-keto ester **9** proceeded to full conversion, affording the desired products **10** and **11** with low enantioselectivity (up to 29% ee). α-Hydroxylation and ring opening of *N*-tosylaziridine **14** gave no product. Finally, the Michael addition of glycine Schiff base **16** to methyl acrylate (**17**) gave the expected product **18** with full conversion and up to 11% ee.

Supplementary Materials: The following supporting information can be downloaded at: https://www.mdpi.com/article/10.3390/molecules28031515/s1. Synthesis and characterization data; HPLC data; copies of ^{1}H- and ^{13}C-NMR spectra; copies of HRMS reports; structure determination by X-ray diffraction analysis. Figure S1. Applied organocatalysts **I–IX**. Figure S2. Molecular structure of compound **VI-Br**. Thermal ellipsoids are shown at 50% probability. Figure S3. Molecular structure of compound **III**. Thermal ellipsoids are shown at 50% probability. Table S1. Evaluation of organocatalysts **I–IX** in the fluorination of β-keto ester **9**. Table S2. Further evaluation of organocatalysts **III**, **VIII**, and **IX** in the fluorination of β-keto ester **9**. Table S3. Evaluation of organocatalysts **II**, **III**, **VI–VIII**, and **IX** in the chlorination of β-keto ester **9**. Table S4. Evaluation of organocatalysts **III**, **IV**, **VII**, **VIII**, and **IX** in the hydroxylation of β-keto ester **9**. Table S5. Evaluation of organocatalysts **IV**, **VIII**, and **IX** in the addition of β-keto ester **9** to tosylaziridine **14**. Table S6. Evaluation of organocatalysts **I**, **III–V**, **VII**, **VIII**, and **IX** in the addition of tert-butyl 2-((diphenylmethylene)amino)acetate (**16**) to methyl acrylate (**17**). Table S7. Correlation between the multiplicity of the H–C(3)-endo proton (He) and the endo absolute configuration at the C-2 chiral center of compounds **1a**, **3b**, **5a**, **7a,b**, and **III–X**. Table S8. Crystal data and structure refinement for compound **VI-Br**. Table S9. Crystal data and structure refinement for compound **III**.

Author Contributions: Conceptualization, L.C., U.G. and J.S.; methodology, L.C. and U.G.; software, L.C., U.G. and J.S.; validation, L.C., U.G., J.S. and M.W.; formal analysis, L.C., H.B. and M.W.; investigation, L.C., M.W. and U.G.; resources, L.C., U.G., M.W. and J.S.; data curation, L.C., U.G., J.S., H.B., M.W. and B.Š.; writing—original draft preparation, L.C. and U.G.; writing—review and editing, L.C., U.G., J.S., M.W., F.P. and B.Š.; visualization, L.C., U.G. and J.S.; supervision, U.G. and M.W.; project administration, U.G., M.W. and J.S.; funding acquisition, U.G. and J.S. All authors have read and agreed to the published version of the manuscript.

Funding: This research was funded by the Slovenian Research Agency through grant P1-0179.

Institutional Review Board Statement: Not applicable.

Informed Consent Statement: Not applicable.

Data Availability Statement: Not applicable.

Conflicts of Interest: The authors declare no conflict of interest.

Sample Availability: Samples of the compounds are available from the authors.

References

1. Helder, R.; Hummelen, J.; Laane, R.; Wiering, J.; Wynberg, H. Catalytic asymmetric induction in oxidation reactions. The synthesis of optically active epoxides. *Tetrahedron Lett.* **1976**, *17*, 1831–1834. [CrossRef]
2. Dolling, U.H.; Davis, P.; Grabowski, E.J.J. Efficient catalytic asymmetric alkylations. 1. Enantioselective synthesis of (+)-indacrinone via chiral phase-transfer catalysis. *J. Am. Chem. Soc.* **1984**, *106*, 446–447. [CrossRef]
3. O'Donnell, M.J.; Bennett, W.D.; Wu, S. The stereoselective synthesis of α-amino acids by phase-transfer catalysis. *J. Am. Chem. Soc.* **1989**, *111*, 2353–2355. [CrossRef]
4. O'Donnell, M.J. *Catalytic Asymmetric Syntheses*, 2nd ed.; Ojima, I., Ed.; Wiley-VCH: New York, NY, USA, 2000; pp. 727–755.
5. Maruoka, K. (Ed.) *Asymmetric Phase Transfer Catalysis*; Wiley-VCH: Weinheim, Germany, 2008.
6. Vitale, M.R.; Oudeyer, S.; Levacher, V.; Briere, J. *Radical and Ion-Pairing Strategies in Asymmetric Organocatalysis*; Elsevier: Amsterdam, The Netherlands, 2017.
7. Shirakawa, S.; Maruoka, K. Recent developments in asymmetric phase-transfer reactions. *Angew. Chem. Int. Ed.* **2013**, *52*, 4312–4348. [CrossRef] [PubMed]
8. Waser, M.; Winter, M.; Mairhofer, C. (Thio)urea containing chiral ammonium salt catalysts. *Chem. Rec.* **2023**, *23*, e202200198. [CrossRef]
9. Ooi, T.; Kameda, M.; Maruoka, K. Molecular Design of a C_2-Symmetric Chiral Phase-Transfer Catalyst for Practical Asymmetric Synthesis of α-Amino Acids. *J. Am. Chem. Soc.* **1999**, *121*, 6519–6520. [CrossRef]
10. Ooi, T.; Takeuchi, M.; Kameda, M.; Maruoka, K. Practical catalytic enantioselective synthesis of α,α-dialkyl-α-amino acids by chiral phase-transfer catalysis. *J. Am. Chem. Soc.* **2000**, *122*, 5228–5229. [CrossRef]
11. Shibuguchi, T.; Fukuta, Y.; Akachi, Y.; Sekine, A.; Ohshima, T.; Shibasaki, M. Development of new asymmetric two-center catalysts in phase-transfer reactions. *Tetrahedron Lett.* **2002**, *43*, 9539–9543. [CrossRef]
12. Waser, M.; Gratzer, K.; Herchl, R.; Müller, N. Design, synthesis, and application of tartaric acid derived N-spiro quaternary ammonium salts as chiral phase-transfer catalysts. *Org. Biomol. Chem.* **2012**, *10*, 251–254. [CrossRef]
13. Wang, H.-Y.; Chai, Z.; Zhao, G. Novel bifunctional thiourea–ammonium salt catalysts derived from amino acids: Application to highly enantio-and diastereoselective aza-Henry reaction. *Tetrahedron* **2013**, *69*, 5104–5111. [CrossRef]
14. Wang, H.-Y.; Zhang, J.-X.; Cao, D.-D.; Zhao, G. Enantioselective addition of thiols to imines catalyzed by thiourea–quaternary ammonium salts. *ACS Catal.* **2013**, *3*, 2218–2221. [CrossRef]
15. Novacek, J.; Waser, M. Syntheses and Applications of (Thio) Urea-Containing Chiral Quaternary Ammonium Salt Catalysts. *Eur. J. Org. Chem.* **2014**, *2014*, 802–809. [CrossRef]
16. Denmark, S.E.; Gould, N.D.; Wolf, L.M. A systematic investigation of quaternary ammonium ions as asymmetric phase-transfer catalysts. Synthesis of catalyst libraries and evaluation of catalyst activity. *J. Org. Chem.* **2011**, *76*, 4260–4336. [CrossRef]

17. Wang, H. Chiral phase-transfer catalysts with hydrogen bond: A powerful tool in the asymmetric synthesis. *Catalysts* **2019**, *9*, 244. [CrossRef]
18. Bernal, P.; Fernández, R.; Lassaletta, J.M. Organocatalytic asymmetric cyanosilylation of nitroalkenes. *Chem. Eur. J.* **2010**, *16*, 7714–7718. [CrossRef] [PubMed]
19. Wang, B.; Liu, Y.; Sun, C.; Wei, Z.; Cao, J.; Liang, D.; Lin, Y.; Duan, H. Asymmetric phase-transfer catalysts bearing multiple hydrogen-bonding donors: Highly efficient catalysts for enantio- and diastereoselective nitro-Mannich reaction of amidosulfones. *Org. Lett.* **2014**, *16*, 6432–6435. [CrossRef] [PubMed]
20. Zhang, J.; Wu, X.Y.; Zhou, Q.-L.; Sun, J. Chiral Camphor Derivatives as New Catalysts for Asymmetric Phase-Transfer Alkylation. *Chin. J. Org. Chem.* **2001**, *19*, 630–633. [CrossRef]
21. Money, T. Remote functionalization of camphor: Application to natural product synthesis. *Org. Synth. Theory Appl.* **1996**, *3*, 1–83.
22. Money, T. Camphor: A chiral starting material in natural product synthesis. *Nat. Prod. Rep.* **1985**, *2*, 253–289. [CrossRef] [PubMed]
23. Holton, R.A.; Somoza, C.; Kim, H.-B.; Liang, F.; Biediger, R.J.; Boatman, P.D.; Shindo, M.; Smith, C.C.; Kim, S.; Nadizadeh, H. The Total Synthesis of Paclitaxel Starting with Camphor. *ACS Symp. Ser.* **1995**, *583*, 288–301. [CrossRef]
24. Nicolaou, K.C.; Yang, Z.; Liu, J.J.; Ueno, H.; Nantermet, P.G.; Guy, R.K.; Claiborne, C.F.; Renaud, J.; Couladouros, E.A.; Paulvannan, K. Total synthesis of taxol. *Nature* **1994**, *367*, 630–634. [CrossRef] [PubMed]
25. Oppolzer, W. Camphor as a natural source of chirality in asymmetric synthesis. *Pure Appl. Chem.* **1990**, *62*, 1241–1250. [CrossRef]
26. Kitamura, M.; Suga, S.; Kawai, K.; Noyori, R. Catalytic asymmetric induction. Highly enantioselective addition of dialkylzincs to aldehydes. *J. Am. Chem. Soc.* **1986**, *108*, 6071–6072. [CrossRef]
27. Mahdy, A.-H.S.; Zayed, S.E.; Abo-Bakr, A.M.; Hassan, E.A. Camphor: Synthesis, reactions and uses as a potential moiety in the development of complexes and organocatalysts. *Tetrahedron* **2022**, *121*, 132913. [CrossRef]
28. Grošelj, U. Camphor-Derivatives in Asymmetric Organocatalysis–Synthesis and Application. *Curr. Org. Chem.* **2015**, *19*, 2048–2074. [CrossRef]
29. Ričko, S.; Požgan, F.; Štefane, B.; Svete, J.; Golobič, A.; Grošelj, U. Stereodivergent Synthesis of Camphor-Derived Diamines and Their Application as Thiourea Organocatalysts. *Molecules* **2020**, *25*, 2978. [CrossRef]
30. Ričko, S.; Svete, J.; Štefane, B.; Perdih, A.; Golobič, A.; Meden, A.; Grošelj, U. 1,3-Diamine-Derived Bifunctional Organocatalyst Prepared from Camphor. *Adv. Synth. Catal.* **2016**, *358*, 3786–3796. [CrossRef]
31. Ričko, S.; Meden, A.; Ivančič, A.; Perdih, A.; Štefane, B.; Svete, J.; Grošelj, U. Organocatalyzed Deracemization of Δ^2-Pyrrolin-4-ones. *Adv. Synth. Catal.* **2017**, *359*, 2288–2296. [CrossRef]
32. Xu, J.; Hu, Y.; Huang, D.; Wang, K.-H.; Xu, C.; Niua, T. Thiourea-Catalyzed Enantioselective Fluorination of β-Keto Esters. *Adv. Synth. Catal.* **2012**, *354*, 515–526. [CrossRef]
33. Novacek, J.; Monkowius, U.; Himmelsbach, M.; Waser, M. Asymmetric α-chlorination of β-ketoesters using bifunctional ammonium salt catalysis. *Monatsh. Chem.* **2016**, *147*, 533–538. [CrossRef]
34. Mairhofer, C.; Novacek, J.; Waser, M. Synergistic Ammonium (Hypo)Iodite/Imine Catalysis for the Asymmetric α-Hydroxylation of β-Ketoesters. *Org. Lett.* **2020**, *22*, 6138–6142. [CrossRef]
35. Haider, V.; Kreuzer, V.; Tiffner, M.; Spingler, B.; Waser, M. Ammonium Salt-Catalyzed Ring-Opening of Aryl-Aziridines with β-Keto Esters. *Eur. J. Org. Chem.* **2020**, *32*, 5173–5177. [CrossRef] [PubMed]
36. Tiffner, M.; Novacek, J.; Busillo, A.; Gratzer, K.; Massa, A.; Waser, M. Design of chiral urea-quaternary ammonium salt hybrid catalysts for asymmetric reactions of glycine Schiff bases. *RSC Adv.* **2015**, *5*, 78941–78945. [CrossRef]
37. 37. In *CrysAlis PRO*; Agilent Technologies UK Ltd.: Oxfordshire, UK, 2011.
38. Dolomanov, O.V.; Bourhis, L.J.; Gildea, R.J.; Howard, J.A.K.; Puschmann, H. *OLEX2*: A complete structure solution, refinement and analysis program. *J. Appl. Cristallogr.* **2009**, *42*, 339–341. [CrossRef]
39. Sheldrick, G.M. A short history of *SHELX*. *Acta Crystallogr. A* **2008**, *64*, 112–122. [CrossRef] [PubMed]
40. Sheldrick, G.M. *SHELXT*-Integrated space-group and crystal-structure determination. *Acta Crystallogr. Sect. A Found. Adv.* **2015**, *71*, 3–8. [CrossRef]
41. Sheldrick, G.M. Crystal structure refinement with *SHELXL*. *Acta Crystallogr. Sect. C Struct. Chem.* **2015**, *71*, 3–8. [CrossRef] [PubMed]
42. Macrae, C.F.; Edgington, P.R.; McCabe, P.; Pidcock, E.; Shields, G.P.; Taylor, R.; Towler, M.; van de Streek, J. Synthesis, *Mercury*: Visualization and analysis of crystal structures. *J. Appl. Crystallogr.* **2006**, *39*, 453–457. [CrossRef]
43. Spek, A.L. Single-crystal structure validation with the program PLATON. *J. Appl. Crystallogr.* **2003**, *36*, 7–13. [CrossRef]

Disclaimer/Publisher's Note: The statements, opinions and data contained in all publications are solely those of the individual author(s) and contributor(s) and not of MDPI and/or the editor(s). MDPI and/or the editor(s) disclaim responsibility for any injury to people or property resulting from any ideas, methods, instructions or products referred to in the content.

Communication

Novel Synthesis of Dihydroisoxazoles by *p*-TsOH-Participated 1,3-Dipolar Cycloaddition of Dipolarophiles with α-Nitroketones

Caiyun Yang [1], Sirou Hu [1], Xinhui Pan [1,2,*], Ke Yang [1], Ke Zhang [1], Qingguang Liu [1], Xiaobing Xin [1], Jie Li [1], Jinhui Wang [1] and Xiaoda Yang [1,2,*]

[1] Key Laboratory of Xinjiang Phytomedicine Resource and Utilisation, Ministry of Education, School of Pharmaceutical Sciences, Shihezi University, Shihezi 832002, China

[2] Stake Key Laboratory of Natural and Biomimetic Drugs, Department of Chemical Biology, School of Pharmaceutical Sciences, Peking University, Beijing 100191, China

* Correspondence: panxhshzu@shzu.edu.cn (X.P.); xyang@bjmu.edu.cn (X.Y.)

Abstract: This article reports in detail a method for the synthesis of 3-benzoxoxazoline by the reaction of alkenes (alkynes) and a variety of α-nitroketones in the presence of *p*-TsOH. The scope of alkenes is broad, including different alkenes and the alkyne. This reaction provides a convenient and efficient synthetic method of 3-benzoylisoxazolines.

Keywords: α-nitroketones; alkenes; alkynes; 1,3-dipolar cycloaddition; *p*-TsOH; isoxazolines

1. Introduction

Heterocycles are important structural elements, which are present in natural products from all classes and in many biologically active synthetic compounds [1]. Heterocyclic compounds perform an important role in chemical industry, e.g., food fragrance and dyes. Amongst these, isoxazole and its derivatives represent a group of five-element heterocyclic compounds containing oxygen and nitrogen atoms of a valuable class [2]. Additionally, they have performed a vital role in the theoretical development of heterocyclic chemistry and are also extensively used in organic synthesis [3]. Isoxazoles have attracted an increasing research interest, and are widely used and studied in the modern drug discoveries as non-classical amide or ester bioisosteres, and potential pharmacophores endowed, and most isoxazoles have strong biological activity [4,5].

Isoxazolines are partially saturated analogs of isoxazoles as important intermediates for synthesis of varieties of fascinating organic molecules applicable to both basic organic synthesis and life sciences [6]. Isoxazolines can be converted into various synthetic units, such as hydroxy ketones [7], amino alcohols [8], β-hydroxynitrile [9], and masked aldols [10], and be used as synthetic equivalent of 1,3-dicarbonyl structure [11]. Isoxazolines can exhibit a variety of bioactivities, such as anti-inflammatory [12], anticancer [13], hypoglycemic [14], antibacterial [15], anti-HIV [16], anti-Alzheimer's [17], antifungal [18], antimalarial [19], antioxidant [20], anti-tuberculosis [21], and antinociceptive [22] activities (Figure 1). Isoxazolines can also be good herbicides [23] and insecticides [24]. Therefore, the development of new methods for more efficient synthesis has been always an attractive task.

Based on the characteristics and wide application of isoxazolines derivatives, the research progress of isoxazolines derivatives have progressed rapidly in recent years, and a large number of synthetic methods for isoxazole derivatives are reported in the literature every year. These approaches can be summarized as the following four major types: (1) 1,3-dipolar cycloaddition between nitrile oxide and unsaturated hydrocarbon [25–29], (2) intramolecular addition cyclization reaction of unsaturated hydroxime [30–33], (3) condensation reactions of 1,3-dicarbonyl derivatives [34], and (4) cycloisomerization [35]. In the past decades, the 1,3-dipole cycloaddition reaction of alkenes with nitrile oxide is the most direct and extensive

Citation: Yang, C.; Hu, S.; Pan, X.; Yang, K.; Zhang, K.; Liu, Q.; Xin, X.; Li, J.; Wang, J.; Yang, X. Novel Synthesis of Dihydroisoxazoles by *p*-TsOH-Participated 1,3-Dipolar Cycloaddition of Dipolarophiles with α-Nitroketones. *Molecules* **2023**, *28*, 2565. https://doi.org/10.3390/molecules28062565

Academic Editors: Alison Rinderspacher, Mircea Darabantu and Gloria Proni

Received: 1 February 2023
Revised: 6 March 2023
Accepted: 6 March 2023
Published: 11 March 2023

Copyright: © 2023 by the authors. Licensee MDPI, Basel, Switzerland. This article is an open access article distributed under the terms and conditions of the Creative Commons Attribution (CC BY) license (https://creativecommons.org/licenses/by/4.0/).

method for the construction of isoxazoline skeletons [36]. Nitrile oxides are usually derived from aldoximes and nitro compounds [37,38], but the common use of transition metal catalysts, such as Cu(I), Cu(II), and Ru(II), in the reaction makes the products residual metal and cytotoxic [39–41], which limits its application in biology and drug development (Scheme 1). Therefore, the development of practical, simple, and cost-effective new methods for synthesis 2-oxazolines would complement current methods.

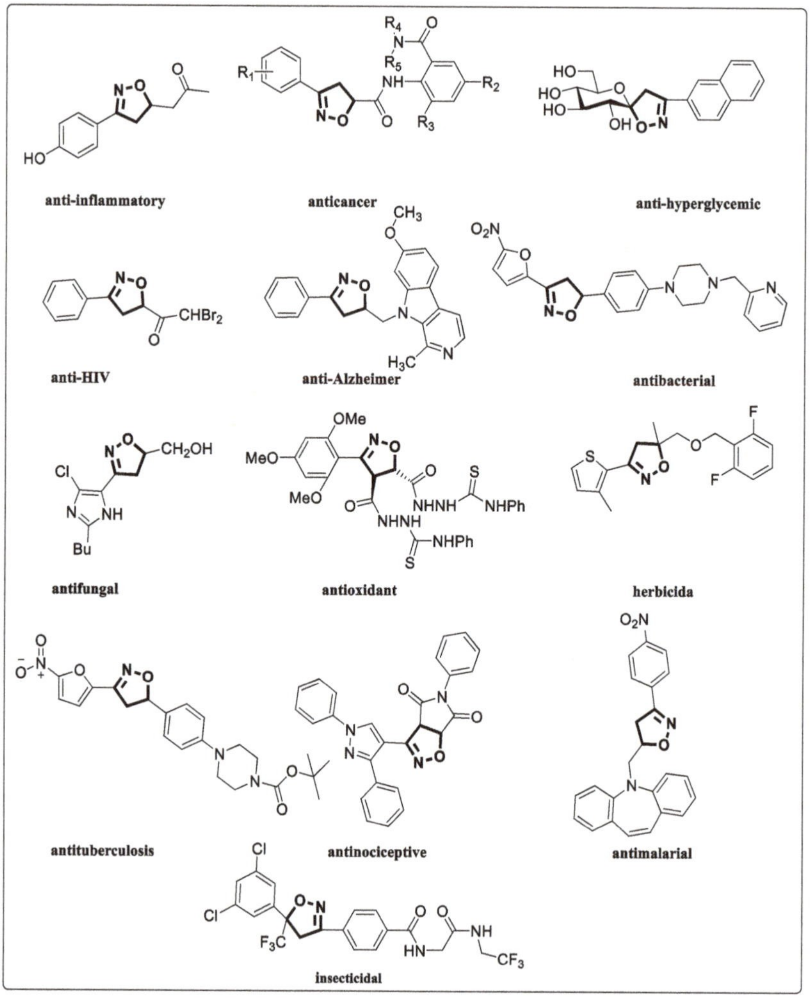

Figure 1. Structures of isooxazolines with bioactivity.

In the previous study, we first used the alkaline catalyst chloramine-T to catalyze the reaction of α-nitroketone and alkene to synthesize isoxazoline with a yield of 77% [42]. In the present work, we report a novel synthesis of dihydroisoxazoles by p-TsOH (anhydrous)-participated 1,3-dipolar cycloaddition of isoxazoline with α-nitroketones. On one hand, compared with the strong acid (H_2SO_4)-catalyzed synthesis of isoxazole [43], p-TsOH gives a milder reaction condition that avoids carbonization of organic substance, and it is low in toxicity and is inexpensive. On the other hand, Natarajan Arumugam and co-workers [44] reported a good, facile, and efficient method for the rapid synthesis of fused pyrrolidine and indolizinoindole heterocycles through 1,3-dipolar cycloaddition in the presence of *p*-TsOH. Additionally, Zhenghui Guan and co-workers [45] also demonstrated a *p*-TsOH mediated 1,3-dipolar cycloaddition approach of nitroolefins and sodium azide for the synthesis of 4-aryl-NH-1,2,3-triazoles, and a slightly higher yield (93%) was isolated. It is an efficient *p*-TsOH-mediated 1,3-dipole cycloaddition reaction that can tolerate a wide range of functional groups, and quickly and easily obtain the target product under mild conditions. *p*-TsOH was discovered as a vital additive in this type of 1,3-dipolar cycloaddition. Herein, isoxazolines, given the importance of preparing biologically active molecules, are chosen for validation of the accessibility, operational simplicity, and atom economy of our method.

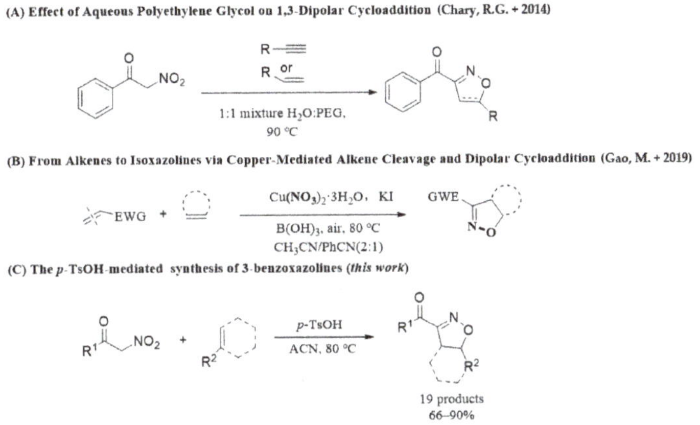

Scheme 1. Overview of the 1,3-dipolar cycloaddition of hydrocarbons and α-nitro ketones [36,39].

2. Results and Discussion

2.1. Optimization of the Reaction Conditions

First, we compared the effects of different acids and solvents. The reaction of benzoyl-nitromethane **1a** with allylbenzene **2a** to form isoxazoline **3a** was performed and the results were summarized in Table 1. It is noted that the reaction without acid did not proceed. In the presence of various acids, i.e., HCl, HNO_3, H_2SO_4, TFA, H_3PO_4, fluoroboric acid, MsOH, and *p*-TsOH, the yield was significantly improved. It appears that oxidative acids produced similar good yield (Table 1, entries 3 and 4), but reagents used there are expensive, toxic, and dangerous. It was found that MsOH in *i*-PrOH at 80 °C effectively promoted the formation desired isooxazoline. However, the yield of **3a** was slightly lower than that of *p*-TsOH (LD_{50}: 1410 mg/kg) (Table 1, entries 8 and 9), and MsOH (LD_{50}: 200 mg/kg) is highly toxic. Therefore, *p*-TsOH, which is non-oxidizing, corrosive, and low toxic was selected to participate in the synthesis of isoxazoline with a yield of 67%. While among the five tested solvents (ACN, *i*-PrOH, DMF, DMSO, and H_2O), it was found that the yield reduced significantly with the relative polarity of the solvent (Table 1, entries 9–13) and can obviously is the best solvent.

The reaction temperature and amount of *p*-TsOH were further optimized. The results were shown in Figure 2. Then, the optimal condition was regarded acanACN as solvent, 4 equiv of *p*-TsOH was involved in the reaction at 80 °C for 22 h, in which a good yield of 90% could be obtained.

Table 1. Reaction condition optimization [a].

[Reaction scheme: 1a (phenacyl nitro compound, PhC(O)CH2NO2) + 2a (allylbenzene) → 3a (isoxazoline with Ph and CH2Ph substituents), acid/solvent]

Entry	Acid	Amount (eq.) of Acid	Solvent (μ/10^{-30}C·m) [b]	Yield [c] (%) of 3a
1	-	-	*i*-PrOH (5.63)	<5
2	Concd. HCl (37%)	5	-//-	31
3	Concd. HNO$_3$ (70%)	5	-//-	83
4	Concd. H$_2$SO$_4$ (98%)	5	-//-	87
5	TFA	5	-//-	<5
6	H$_3$PO$_4$ (85%)	5	-//-	<5
7	Fluoroboric acid	5	-//-	33
8	MsOH	5	-//-	60
9	*p*-TsOH	5	-//-	67
10	-//-	5	ACN (10.68)	90
11	-//-	5	DMF (7.10)	<5
12	-//-	5	DMSO (14.00)	<5
13	-//-	5	H$_2$O (6.24)	<5

[a] General conditions: **1a** (0.125 mmol), **2a** (0.625 mmol), solvent (0.2 mL) at 80 °C for 22 h. [b] Data from PubChem 25 °C. [c] Isolated yield.

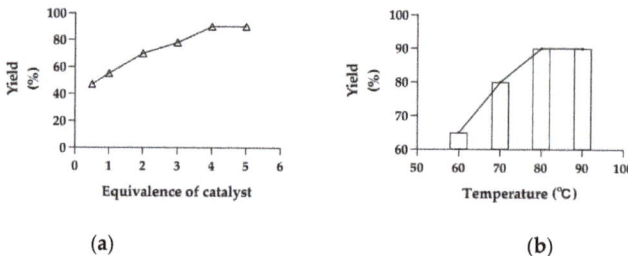

(a)   (b)

Figure 2. Optimization of the reaction conditions. (**a**) The effect of various equivalence of *p*-TsOH; (**b**) The effect of various temperature. The amount of *p*-TsOH was 0.5 mmol. General conditions: **1a** (0.125 mmol), **2a** (0.625 mmol), ACN (0.2 mL) at 80 °C or indicated temperature for 22 h.

2.2. Substrate Scope Studies

With the optimal reaction conditions, we first tested the 1,3-dipolar addition reaction for benzoyl nitromethane and allylbenzene and had a yield of product of 90%. Then, to examine the generality and scopes of this methodology, we took a variety of benzoylnitromethane derivatives **1** (Scheme 2) and allylbenzene **2a** as substrates and representative results were shown in Scheme 2. These results showed that a variety of electronically varied aromatic α-nitroketones were well compatible with the cycloaddition in all the reactions, and reaction generally obtains in moderate to good yields for the synthesis of isoxazoline derivatives. Moreover, in this reaction, we found a good regioselectivity, which was consistent with the work of Ken-ichi Itoh [45].

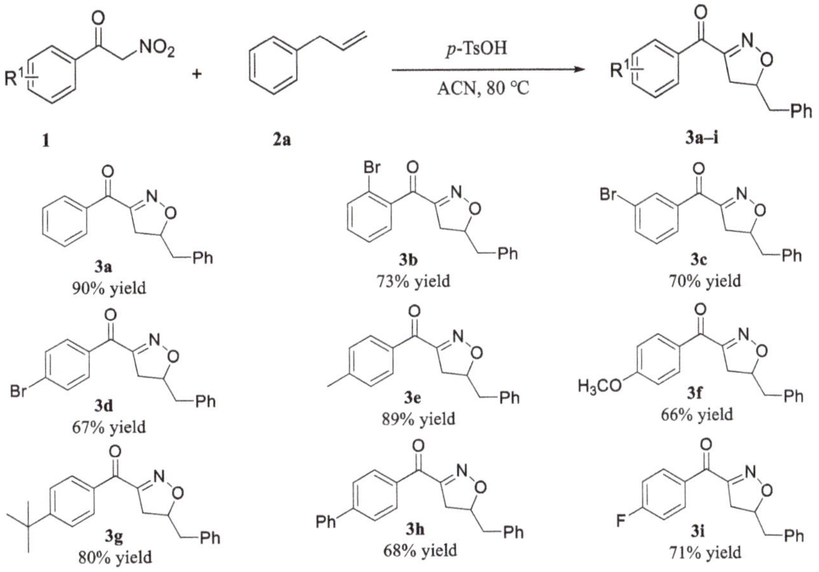

Scheme 2. Scopes of the substituted phenylnitroketones.

At first, it was found that the R_1 substituents would affect the cycloaddition efficiency in these reactions. The electron-rich α-nitroketones (**1e**, **1g**, Scheme 2) provided products (**3e**, **3g**, Scheme 2) in slightly better yields in comparison to the electron-deficient ones (**1d**, **1f**, **1i**, Scheme 2). Different electron-withdrawing substituents at the same position of phenyl-α-nitroketone resulted in similar yields (**3d**, **3i**, Scheme 2). Additionally, surprisingly, isoxazoline derivative (**3f**, Scheme 2) were obtained in moderate yields when electron-donating group(**-OMe**) were used with a yield of 66%, which was close to the yield of isoxazoline obtained by electron-deficient α-nitroketone. In the case of electron-deficient α-nitroketone, the corresponding were obtained in good yields (**3b–3d**, Scheme 2), respectively 73%, 70%, and 67%. The results show that the position of the substituent has no effect on the reaction results. Additionally, aromatic substrate, such as benzene, behaved similar to an electron-withdrawing substituents and gave a yield of 68% for product **3h**. However, the reaction rate was slower than that of aliphatic substituted substrates.

Next, we also investigated the scope of the alkenes (Table 2). The reaction of **1a** with alkenes derivatives **4** was carried out under the optimum reaction conditions whose results are shown in Table 2. All the reactions gave **5a–5f** as product, respectively, in good to excellent yield, except **5f**. The type of reaction substrate alkenes was modified, and it was found that the reaction proceeded well with both aliphatic alkenes and aromatic alkenes affording isoxazolines in good yields from the same α-nitroketone (entries 1–5, Table 2). In addition, cycloaddition of cyclohexene (**4f**) with benzoylnitromethane (**1a**) could also be achieved in fairly good yields, the corresponding isoxazoline **5f** was obtained in 69% yield (entry 6, **5f**, **Table** 2).

In addition, in order to expand the applicability of the reaction, we further examined the types of reaction materials. Isoxazolines were synthesized using the dipolarophiles **2a** (allylbenzene) and alkyl nitroketones **6** (Scheme 3). The results showed that in the presence of *p*-TsOH, alkyl nitroketones were also able to react with dipolar reagents to obtain isoxazoline derivatives; unfortunately, compared with **3** and **5** phenylisoxazoline, **7a** and **7b** yields were lower, 23% and 20%, respectively.

Table 2. Scopes of alkenes.

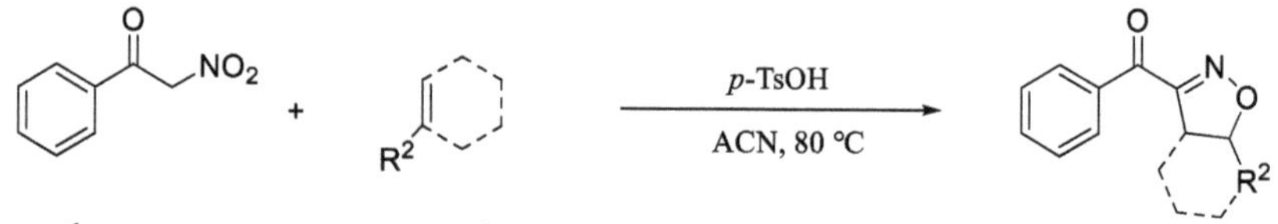

Entry [a]	Series	Products	Yields [b]
1	5a	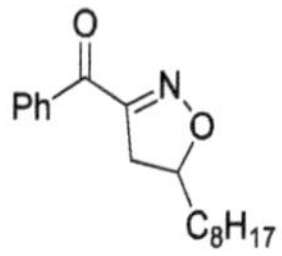 Ph-C(O)-isoxazoline-C$_8$H$_{17}$	86%
2	5b	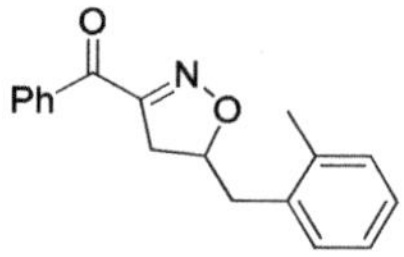 Ph-C(O)-isoxazoline-CH$_2$-(2-methylphenyl)	87%
3	5c	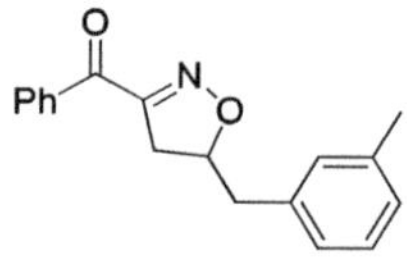 Ph-C(O)-isoxazoline-CH$_2$-(3-methylphenyl)	85%
4	5d	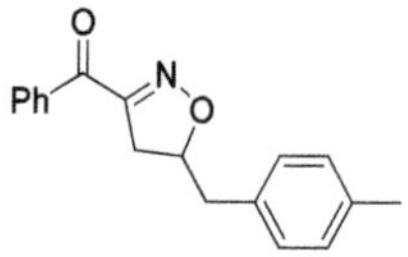 Ph-C(O)-isoxazoline-CH$_2$-(4-methylphenyl)	85%
5	5e	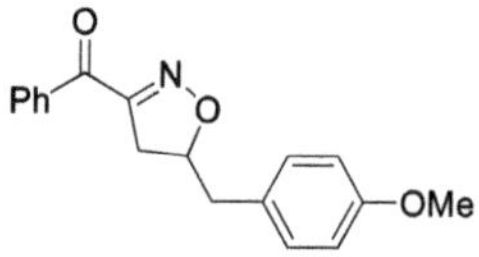 Ph-C(O)-isoxazoline-CH$_2$-(4-OMe-phenyl)	89%
6	5f	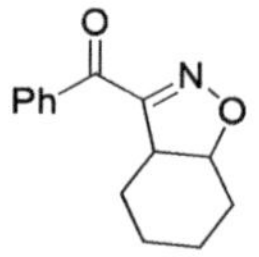 Ph-C(O)-bicyclic isoxazoline	69%

[a] General conditions: **1a** (0.125 mmol), **4** (0.625 mmol), *p*-TsOH (0.5 mmol), ACN (0.2 mL) at 80 °C for 22 h, unless stated otherwise. [b] Isolated yield.

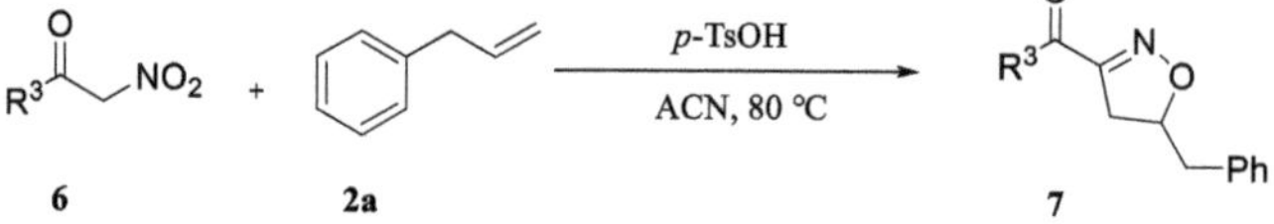

Scheme 3. Variation of alkyl nitroketones.

R^3-C(O)-CH$_2$-NO$_2$ (**6**) + styrene (**2a**) →(*p*-TsOH, ACN, 80 °C)→ **7**

7a R^3 = Me, 23% yield
7b R^3 = *i*-Bu, 20% yield

Finally, the alkyen **8** was used for the reaction with α-nitroketone (**1a**) under the optimized reaction conditions, which obtained in excellent yields Isoxazoles. Under the same conditions, the reaction rate with alkyen was quicker than with alkenes. Nevertheless, **9a** and **9b** were obtained in 85% and 88% yield, respectively (Scheme 4).

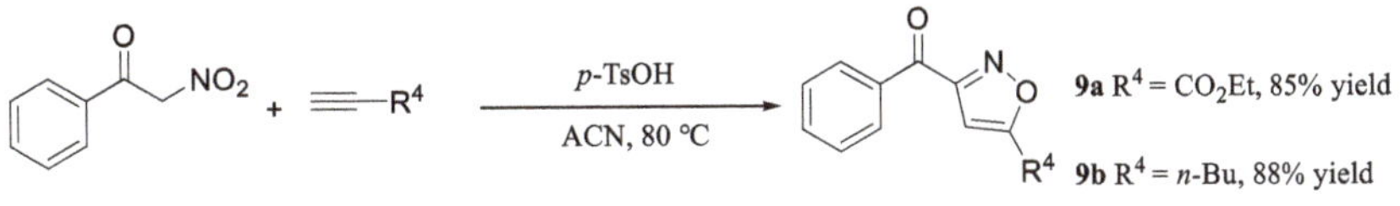

Scheme 4. Variation of alkyne.

2.3. Mechanistic Studies

After screening the reaction conditions and studying the application of the products, the reaction mechanism was also studied. On the basis of the reaction mechanism reported by Ken-ichi Itoh et al. [46,47]. We proposed the theory of 1,3-dipolar cycloaddition of benzoylnitromethane with allylbenzene was deduced as follows (Scheme 5): In this reaction, α-nitroketones are converted to nitroso cations in the presence of non-aqueous phase protons, then nitrile oxides are formed from nitroso cations. Finally, isoxazolines and their derivatives are obtained by intermolecular the 1,3-dipolar cycloaddition cyclization of dipolarophiles (alkenes or alkynes) and nitrile oxide.

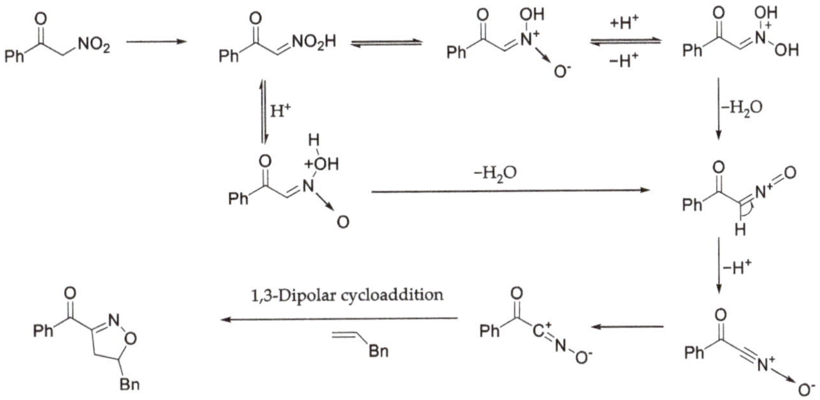

Scheme 5. Reaction mechanism.

3. Materials and Methods

3.1. General Experimental Methods

The structures of produced compounds were firmly confirmed by ^{13}C NMR and ^{1}HNMR spectra, and supported by HRMS, and IR data (see the Supplementary Materials).

^{1}H NMR (400 MHz) and ^{13}C NMR (101 MHz) were recorded at room temperature on DRX-400 spectrometer (Bruker, Saarbrücken, Saarland, Germany) in CDCl$_3$. The chemical shifts are given in parts per million (ppm) on the delta (δ) scale. The solvent peak was used as a reference value, for ^{1}H NMR: CDCl$_3$ δ$_H$ 7.26; for ^{13}C NMR: CDCl$_3$ δ$_C$ 77.16 ppm. IR spectra were recorded using an Avatar 360 FT-IR ESP spectrometer (Nicolet, Madison, Wisconsin, USA) at room temperature. HR-ESI-MS spectra were acquired using an Agilent 6210 ESI/TOF mass spectrometer (Agilent Technologies, Santa Clara, CA, USA). Analytical TLC was conducted on silica gel plates (GF254, Yantai Institute of Chemical Technology, Yantai, China). Spots on the plates were observed under UV light. Column chromatography was performed on silica gel (200~300 mesh and 300~400 mesh; Qingdao Marine Chemical Factory, Qingdao, China). Super-dry solvent i-PrOH, ACN, DMSO and DMF were purchased from Aldrich and used as supplied. The α-nitroketones were synthesized using the same method as reported in the literature [48,49].

3.2. General Procedure for the Cycloaddition of Alkenes and α-Nitroketones

p-TsOH (0.500 mmol, 4 equiv) was added to a solution of **1** (0.125 mmol, 1 equiv) or **6** (0.125 mmol, 1 equiv) and **2** (0.625 mmol, 5 equiv) (or **4** (0.625 mmol, 5 equiv) or **8** (0.625 mmol, 5 equiv)) in ACN (0.2 mL). The mixture was then stirred at 80 °C until the starting material disappeared as monitored by TLC. Subsequently, the mixture was directly purified by flash chromatography (with ethyl acetate/petroleum ether as the eluent) to obtain the desired product (**3**, **5**, **7** or **9**).

4. Conclusions

In conclusion, we have developed an efficient cycloaddition of a variety of α-nitroketones with alkenes or alkyne using inexpensive and gentle acid. Among the synthesized compounds, the yield of cycloaddition products of substituted phenylnitroketone is high (66–90 %), while the yield of cycloaddition products of alkyl nitroketones (such as **7a–b**) is low (20–23 %). This synthesis is based on cycloaddition 1,3-dipolar in presence of *p*-TsOH, which is attractive that the low cost, simple synthetic route, and ease of handling of the gentle acid. It is particularly noteworthy that the reaction provides an effective synthesis method for 3-carbonylisoxazolines. The development of other methods for the synthesis of 3-carbonylisoxazoline is currently under investigation and will be disclosed in due course.

Supplementary Materials: The following supporting information can be downloaded at: https://www.mdpi.com/article/10.3390/molecules28062565/s1, The Supplementary Materials contain experimental protocols, analytical data for products and NMR spectra [29,42,46,48–50].

Author Contributions: Conceptualization, X.P., X.Y., and J.W.; methodology, X.P. and X.X., formal analysis, C.Y., X.P.; K.Y.; S.H., K.Z., Q.L., X.X., and J.L.; original manuscript writing, C.Y. and X.P.; revision and supervision, X.P., X.Y., and J.W.; funding acquisition, X.P., X.Y., and J.W. All authors have read and agreed to the published version of the manuscript.

Funding: This research was by National Natural Science Foundation of China (82160651); The Open Project of Stake Key Laboratory of Natural and Biomimetic Drugs, Peking University (K202103); The Open Project of Key Laboratory of Xinjiang Phytomedicine Resource and Utilization, Ministry of Education (XPRU202004); Youth Innovative Talent Cultivation Projects of Shihezi University (CXPY202005); The National Innovation and Entrepreneurship Training Program for Undergraduate (202210759029); and The Open Sharing Fund for the Large-scale Instruments and Equipments of Shihezi University.

Institutional Review Board Statement: Not applicable.

Informed Consent Statement: Not applicable.

Data Availability Statement: The data presented in this study are available in the article and Supplementary Materials.

Acknowledgments: The authors thank Liang Guo for assistance with NMR.

Conflicts of Interest: The authors declare no conflict of interest.

Sample Availability: Samples of the compounds **3a-i**, **5a-f**, **7a-b** and **9a-b** are available from the authors.

References

1. Hemmerling, F.; Hahn, F. Biosynthesis of oxygen and nitrogen-containing heterocycles in polyketides. *Beilstein J. Org. Chem.* **2016**, *12*, 1512–1550. [CrossRef] [PubMed]
2. Sysak, A.; Obminska-Mrukowicz, B. Isoxazole ring as a useful scaffold in a search for new therapeutic agents. *Eur. J. Med. Chem.* **2017**, *137*, 292–309. [CrossRef] [PubMed]
3. Gutierrez, M.; Matus, M.F.; Poblete, T.; Amigo, J.; Vallejos, G.; Astudillo, L. Isoxazoles: Synthesis, evaluation and bioinformatic design as acetylcholinesterase inhibitors. *J. Pharm. Pharmacol.* **2013**, *65*, 1796–1804. [CrossRef]
4. Koufaki, M.; Fotopoulou, T.; Kapetanou, M.; Heropoulos, G.A.; Gonos, E.S.; Chondrogianni, N. Microwave-assisted synthesis of 3,5-disubstituted isoxazoles and evaluation of their anti-ageing activity. *Eur. J. Med. Chem.* **2014**, *83*, 508–515. [CrossRef]

5. Bano, S.; Alam, M.S.; Javed, K.; Dudeja, M.; Das, A.K.; Dhulap, A. Synthesis, biological evaluation and molecular docking of some substituted pyrazolines and isoxazolines as potential antimicrobial agents. *Eur. J. Med. Chem.* **2015**, *95*, 96–103. [CrossRef] [PubMed]
6. Liao, J.; Ouyang, L.; Jin, Q.; Zhang, J.; Luo, R. Recent advances in the oxime-participating synthesis of isoxazolines. *Org. Biomol. Chem.* **2020**, *18*, 4709–4716. [CrossRef] [PubMed]
7. Hyean Kim, B.; Jun Chung, Y.; Jung Ryu, E. Synthesis of α-hydroxy ketomethylene dipeptide isosteres. *Tetrahedron Lett.* **1993**, *34*, 8465–8468. [CrossRef]
8. Goncalves, I.L.; Machado das Neves, G.; Porto Kagami, L.; Eifler-Lima, V.L.; Merlo, A.A. Discovery, development, chemical diversity and design of isoxazoline-based insecticides. *Bioorg. Med. Chem.* **2021**, *30*, 115934. [CrossRef]
9. Pohjakallio, A.; Pihko, P.M.; Laitinen, U.M. Synthesis of 2-isoxazolines: Enantioselective and racemic methods based on conjugate additions of oximes. *Chemistry* **2010**, *16*, 11325–11339. [CrossRef]
10. Liu, X.; Ma, X.; Feng, Y. Introduction of an isoxazoline unit to the β-position of porphyrin via regioselective 1,3-dipolar cycloaddition reaction. *Beilstein J. Org. Chem.* **2019**, *15*, 1434–1440. [CrossRef]
11. Akagawa, K.; Kudo, K. Iterative Polyketide Synthesis via a Consecutive Carbonyl-Protecting Strategy. *J. Org. Chem.* **2018**, *83*, 4279–4285. [CrossRef]
12. Lopes, E.F.; Penteado, F.; Thurow, S.; Pinz, M.; Reis, A.S.; Wilhelm, E.A.; Luchese, C.; Barcellos, T.; Dalberto, B.; Alves, D.; et al. Synthesis of Isoxazolines by the Electrophilic Chalcogenation of beta, gamma-Unsaturated Oximes: Fishing Novel Anti-Inflammatory Agents. *J. Org. Chem.* **2019**, *84*, 12452–12462. [CrossRef]
13. Kaur, K.; Kumar, V.; Sharma, A.K.; Gupta, G.K. Isoxazoline containing natural products as anticancer agents: A review. *Eur. J. Med. Chem.* **2014**, *77*, 121–133. [CrossRef]
14. Goyard, D.; Konya, B.; Chajistamatiou, A.S.; Chrysina, E.D.; Leroy, J.; Balzarin, S.; Tournier, M.; Tousch, D.; Petit, P.; Duret, C.; et al. Glucose-derived spiro-isoxazolines are anti-hyperglycemic agents against type 2 diabetes through glycogen phosphorylase inhibition. *Eur. J. Med. Chem.* **2016**, *108*, 444–454. [CrossRef]
15. Picconi, P.; Prabaharan, P.; Auer, J.L.; Sandiford, S.; Cascio, F.; Chowdhury, M.; Hind, C.; Wand, M.E.; Sutton, J.M.; Rahman, K.M. Novel pyridyl nitrofuranyl isoxazolines show antibacterial activity against multiple drug resistant Staphylococcus species. *Bioorg. Med. Chem.* **2017**, *25*, 3971–3979. [CrossRef] [PubMed]
16. Dallanoce, C.; Meroni, G.; De Amici, M.; Hoffmann, C.; Klotz, K.N.; De Micheli, C. Synthesis of enantiopure Delta2-isoxazoline derivatives and evaluation of their affinity and efficacy profiles at human beta-adrenergic receptor subtypes. *Bioorg. Med. Chem.* **2006**, *14*, 4393–4401. [CrossRef]
17. Filali, I.; Bouajila, J.; Znati, M.; Bousejra-El Garah, F.; Ben Jannet, H. Synthesis of new isoxazoline derivatives from harmine and evaluation of their anti-Alzheimer, anti-cancer and anti-inflammatory activities. *J. Enzyme Inhib. Med. Chem.* **2015**, *30*, 371–376. [CrossRef]
18. Basappa; Sadashiva, M.P.; Mantelingu, K.; Swamy, S.N.; Rangappa, K.S. Solution-phase synthesis of novel delta2-isoxazoline libraries via 1,3-dipolar cycloaddition and their antifungal properties. *Bioorg. Med. Chem.* **2003**, *11*, 4539–4544. [CrossRef]
19. Vinay Kumar, K.S.; Lingaraju, G.S.; Bommegowda, Y.K.; Vinayaka, A.C.; Bhat, P.; Pradeepa Kumara, C.S.; Rangappa, K.S.; Gowda, D.C.; Sadashiva, M.P. Synthesis, antimalarial activity, and target binding of dibenzazepine-tethered isoxazolines. *RSC Adv.* **2015**, *5*, 90408–90421. [CrossRef]
20. Alshamari, A.; Al-Qudah, M.; Hamadeh, F.; Al-Momani, L.; Abu-Orabi, S. Synthesis, Antimicrobial and Antioxidant Activities of 2-Isoxazoline Derivatives. *Molecules* **2020**, *25*, 4271. [CrossRef] [PubMed]
21. Tangallapally, R.P.; Sun, D.; Rakesh; Budha, N.; Lee, R.E.; Lenaerts, A.J.; Meibohm, B.; Lee, R.E. Discovery of novel isoxazolines as anti-tuberculosis agents. *Bioorg. Med. Chem. Lett.* **2007**, *17*, 6638–6642. [CrossRef]
22. Karthikeyan, K.; Veenus Seelan, T.; Lalitha, K.G.; Perumal, P.T. Synthesis and antinociceptive activity of pyrazolyl isoxazolines and pyrazolyl isoxazoles. *Bioorg. Med. Chem. Lett.* **2009**, *19*, 3370–3373. [CrossRef]
23. Yang, J.; Guan, A.; Wu, Q.; Cui, D.; Liu, C. Design, synthesis and herbicidal evaluation of novel uracil derivatives containing an isoxazoline moiety. *Pest Manag. Sci.* **2020**, *76*, 3395–3402. [CrossRef] [PubMed]
24. Mita, T.; Kikuchi, T.; Mizukoshi, T.; Yaosaka, M.; Komoda, M. Preparation of Isoxazoline-Substituted Benzamide Derivatives as Insecticides, Acaricides, and Parasiticides. CN1930136; Nissan Chemical Industries Co., Ltd.: Tokyo, Japan, 2007; 2007-03-14.
25. Slagbrand, T.; Kervefors, G.; Tinnis, F.; Adolfsson, H. An Efficient One-pot Procedure for the Direct Preparation of 4,5-Dihydroisoxazoles from Amides. *Adv. Synth. Catal.* **2017**, *359*, 1990–1995. [CrossRef]
26. Boruah, M.; Konwar, D. KF/Al$_2$O$_3$: Solid-Supported Reagent Used in 1,3-Dipolar Cycloaddition Reaction of Nitrile Oxide. *Synth. Commun.* **2012**, *42*, 3261–3268. [CrossRef]
27. Kesornpun, C.; Aree, T.; Mahidol, C.; Ruchirawat, S.; Kittakoop, P. Water-Assisted Nitrile Oxide Cycloadditions: Synthesis of Isoxazoles and Stereoselective Syntheses of Isoxazolines and 1,2,4-Oxadiazoles. *Angew. Chem. Int. Ed. Engl.* **2016**, *55*, 3997–4001. [CrossRef]
28. Svejstrup, T.D.; Zawodny, W.; Douglas, J.J.; Bidgeli, D.; Sheikh, N.S.; Leonori, D. Visible-light-mediated generation of nitrile oxides for the photoredox synthesis of isoxazolines and isoxazoles. *Chem. Commun.* **2016**, *52*, 12302–12305. [CrossRef] [PubMed]
29. Dai, P.; Tan, X.; Luo, Q.; Yu, X.; Zhang, S.-G.; Liu, F.; Zhang, H.-W. Synthesis of 3-Acyl-isoxazoles and Δ2-Isoxazolines from Methyl Ketones, Alkynes or Alkenes, and tert-Butyl Nitrite via a Csp^3-H Radical Functionalization/Cycloaddition Cascade. *Org Lett.* **2019**, *21*, 5096–5100. [CrossRef] [PubMed]

30. Zhang, X.-W.; Xiao, Z.-F.; Zhuang, Y.-J.; Wang, M.-M.; Kang, Y.-B. Metal-Free Autoxidative Nitrooxylation of Alkenyl Oximes with Molecular Oxygen. *Adv. Synth. Catal.* **2016**, *358*, 1942–1945. [CrossRef]
31. Liu, R.-H.; Wei, D.; Han, B.; Yu, W. Copper-Catalyzed Oxidative Oxyamination/Diamination of Internal Alkenes of Unsaturated Oximes with Simple Amines. *ACS Catal.* **2016**, *6*, 6525–6530. [CrossRef]
32. Wang, L.-J.; Chen, M.; Qi, L.; Xu, Z.; Li, W. Copper-mediated oxysulfonylation of alkenyl oximes with sodium sulfinates: A facile synthesis of isoxazolines featuring a sulfone substituent. *Chem. Commun. (Camb).* **2017**, *53*, 2056–2059. [CrossRef]
33. Triandafillidi, I.; Kokotos, C.G. Green Organocatalytic Synthesis of Isoxazolines via a One-Pot Oxidation of Allyloximes. *Org. Lett.* **2017**, *19*, 106–109. [CrossRef] [PubMed]
34. Wei, W.; Tang, Y.; Zhou, Y.; Deng, G.; Liu, Z.; Wu, J.; Li, Y.; Zhang, J.; Xu, S. Recycling Catalyst as Reactant: A Sustainable Strategy To Improve Atom Efficiency of Organocatalytic Tandem Reactions. *Org. Lett.* **2018**, *20*, 6559–6563. [CrossRef] [PubMed]
35. Tu, K.N.; Hirner, J.J.; Blum, S.A. Oxyboration with and without a Catalyst: Borylated Isoxazoles via B-O sigma-Bond Addition. *Org. Lett.* **2016**, *18*, 480–483. [CrossRef] [PubMed]
36. Chary, R.G.; Reddy, G.R.; Ganesh, Y.S.S.; Prasad, K.V.; Raghunadh, A.; Krishna, T.; Mukherjee, S.; Pal, M. Effect of Aqueous Polyethylene Glycol on 1,3-Dipolar Cycloaddition of Benzoylnitromethane/Ethyl 2-Nitroacetate with Dipolarophiles: Green Synthesis of Isoxazoles and Isoxazolines. *Adv. Synth. Catal.* **2014**, *356*, 160–164. [CrossRef]
37. Zhang, X.-W.; He, X.-L.; Yan, N.; Zheng, H.X.; Hu, X.-G. Oxidize Amines to Nitrile Oxides: One Type of Amine Oxidation and Its Application to Directly Construct Isoxazoles and Isoxazolines. *J. Org. Chem.* **2020**, *85*, 15726–15735. [CrossRef]
38. Li, B.-B.; Jiang, Q. Preparation of a novel isoxazoline compound, fluoreamine. *Guangzhou Chem. Ind.* **2020**, *47*, 215–216.
39. Gao, M.; Gan, Y.; Xu, B. From Alkenes to Isoxazolines via Copper-Mediated Alkene Cleavage and Dipolar Cycloaddition. *Org. Lett.* **2019**, *21*, 7435–7439. [CrossRef]
40. Morita, T.; Fukuhara, S.; Fuse, S.; Nakamura, H. Gold(I)-Catalyzed Intramolecular S(E)Ar Reaction: Efficient Synthesis of Isoxazole-Containing Fused Heterocycles. *Org. Lett.* **2018**, *20*, 433–436. [CrossRef]
41. Ueda, M.; Ikeda, Y.; Sato, A.; Ito, Y.; Kakiuchi, M.; Shono, H.; Miyoshi, T.; Naito, T.; Miyata, O. Silver-catalyzed synthesis of disubstituted isoxazoles by cyclization of alkynyl oxime ethers. *Tetrahedron* **2011**, *67*, 4612–4615. [CrossRef]
42. Pan, X.-H.; Xin, X.-B.; Mao, Y.; Li, X.; Zhao, Y.-N.; Liu, Y.-D.; Zhang, K.; Yang, X.-D.; Wang, J.-H. 3-Benzoylisoxazolines by 1,3-Dipolar Cycloaddition: Chloramine-T-Catalyzed Condensation of alpha-Nitroketones with Dipolarophiles. *Molecules* **2021**, *26*, 3491. [CrossRef] [PubMed]
43. Nazarenko, K.G.; Shvidenko, K.V.; Pinchuk, A.M.; Tolmachev, A.A. Synthesis of 7-Amino-1-nitro-2-heptanone Derivatives. *Synth. Commun.* **2003**, *33*, 4241–4252. [CrossRef]
44. Arumugam, N.; Raghunathan, R.; Almansour, A.I.; Karama, U. An efficient synthesis of highly functionalized novel chromeno [4,3-b] pyrroles and indolizino [6,7-b] indoles as potent antimicrobial and antioxidant agents. *Bioorg. Med. Chem. Lett.* **2012**, *22*, 1375–1379. [CrossRef]
45. Quan, X.-J.; Ren, Z.-H.; Wang, Y.-Y.; Guan, Z.-H. p-Toluenesulfonic acid mediated 1,3-dipolar cycloaddition of nitroolefins with NaN3 for synthesis of 4-aryl-NH-1,2,3-triazoles. *Org. Lett.* **2014**, *16*, 5728–5731. [CrossRef]
46. Itoh, K.-i.; Aoyama, T.; Satoh, H.; Fujii, Y.; Sakamaki, H.; Takido, T.; Kodomari, M. Application of silica gel-supported polyphosphoric acid (PPA/SiO2) as a reusable solid acid catalyst to the synthesis of 3-benzoylisoxazoles and isoxazolines. *Tetrahedron Lett.* **2011**, *52*, 6892–6895. [CrossRef]
47. Itoh, K.; Horiuchi, C.A. Formation of isoxazole derivatives via nitrile oxide using ammonium cerium nitrate (CAN): A novel one-pot synthesis of 3-acetyl- and 3-benzoylisoxazole derivatives. *Tetrahedron* **2004**, *60*, 1671–1681. [CrossRef]
48. June, L.J.; Jihye, K.; Moo, J.Y.; Min, L.B.; Hyo, K.B. Indium-mediated one-pot synthesis of benzoxazoles or oxazoles from 2-nitrophenols or 1-aryl-2-nitroethanones. *Tetrahedron* **2009**, *65*, 8821–8831.
49. Zhang, H.Q.; Pan, X.H.; Zhang, K.; Wang, H.Y.; Wang, J.H. Synthesis of α-nitroketones. *J. Shihezi Univ. (Nat. Sci. Ed.)* **2017**, *35*, 602–605.
50. Lindsay, A.C.; Kilmartin, P.A.; Organic, J.S.J.; Sperry, J. Synthesis of 3-nitroindoles by sequential paired electrolysis. *Org. Biomol. Chem.* **2021**, *19*, 7903–7913. [CrossRef]

Disclaimer/Publisher's Note: The statements, opinions and data contained in all publications are solely those of the individual author(s) and contributor(s) and not of MDPI and/or the editor(s). MDPI and/or the editor(s) disclaim responsibility for any injury to people or property resulting from any ideas, methods, instructions or products referred to in the content.

Article

Synthesis of Tetrasubstituted Nitroalkenes and Preliminary Studies of Their Enantioselective Organocatalytic Reduction

Patricia Camarero González [1], Sergio Rossi [1], Miguel Sanz [2], Francesca Vasile [1] and Maurizio Benaglia [1,*]

[1] Dipartimento di Chimica, Università degli Studi di Milano, Via Golgi 19, 20133 Milano, Italy
[2] Taros Chemicals GmbH & Co. KG, Emil-Figge-Str. 76a, 44227 Dortmund, Germany
* Correspondence: maurizio.benaglia@unimi.it; Tel.: +02-5031-4171

Abstract: Starting from commercially available ketones, a reproducible and reliable strategy for the synthesis of tetrasubstituted nitroalkenes was successfully developed, using a two-step procedure; the HWE olefination of the ketone to form the corresponding α,β-unsaturated esters is followed by a nitration reaction to introduce the nitro group in the α position of the ester group. The enantioselective organocatalytic reduction of these compounds has also been preliminarily studied, to access the functionalized enantioenriched nitroalkanes, which are useful starting materials for further synthetic elaborations. The absolute configuration of the reduction product was established by chemical correlation of the chiral nitroalkane with a known product; preliminary DFT calculations were also conducted to rationalize the stereochemical outcome of the organocatalytic enantioselective reduction.

Keywords: nitroacrylates; organocatalysis; stereoselective synthesis; reduction; chiral nitro derivatives

1. Introduction

Among unnatural α-amino acids, α,α-disubstituted amino acids are key biological scaffolds with many specific roles and properties that have made them increasingly attractive in the fields of organic chemistry, biochemical research and drug discovery [1–3]. They have unique structural properties, which make them ideal candidates to be included in the design of new pharmaceutically active compounds, as well as intermediates for the study of pathological pathways [4–6].

Synthetic approaches for the synthesis of di- or trisubstituted nitroalkenes, valuable intermediates for the synthesis of α,α-disubstituted amino acids, are abundant in the literature, but there are very scarce data reporting synthetic routes for tetrasubstituted nitroalkenes [7,8]. Herein we describe a reproducible methodology for the synthesis of these compounds (Figure 1) that started from easily available ketones (**1**), which are converted to tetrasubstituted nitroolefins (**3**), by reaction of an acrylate intermediate (**2**), with an appropriate nitration reagent. Nitroalkenes will be organocatalytically reduced to afford enantioenriched nitroalkanes (**4**), highly functionalized chiral starting materials for further transformations (Figure 1).

Citation: González, P.C.; Rossi, S.; Sanz, M.; Vasile, F.; Benaglia, M. Synthesis of Tetrasubstituted Nitroalkenes and Preliminary Studies of Their Enantioselective Organocatalytic Reduction. *Molecules* 2023, 28, 3156. https://doi.org/10.3390/molecules28073156

Academic Editor: Renata Riva

Received: 26 February 2023
Revised: 28 March 2023
Accepted: 29 March 2023
Published: 1 April 2023

Copyright: © 2023 by the authors. Licensee MDPI, Basel, Switzerland. This article is an open access article distributed under the terms and conditions of the Creative Commons Attribution (CC BY) license (https:// creativecommons.org/licenses/by/ 4.0/).

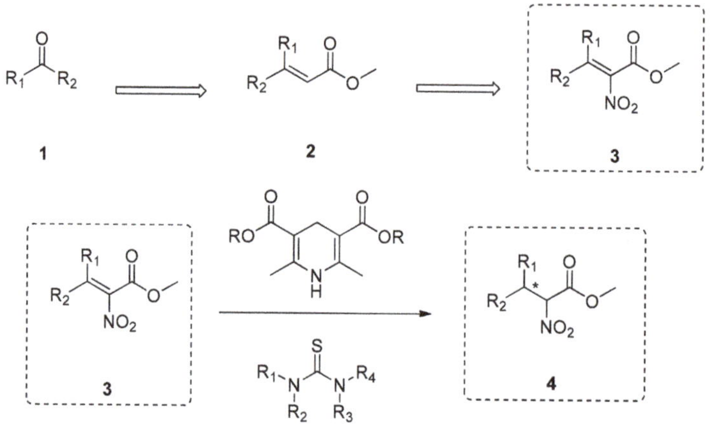

Figure 1. Synthetic approach for the synthesis of tetrasubstituted nitroalkenes and their enantioselective catalytic reduction.

2. Results and Discussion

2.1. Synthesis of Tetrasubstituted Nitroalkenes

We started the investigations by exploring a two-step synthetic strategy for the synthesis of tetrasubstituted nitroalkenes (Scheme 1). The first reaction involves the formation of acrylates **2a–f**, by reaction of different commercially available ketones **1a–f**, using Horner–Wadstworth–Emmons reaction conditions [9]. The results are summarized in Table 1.

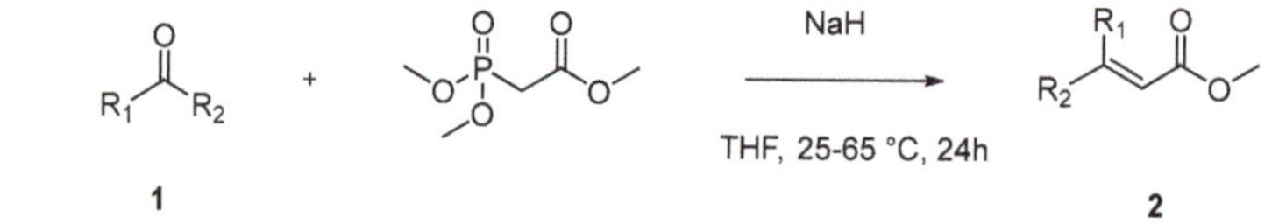

R_1 = Me; R_2 = Ph **1a**
R_1 = CF$_3$; R_2 = Ph **1b**
R_1 = Me; R_2 = ^{t}Bu **1c**
R_1 = ^{i}Pr; R_2 = Ph **1d**
R_1 = Me; R_2 = 4-BrPh **1e**
R_1 = Me; R_2 = 4-MePh **1f**

R_1 = Me; R_2 = Ph **2a**
R_1 = CF$_3$; R_2 = Ph **2b**
R_1 = Me; R_2 = ^{t}Bu **2c**
R_1 = ^{i}Pr; R_2 = Ph **2d**
R_1 = Me; R_2 = 4-BrPh **2e**
R_1 = Me; R_2 = 4-MePh **2f**

Scheme 1. Synthesis of acrylates intermediates (2).

Table 1. Synthesis of acrylate intermediates **2a–f**.

Entry	R_1	R_2	T	Product	E/Z Ratio [a]	Yield [b]
1	Me	Ph	RT	2a	80:20	52%
2	CF$_3$	Ph	RT	2b	75:25	79%
3	Me	^{t}Bu	66 °C	2c	90:10	14%
4	^{i}Pr	Ph	66 °C	2d	67:33	33%
5	Me	4-BrPh	RT	2e	65:35	76%
6	Me	4-MePh	66 °C	2f	75:25	86%

[a] Determined by ^{1}H-NMR. [b] Yield was determined after purification using column chromatography.

Compounds **2a–f** were obtained after reaction of the corresponding commercially available ketones with trimethylphosphonoacetate and sodium hydride in THF for 24 h with good to moderate yields and excellent diastereoselectivities after purification with column chromatography. The reactions were performed at room temperature, but with compounds **2c**, **2d** and **2f** (Table 1, entries 3,4,6) it was necessary to heat the reaction to 66 °C. Their E/Z ratio was checked by ^{1}H-NMR of reaction crude. The low yield observed for the isolation of pure product **2c** (Table 1, entry 3) can be explained by its high volatility, and optimization of the product isolation is underway.

Then, our efforts were concentrated on the nitration reaction of these acrylate intermediates using common nitration reagents such as HNO$_3$ [10,11]. However, the use of nitric acid mainly led to the formation of products with the nitro group on the aromatic ring, and the yields after purification were very low (<10%). Furthermore, several problems were also detected during the isolation process. Selective nitration conditions for double bonds were also considered, but no formation of the desired nitroalkane was observed. Thus, alternative strategies involving the condensation between acetophenone and ethyl nitroacetate or the reaction between phenylacetylene with ethyl nitroacetate catalyzed by indium salts were also explored, [12–15] but without any satisfactory results (see Supplementary Materials).

Finally, Buevich and co-workers reported the first example of a α-nitro addition to a cinammic ester for the synthesis of dehydrophenylalanine derivatives, which are precursors of α-amino acids, by utilizing a CAN-NaNO$_2$ system [16]. Therefore, we decided to investigate the methodology for the synthesis of target tetrasubstituted nitroalkenes **3a–f** starting from acrylates **2a–f** (Scheme 2).

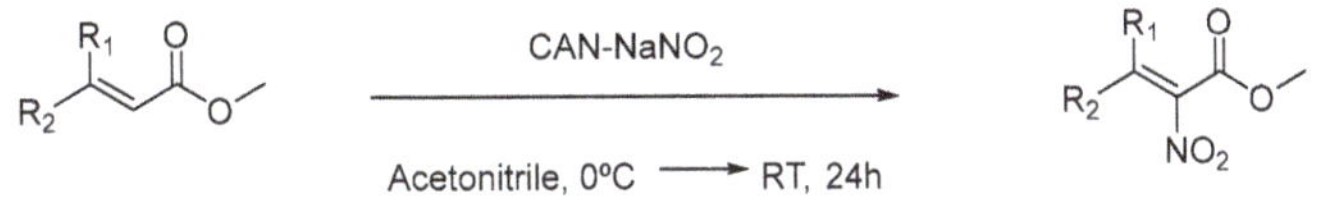

R$_1$ = Me; R$_2$ = Ph **2a**
R$_1$ = CF$_3$; R$_2$ = Ph **2b**
R$_1$ = Me; R$_2$ = ^{t}Bu **2c**
R$_1$ = ^{i}Pr; R$_2$ = Ph **2d**
R$_1$ = Me; R$_2$ = 4-BrPh **2e**
R$_1$ = Me; R$_2$ = 4-MePh **2f**

R$_1$ = Me; R$_2$ = Ph **3a**
R$_1$ = CF$_3$; R$_2$ = Ph **3b**
R$_1$ = Me; R$_2$ = ^{t}Bu **3c**
R$_1$ = ^{i}Pr; R$_2$ = Ph **3d**
R$_1$ = Me; R$_2$ = 4-BrPh **3e**
R$_1$ = Me; R$_2$ = 4-MePh **3f**

Scheme 2. Synthesis of tetrasubstituted nitroalkenes **3a–f**.

The results are summarized in Table 2.

Table 2. Synthesis of tetrasubstituted nitroalkenes **3a–f**.

Entry	R$_1$	R$_2$	Acrylate	Nitroacrylate	Yield [a]	Isomers Ratio
1	Me	Ph	2a	3a	46%	60/40
2	CF$_3$	Ph	2b	-	[b]	-
3	Me	^{t}Bu	2c	3c	—[c]	-
4	^{i}Pr	Ph	2d	3d	25%	99/1
5	Me	4-BrPh	2e	3e	47%	62/38
6	Me	4-MePh	2f	3f	48%	57/43

[a] Isolated yield after purification with column chromatography. [b] Side-product (Z)-methyl 4,4,4-trifluoro-3-(4-nitrophenyl)but-2-enoate was obtained in 76% yield. [c] Reaction did not occur, unreacted starting material was recovered.

The nitration reaction of acrylates led to the formation of the corresponding tetrasubstituted nitroacrylates **3a–f** in low to moderate yields after chromatographic purification, and typically in a 6/4 diasteoisomeric ratio, except in the case of **3d**, when a single isomer was isolated. In case of nitroacrylates **3a,e** (Table 2, entries 1 and 5), two different fractions corresponding to the two separated diastereoisomers were obtained after purification., while for product **3f** (Table 2, entry 6) a single fraction containing both, non-separable, isomers, was obtained after chromatography.

The nitration of acrylate **2b** (Table 2, entry 2), did not lead to the formation of the corresponding nitroacrylate **3b**, but (Z)-methyl 4,4,4-trifluoro-3-(4-nitrophenyl)but-2-enoate was obtained as major compound in this reaction in 76% yield. The nitration of acrylate **2c** (Table 2, entry 3) did not afford any product.

As previously mentioned, in the nitration of acrylates **2a,e** (Table 2, entries 1 and 5), it was possible to separate the two diastereoisomers. In order to clarify the configuration of the two products, additional NMR experiments were conducted on the isomers of compound **3a** (Figure 2).

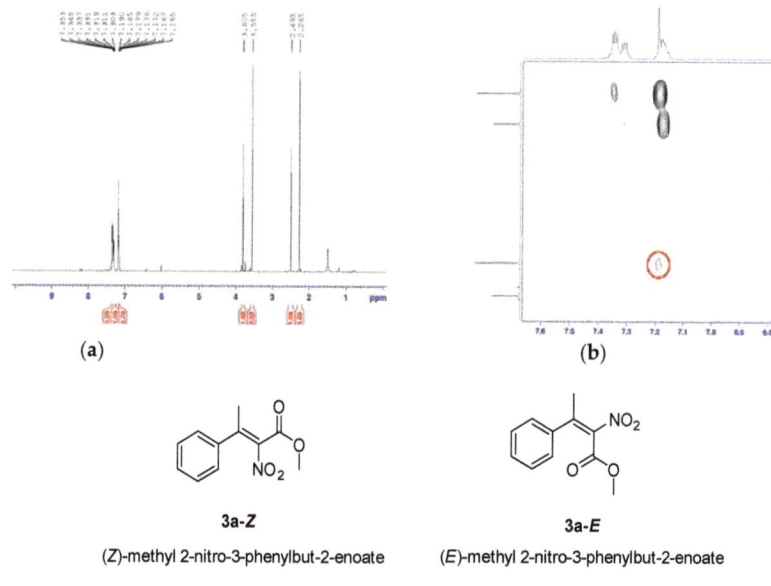

Figure 2. Further NMR experiments performed for tetrasubstituted nitroacrylate **3a**. (**a**): NMR spectra of a mixture fraction of compound **3a**. (**b**): NOE contact observed between OMe and Ph groups in the more abundant form of the tetrasubstituted nitroacrylate **3a**. (**c**): structures of nitroacrylates **3a–Z** and **3a–E**.

A chemical shift study on the methoxy group signal of both Z/E forms of the tetrasubstituted nitroalkene **3a** has been performed, using a NOESY experiment. As illustrated in Figure 2a, the OMe group of the molecule is more shielded (3.55 ppm) in the more abundant form, suggesting that it can be near to the shielding cone of the aromatic ring (corresponding to the (E) isomer), while it resonates at 3.8 ppm in the less abundant isomer of Figure 2a.

NOESY experimental results, showing through-space correlation within the molecule, were also acquired to predict if the more abundant isomer of the synthesized tetrasubstituted nitroalkene corresponds to (E) or (Z), considering that the NOE contact between the methoxy group and the phenyl ring can only be observed in the (E) isomer. The analysis of NOE contacts suggested that the significant cross peak between the OMe group and phenyl ring (Figure 2b) is present only in the major isomer that can be determined to have the (E) configuration (product **3a–E**).

2.2. Enantioselective Reduction of Tetrasubstituted Nitroalkenes (3)

The use of Hantzsch esters as biomimetic reducing agents [17] has been reported in different organocatalytic reductions of nitroalkenes, but also ketoimines and ketoesters [18–20].

In addition, Hantzsch esters are easily synthetized, and their structure readily tuned in order to maximize the efficiency of enantioselective reactions [21].

The enantioselective reduction of the synthetized tetrasubstituted nitroalkenes (**3**) using Hantzsch esters as a reductive agent and a thiourea based chiral catalyst was investigated to obtain the corresponding functionalized nitroalkanes (**4**) (Scheme 3), in the presence of a few chiral bifunctional catalysts **A–D**, representative of different classes of the most popular organocatalysts for this transformation. The results are reported in Table 3.

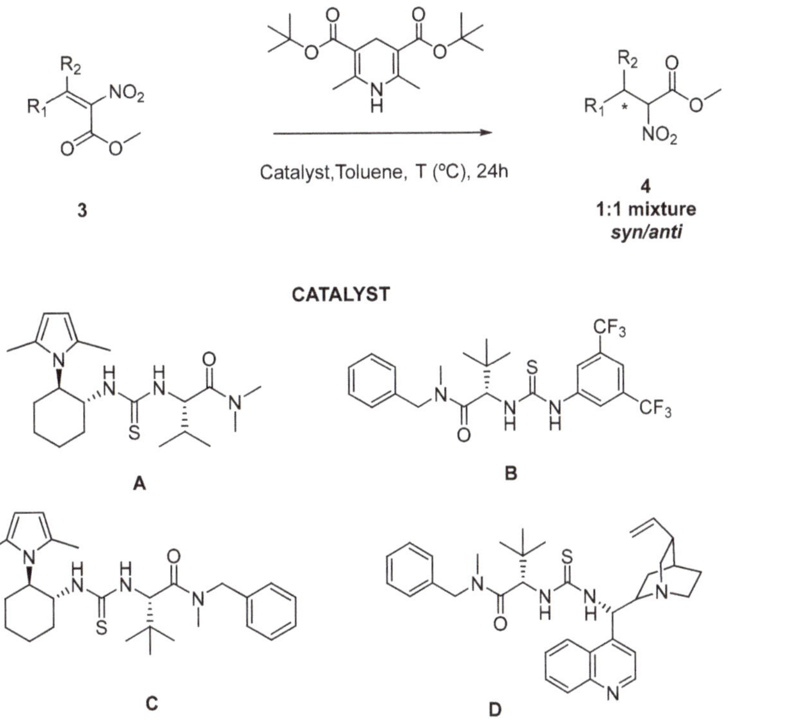

Scheme 3. Organocatalytic reduction of tetrasubstituted nitroalkenes.

In general, compounds **4** were obtained in a 1:1 mixture of *syn/anti* products after 24 h of reaction time. Initial experiments were conducted using nitroalkane **3a** (Table 3, entries 1–8) as model substrate. The reactions performed in the presence of catalyst **A** starting from different mixtures of nitroalkene **3a** (Table 3, entries 1–3), demonstrate a different reactivity of the *E–Z* isomers of this compound, showing that one isomer reacted more quickly than the other one. This interesting discovery was confirmed when the reaction was performed with a pure fraction of the less reactive isomer (which was previously assigned as having a *Z* configuration) and no reaction was observed (Table 3, entry 1). Starting from differently enriched mixtures in the *E* isomer led to similar results (entries 2–3), leading to the chiral alkanes in up to 67% enantiomeric excess (entries 2–3). Attempts to increase the yield by operating at a higher temperature or to improve the enantioselectivity by running the reaction at a lower temperature did not lead to any significant results. The enantioselectivity of the reaction was measured by HPLC analysis of the pure samples on the chiral stationary phase, and two pairs of enantiomers, corresponding to *syn/anti* products, were found (see experimental section).

Table 3. Enantioselective reduction of tetrasubstituted nitroalkenes.

Entry	T [a]	Nitroalkene	Nitroalkane	Catalyst	Yield [d] 4	ee% Syn–ee% Anti [e]
1 [a]	60 °C	3a–Z	4a	A	trace	n.d.
2 [b]	60 °C	3a–E	4a	A	51%	67–63
3 [c]	60 °C	3a (E+Z)	4a	A	53%	63–52
4 [c]	100 °C	3a (E+Z)	4a	A	37%	27–22
5 [c]	25 °C	3a (E+Z)	4a	A	15%	n.d.
6 [c]	60 °C	3a (E+Z)	4a	B	69%	33–32
7	60 °C	3a (E+Z)	4a	C	19%	n.d.
8	60 °C	3a (E+Z)	4a	D	10%	n.d
9	60 °C	3d	4d	A	32%	12–13
10	60 °C	3e	4e	A	60%	24–4
11	60 °C	3f	4f	A	51%	50–46

[a] Reaction was performed using a 4:96 E–Z mixture of the nitroalkene; [b] Reaction was performed using an 90:10 E–Z mixture of the nitroalkene; [c] Reaction was performed using a 70:30 E–Z mixture of the nitroalkene; [d] Yield was determined after purification with column chromatography; [e] Determined using chiral HPLC column Phenomenex-Lux-Cellulose 5, Hexane/IPA 98:2 for compound 4a and Hexane/IPA 95:5 for compounds 4d–f; n.d. = not determined.

When thiourea catalyst **B** was used (Table 3, entry 6), the corresponding nitroalkane **3a** was obtained with good yield, but the enantioselectivity observed was lower than with catalyst **A** (Table 3, entry 2), which was the best catalyst for this transformation. When the thiourea catalysts **C** and **D** were tested (Table 3, entries 7,8), the yields were drastically reduced, with a very low quantity of product obtained after purification. With the optimized conditions in hand, the enantioselective reduction of nitroalkenes **3d–f** (Table 3, entries 9–11) was carried out using Hanztsch ester and thiourea catalyst **A**. Compounds **4d–f** were obtained with good yields after purification, but generally low or modest enantioselectivities.

2.3. Determination of the Absolute Configuration of Nitroalkanes (4)

The absolute configuration of the reduction product, the nitroalkane (**4**), was experimentally established by converting **4a** into a known compound. To achieve this goal, two approaches were attempted (Scheme 4).

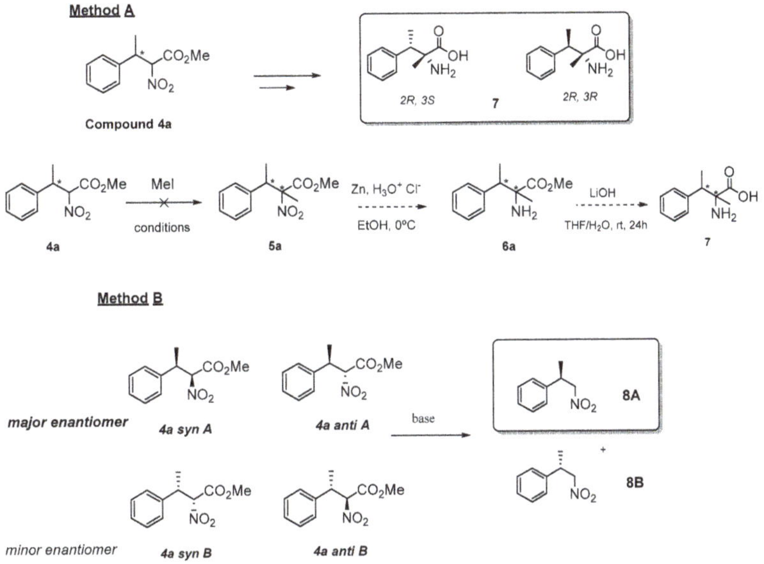

Scheme 4. Strategies for the experimental determination of the absolute configuration of nitroalkane **4a**.

The first strategy (Scheme 4, method A) explored the transformation of **4a** into the known compound **7**, by a three-step procedure involving the alkylation of nitroalkane **4a**, followed by the reduction of the nitro group and hydrolysis of the ester moiety as depicted in Scheme 4 [22–24]. However, despite several conditions being attempted, the alkylation did not lead to the desired product, but mixtures of unreacted starting material and O-alkylated products were obtained (In an additional experimental effort, we have observed that nitroalkane **4a** reacted with "softer" electrophiles, such as methyl vinyl ketone in a Michael reaction, additional studies are underway to further explore these transformations).

Thus, our efforts were focused on an alternative approach (Scheme 4, Method B). The decarboxylation of the ester moiety of **4a** (as 50:50 mixture of *syn/anti* isomers, enantioenriched) to afford the corresponding trisubstituted nitroalkane **8** was then explored [25]. This simple approach was found to be effective and the desired decarboxylated nitroalkane **8** was finally synthetized. The experimental optical rotation was measured using a polarimeter and by comparison with the literature data, the compound **8**, derived from the major enantiomer of product **4a** (**4a** *syn* **A** and **4a** *anti* **A**), was established to have the *R*-configuration [26].

DFT computational studies on the enantioselective reduction of tetrasubstituted nitroalkanes were preliminarily performed to rationalize the stereochemical outcome of the reaction, using the Gaussian g16 package using Catalyst **A** as the model catalyst. All geometries of reactants and products (ground states and transition states) were located at a B3LYP/6-31G (d,p) level of theory and finer electronic energies were successively obtained, increasing the basis set up to 6/311 + (2df,2pd) with B3LYP functional [27]. In Figure 3 four possible complexes of nitroalkene **3a** with catalyst **A** and the geometries of the TS leading to the formation of the four stereoisomers of nitroalkane **4a** are represented.

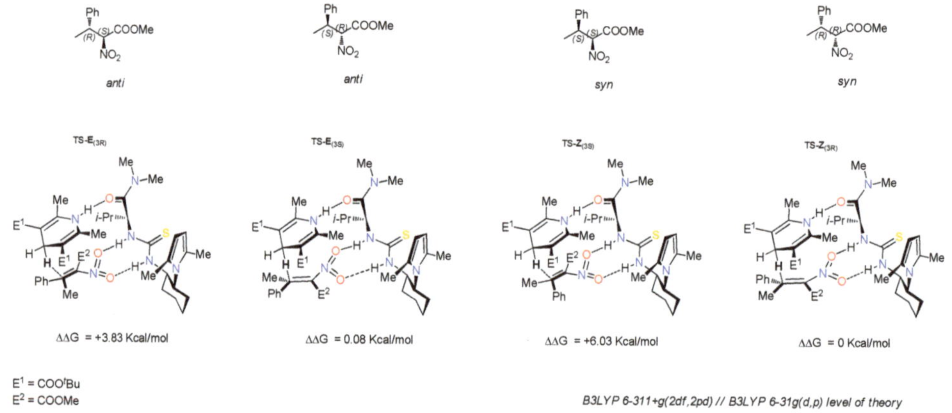

Figure 3. The four possible TS geometries for nitroalkene **3a** complexes with catalyst **A**.

Transition states responsible for the hydride transfer were located assuming the coordination of the nitro group of the nitroalkane **3a** to the thiourea moiety and of the Hantzsch ester NH group with the catalyst carboxyamide group, according to the so-called Takemoto model. The energy profile is depicted in Figure 4.

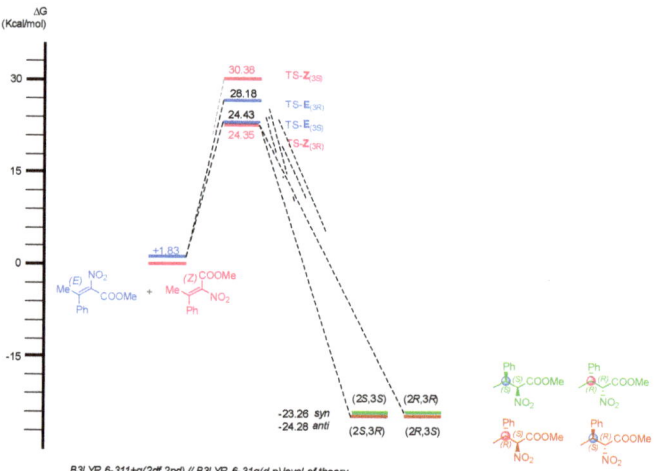

Figure 4. DFT calculations performed for the enantioselective reduction of tetrasubstituted nitroal-kene **3a**.

In Figure 5 the geometries of the transition states originated by the *E*-isomer of compound **3a** are illustrated.

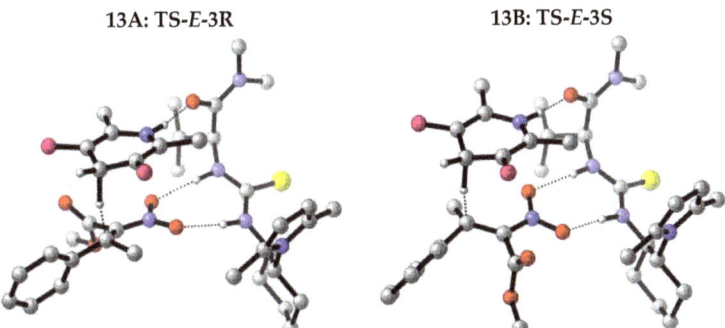

Figure 5. Transition states formed by the *E*- isomer of nitroacrylate **3a**.

The blue spheres represent the nitrogen atom of the nitro group of nitroalkane **3a** and of the thiourea catalyst **A**, the yellow one represents the sulfur atom of the thiourea moiety, and the red spheres represent the oxygen atoms of nitro and ester groups of compound **3a**, and of the carboxyamide group of the catalyst. The pink sphere represents the *tert*-butyl groups of the Hantzsch ester, carbon atoms are grey and hydrogen atoms are white. The broken lines showed the H-bonding interaction between the nitro group of **3a** and the thiourea moiety of the catalyst, and the transfer hydride between Hantzsch ester and carbon C_3 of the nitroolefin.

According to the calculations, among the two transition states originating from Z-olefin, TS-Z-(3R) is the lowest in energy and would lead to the formation of the final product with a *R*-configuration at the C_3 carbon, in agreement with the experimental data. However, the (*Z*) isomer was found to be very poorly reactive, while, as established in NMR analysis, the more reactive isomer is the (*E*)-nitroacrylate, that, according to the calculations should preferably afford the (*S*) enantiomer at C_3 carbon of compound **4a**. These findings are in contrast with the experimentally established absolute configuration (*R*) for the major enantiomer derived from the reaction of the E isomer of **3a**. Therefore, we can conclude that, at the moment, the proposed TS according to the Takemoto model, is not able to explain why the *E* isomer should be more reactive than the *Z* isomer, and, furthermore, cannot predict the correct configuration at the C_3 of the nitroalkane. Those results are probably an indication that other coordination modes are active in the TS of the reactions, and other models need to be taken into consideration to rationalize the stereochemical outcome of the reaction.

3. Materials and Methods

Reactions were monitored by analytical thin-layer chromatography (TLC) using silica gel 60 F_{254} pre-coated glass plates (0.25 mm thickness) and visualized using UV light. Flash chromatography was carried out on silica gel (230–400 mesh). Proton NMR spectra were recorded on spectrometers operating at 300 MHz (Bruker Fourier 300); proton chemical shifts are reported in ppm (δ) with the solvent reference relative to tetramethylsilane (TMS) employed as the internal standard ($CDCl_3$: δ = 7.26 ppm). ^{13}C-NMR spectra were recorded on 300 MHz spectrometers (Bruker Fourier 300) operating at 75 MHz, with complete proton decoupling; carbon chemical shifts are reported in ppm (δ) relative to TMS with the respective solvent resonance as the internal standard ($CDCl_3$: δ = 77.0 ppm). Mass spectra and accurate mass analysis were carried out on a VG AUTOSPEC- M246 spectrometer (double-focusing magnetic sector instrument with EBE geometry) equipped with EI source or with LCQ Fleet ion trap mass spectrometer, ESI source, with acquisition in positive ionization mode in the mass range of 50–2000 m/z. Dry solvents were purchased and stored under nitrogen over molecular sieves (bottles with crown caps). All chemicals were purchased from commercial suppliers and used without further purification unless otherwise specified.

3.1. General Procedure for the Synthesis of Tetrasubstituted Nitroalkenes (3)

Tetrasubstituted nitroalkenes (3) were synthetized using a two-step procedure: Firstly the formation of an acrylate intermediate (2) by a Horner–Wasdforth–Emmons reaction of an appropriate ketone (1) with trimethylphosphonoacetate and sodium hydride, following by a nitration reaction of this intermediate with a mixture of CAN-NaNO$_2$ as an effective nitration reagent.

Compounds 2a–f were synthetized using conditions reported in the literature [9]. First, a solution of trimethyl phosphonoacetate (5.21 mmol) in 20 mL of THF was cooled to 0 °C. Then, sodium hydride (5.21 mmol) was added portion-wise and the mixture was stirred for 30 min. After this time, the appropriate ketone (4.17 mmol) was added at the same temperature and the reaction mixture was allowed to warm to room temperature and stirred for 24 h at the right temperature. Then, 20 mL of saturated solution of ammonium chloride was added dropwise and the mixture was extracted with Et2O.

The combined organic phases were dried using MgSO$_4$, filtered and concentrated in vacuo. The solvent was eliminated under reduced pressure and the crude product was purified using column chromatography and hexanes/EtOAc as eluent. The ^{1}H-NMR of compounds 2a–f were in agreement with the published ones. Compounds 2a–f were directly used in the next step after purification.

Acrylates 2a, 2b, 2c, 2d, 2e and 2f (5.68 mmol) were dissolved in 50 mL of acetonitrile and cooled to 0 °C. Then, sodium nitrite (17 mmol) and cerium ammonium nitrate (17 mmol) were added at the same temperature, and the reaction mixture was allowed to warm to room temperature and stirred for 24 h. After this time, the reaction was filtered through a pad of celite, and the filtrate was concentrated under reduced pressure. The residue was poured into cold water and extracted with DCM (3 × 50 mL). The combined organic layers were dried using MgSO$_4$, filtered and concentrated in vacuo. The crude product was purified by column chromatography using an appropriate mixture of solvents to afford nitroacrylates. For further details see the Supporting Information.

3.2. General Procedure for the Enantioselective Synthesis of Nitroalkanes (4)

To a stirred solution of nitroalkenes (3) in toluene (0.3 mmol 0.3M), catalyst A (10 mol%) and Hanztsch ester (1.2 eq, 0.36 mmol) were added. The reaction mixture was heated at 60 °C for 24 h. Then, the mixture was allowed to warm to room temperature and the solvent was eliminated under reduced pressure, and the crude product was purified using column chromatography and an appropriate mixture of eluents.

4. Conclusions

Although the preparation of tetrasubstituted nitroacrylates proved to be very challenging, in this work a reproducible strategy for the synthesis of tetrasubstituted nitroalkenes was successfully developed using a two-step procedure; the HWE olefination of the ketone followed by the reaction of nitration affords the desired tetrasubstituted nitroalkenes (3).

The enantioselective reduction of these synthetized tetrasubstituted nitroalkenes (3) to access the functionalized nitroalkanes (4) was also performed, using a Hantzsch ester as the reductive agent and a thiourea based chiral catalyst, to afford the products with good to moderate yields, in a 1:1 mixture of syn/anti isomers, and up to 67% e.e. Although the level of enantioselectivity could not be considered satisfactory yet, it should be noted that the enantioselective organocatalytic reduction of tetrasubsituted alkenes was almost completely unknown. Even if the poor reactivity of the substrates represents a major problem, the present work demonstrates that the asymmetric catalytic reduction of functionalized nitroacrylates may offer a viable strategy for the synthesis of chiral amino ester derivatives.

The absolute configuration of the major enantiomer obtained in the enantioselective reduction was established by converting the nitroalkane 4a into a known product. DFT calculations, performed in order to rationalize the stereochemical outcome of the reaction did not lead to satisfactory results. Further studies, considering other alternative coordination modes between the catalyst and the substrate, will be necessary in order to understand the origins of the stereocontrol of the reaction.

Supplementary Materials: The following supporting information can be downloaded at: https://www.mdpi.com/article/10.3390/molecules28073156/s1.

Author Contributions: Conceptualization, M.B.; methodology, M.S., P.C.G. and S.R.; validation, M.S., P.C.G. and S.R.; formal analysis, F.V.; investigation, P.C.G.; data curation, S.R., F.V. and P.C.G.; writing—original draft preparation, M.B. All authors have read and agreed to the published version of the manuscript.

Funding: M.B. and P.C.G. thank ITN-EID project Marie Sklodowska-Curie Actions Innovative Training Network—TECHNOTRAIN H2020-MSCA-ITN-2018 Grant Agreement 812944. www.technotrain-ITN.eu. M.B. and M.S. thank Taros Chemicals.

Institutional Review Board Statement: Not applicable.

Informed Consent Statement: Not applicable.

Data Availability Statement: Data available on request from the authors.

Conflicts of Interest: The authors declare no conflict of interest.

Sample Availability: Not applicable.

References

1. Narancic, T.; Almahboub, S.A.; O'Connor, K.E. Unnatural amino acids: Production and biotechnological potential. *World J. Microbiol. Biotechnol.* **2019**, *35*, 67. [CrossRef]
2. Osberger, T.J.; Rogness, D.C.; Kohrt, J.T.; Stepan, A.F.; White, M.C. Oxidative diversification of amino acids and peptides by small-molecule iron catalysis. *Nature* **2016**, *537*, 214–219. [CrossRef]
3. Bisht, A.; Juyal, D. Unnatural amino acids (UAA'S): A trendy scaffold for pharmaceutical research. *J. Drug Deliv. Ther.* **2019**, *9*, 601–609.
4. Ohfune, Y.; Shinada, T. Enantio- and Diastereoselective Construction of α,α-Disubstituted α-Amino Acids for the Synthesis of Biologically Active Compounds. *Eur. J. Org. Chem.* **2005**, *2005*, 5127–5143. [CrossRef]
5. Li, B.; Zhang, J.; Xu, Y.; Yang, X.; Li, L. Improved synthesis of unnatural amino acids for peptide stapling. *Tetrahedron Lett.* **2017**, *58*, 24–28. [CrossRef]
6. Vogt, H.; Bräse, S. Recent approaches towards the asymmetric synthesis of α,α-disubstituted α-amino acids. *Org. Biomol. Chem.* **2007**, *5*, 406–430. [CrossRef]
7. Noboru, O. Chapter 2: Preparation of Nitro Compounds. In *The Nitro Group in Organic Synthesis*; Willey-VCH: Hoboken, NJ, USA, 2001; pp. 3–29.
8. Gabrielli, S.; Chiurchiù, E.; Palmieri, A. β-Nitroacrylates: A Versatile and Growing Class of Functionalized Nitroalkenes. *Adv. Synth. Catal.* **2019**, *361*, 630–651. [CrossRef]
9. Wen, J.; Jiang, J.; Zhang, X. Rhodium-Catalyzed Asymmetric Hydrogenation of α,β-Unsaturated Carbonyl Compounds via Thiourea Hydrogen Bonding. *Org. Lett.* **2016**, *18*, 4451–4453. [CrossRef]
10. Sharma, V.; Kelly, G.T.; Watanabe, C.M.H. Exploration of the Molecular Origin of the Azinomycin Epoxide: Timing of the Biosynthesis Revealed. *Org. Lett.* **2008**, *10*, 4815–4818. [CrossRef]
11. Maity, S.; Manna, S.; Rana, S.; Naveen, T.; Mallick, A.; Maiti, D. Efficient and Stereoselective Nitration of Mono- and Disubstituted Olefins with AgNO$_2$ and TEMPO. *J. Am. Chem. Soc.* **2013**, *135*, 3355–3358. [CrossRef]
12. Rodríguez, J.M.; Pujol, M.D. Straightforward synthesis of nitroolefins by microwave- or ultrasound-assisted Henry reaction. *Tetrahedron Lett.* **2011**, *52*, 2629–2632. [CrossRef]
13. Zhu, L.; Yan, P.; Zhang, L.; Chen, Z.; Zeng, X.; Zhong, G. TiCl$_4$/DMAP mediated Z-selective knovenagel condensation of isatins with nitroacetates and related compounds. *RSC Adv.* **2017**, *7*, 51352–51358. [CrossRef]
14. Zhang, J.; Blazecka, P.G.; Mark, P.A.; Lovdahl, T.; Curran, T. Indium (III) mediated Markovnikov addition of malonates and β-ketoesters to terminal alkynes and the formation of Knoevenagel condensation products. *Tetrahedron* **2005**, *61*, 7807–7813. [CrossRef]
15. Nakamura, M.; Fujimoto, T.; Endo, K.; Nakamura, E. Stereoselective Synthesis of Tetra-Substituted Olefins via Addition of Zinc Enolates to Unactivated Alkynes. *Org. Lett.* **2004**, *6*, 4837–4840. [CrossRef]
16. Buevich, A.V.; Wu, Y.; Chan, T.M.; Stamford, A. An unusual contra-Michael addition of NaNO2–ceric ammonium nitrate to acrylic esters. *Tetrahedron Lett.* **2008**, *49*, 2132–2135. [CrossRef]
17. Zheng, C.; You, S.-L. Transfer hydrogenation with Hantzsch esters and related organic hydride donors. *Chem. Soc. Rev.* **2012**, *41*, 2498–2518. [CrossRef]

18. Nolwenn, J.A.; Ozores, M.L.; List, B. Organocatalytic Asymmetric Transfer Hydrogenation of Nitroolefins. *J. Am. Chem. Soc.* **2007**, *129*, 8976–8977.
19. Bernardi, L.; Fochi, M. A General Catalytic Enantioselective Transfer Hydrogenation Reaction of β,β-Disubstituted Nitroalkenes Promoted by a Simple Organocatalyst. *Molecules* **2016**, *21*, 1000. [CrossRef]
20. Massolo, E.; Benaglia, M.; Orlandi, M.; Rossi, S.; Celentano, G. Enantioselective Organocatalytic Reduction of β-Trifluoromethyl Nitroalkenes: An Efficient Strategy for the Synthesis of Chiral β-Trifluoromethyl Amines. *Chem. Eur. J.* **2015**, *21*, 3589–3595. [CrossRef]
21. Barrios-Rivera, J.; Xu, Y.; Willis, M.; Vyas, V.K. A diversity of recently reported methodology for asymmetric imine reduction. *Org. Chem. Front.* **2020**, *7*, 3312–3342. [CrossRef]
22. Davis, F.A.; Liang, C.H.; Liu, H. Asymmetric Synthesis of β-Substituted α-Amino Acids Using 2H-Azirine-2-carboxylate Esters. Synthesis of 3,3-Disubstituted Aziridine-2-carboxylate Esters. *J. Org. Chem.* **1997**, *62*, 3796–3797. [CrossRef]
23. Gagnot, G.; Hervin, V.; Coutant, E.P.; Desmons, S.; Baatallah, R.; Monnot, V.; Janin, Y.L. Synthesis of unnatural α-amino esters using ethyl nitroacetate and condensation or cycloaddition reactions. *Beilstein J. Org. Chem.* **2018**, *14*, 2846–2852. [CrossRef] [PubMed]
24. Harel, T.; Rozen, S. Transforming Natural Amino Acids into α-Alkyl-Substituted Amino Acids with the Help of the HOF·CH3CN Complex. *J. Org. Chem.* **2007**, *72*, 6500–6503. [CrossRef]
25. Metz, A.E.; Kozlowski, M.C. 2-Aryl-2-nitroacetates as Central Precursors to Aryl Nitromethanes, α-Ketoesters, and α-Amino Acids. *J. Org. Chem.* **2013**, *78*, 717–722. [CrossRef] [PubMed]
26. Hostmann, T.; Molloy, J.J.; Bussmann, K.; Gilmour, R. Light-Enabled Enantiodivergence: Stereospecific Reduction of Activated Alkenes Using a Single Organocatalyst Enantiomer. *Org. Lett.* **2019**, *21*, 10164–10168. [CrossRef] [PubMed]
27. Frisch, M.J.; Trucks, G.W.; Schlegel, H.B.; Scuseria, G.E.; Robb, M.A.; Cheeseman, J.R.; Scalmani, G.; Barone, V.; Petersson, G.A.; Nakatsuji, H.; et al. *Gaussian 16, Revision C.01*; Gaussian, Inc.: Wallingford, CT, USA, 2016; Available online: https://gaussian.com/citation/ (accessed on 13 December 2022).

Disclaimer/Publisher's Note: The statements, opinions and data contained in all publications are solely those of the individual author(s) and contributor(s) and not of MDPI and/or the editor(s). MDPI and/or the editor(s) disclaim responsibility for any injury to people or property resulting from any ideas, methods, instructions or products referred to in the content.

MDPI
St. Alban-Anlage 66
4052 Basel
Switzerland
www.mdpi.com

Molecules Editorial Office
E-mail: molecules@mdpi.com
www.mdpi.com/journal/molecules

Disclaimer/Publisher's Note: The statements, opinions and data contained in all publications are solely those of the individual author(s) and contributor(s) and not of MDPI and/or the editor(s). MDPI and/or the editor(s) disclaim responsibility for any injury to people or property resulting from any ideas, methods, instructions or products referred to in the content.

www.ingramcontent.com/pod-product-compliance
Lightning Source LLC
LaVergne TN
LVHW070652100526
838202LV00013B/948